Enzymes
and
Their Inhibition

Drug Development

CRC Enzyme Inhibitors Series

Series Editors
H. John Smith and Claire Simons
Cardiff Univeristy
Cardiff, UK

Carbonic Anhydrase: Its Inhibitors and Activators
Edited by Claudiu T. Supuran, Andrea Scozzafava and Janet Conway

Enzymes and Their Inhibitors: Drug Development
Edited by H. John Smith and Claire Simons

Enzymes and Their Inhibition

Drug Development

Edited by

H. John Smith
and
Claire Simons

CRC Press
Taylor & Francis Group
Boca Raton London New York

CRC Press is an imprint of the
Taylor & Francis Group, an **informa** business

CRC Press
Taylor & Francis Group
6000 Broken Sound Parkway NW, Suite 300
Boca Raton, FL 33487-2742

First issued in paperback 2019

ISBN-13: 978-0-415-33402-0 (hbk)
ISBN-13: 978-0-367-39357-1 (pbk)

Library of Congress Cataloging-in-Publication Data

Enzymes and their inhibition: drug development / edited by H. John Smith and Claire Simons.
      p.; cm. — (Enzyme inhibitor)
    Includes bibliographical references and index.
    ISBN 0-415-33402-0 (alk. paper)
    1. Enzyme inhibitors. 2. Drug development. I. Smith, H. J., 1930- II. Simons, Claire.
III. CRC enzyme inhibitors series.
    [DNLM: 1. Enzyme Inhibitors. 2. Chemistry, Pharmaceutical. 3. Drug Design. QU 143
E6154 2004]
    QP601.5.E5945 2004
    615'.35—dc22   2004055375

Library of Congress Card Number 2004055375

Visit the Taylor & Francis Web site at
http://www.taylorandfrancis.com

and the CRC Press Web site at
http://www.crcpress.com

Series Preface

One approach to the development of drugs as medicines, which has gained considerable success over the past two decades, involves inhibition of the activity of a target enzyme in the body or invading parasite by a small molecule inhibitor, leading to a useful clinical effect.

The CRC Enzyme Inhibitor Series consists of an expanding series of monographs on this aspect of drug development, providing timely and in-depth accounts of developing and future targets that collectively embrace the contributions of medicinal chemistry (synthesis, design), pharmacology and toxicology, biochemistry, physiology, and biopharmaceutics necessary in the development of novel pharmaceutics.

H. John Smith
Claire Simons

Preface

The majority of drugs used clinically exert their action in one of two ways: (1) by interfering with a component (agonist) in the body, preventing interaction with its site of action (receptor), i.e., receptor antagonist, or (2) by interfering with an enzyme normally essential for the well-being of the body or involved in bacterial or parasitic or fungal growth causing disease and infectious states, where the removal of its activity by treatment is necessary, i.e., enzyme inhibitors. In recent years the proportion of current drugs described as enzyme inhibitors has increased, and this book gives an account of the steps taken for designing and developing such inhibitors — from identification of the target enzyme to be blocked in a particular disease or infection to their introduction in the marketplace.

Once the enzyme target is selected or discovered, a knowledge of the structure, substrates, kinetics, and mechanism of the enzyme can be brought together in the rational design of an inhibitor. However, the transfer of a prospective drug candidate from the laboratory bench to the marketplace follows a prolonged and difficult pathway due to the body's requirements for suitable absorption, distribution, metabolism, and selectivity characteristics so as to arrive at its site of action at a satisfactory concentration level devoid of unnecessary side effects.

Editors

H. John Smith is a former reader in medicinal chemistry at the Welsh School of Pharmacy, Cardiff University, U.K. He obtained his Ph.D. in medicinal chemistry at the University of London and received his D.Sc. in 1995. A fellow of the Royal Society of Chemistry and the Royal Pharmaceutical Society of Great Britain, he has spent much of his career studying enzyme inhibitors and their potential use in drugs. John Smith has coauthored several texts on this subject and is editor-in-chief of the *Journal of Enzyme Inhibition and Medicinal Chemistry.*

Claire Simons is a lecturer in medicinal chemistry at the Welsh School of Pharmacy, Cardiff University, U.K. She obtained her Ph.D. in organic chemistry at King's College, University of London, and is a member of the Royal Society of Chemistry. Her main research interests are the design, synthesis, and computational analysis of novel heterocyclic and nucleoside compounds as enzyme inhibitors. Claire Simons has authored a textbook on nucleoside chemistry and its therapeutic application and coedited (with John Smith) a textbook, *Proteinase and Peptidase Inhibition.*

Contributors

Paul J. Ala
Incyte Corporation
Wilmington, Delaware

Anthony J. Berdis
Department of Pharmacology
Case Western Reserve University
Cleveland, Ohio

Angela Casini
Department of Chemistry
University of Florence
Florence, Italy

Chong-Hwan Chang
Bristol-Myers Squibb Company
Princeton, New Jersey

Hans-Ulrich Demuth
Probiodrug Research Ltd
Weinbergweg Research Ltd
Halle, Germany

Samer Haidar
Faculty of Pharmaceutical and
   Medicinal Chemistry
University of the Saar
Saarbrucken, Germany

Rolf W. Hartmann
Faculty of Pharmaceutical and
   Medicinal Chemistry
University of the Saar
Saarbrucken, Germany

Neil R. Kitteringham
Department of Pharmacology and
   Therapeutics
University of Liverpool
Liverpool, United Kingdom

Irene Lee
Department of Chemistry
Case Western Reserve University
Cleveland, Ohio

W. Edward Lindup
Department of Pharmacology and
   Therapeutics
University of Liverpool
Liverpool, United Kingdom

Edward A. Meighen
Department of Biochemistry
McGill University
Montreal, Quebec, Canada

André J. Neistroj
Probiodrug Research Ltd
Weinbergweg Research Ltd
Halle, Germany

Bruce A. Palfey
Department of Biological Chemistry
University of Michigan Medical School
Ann Arbor, Michigan

Andrea Scozzafava
Department of Chemistry
University of Florence
Florence, Italy

Claire Simons
Welsh School of Pharmacy
Cardiff University
Cardiff, United Kingdom

H. John Smith
Welsh School of Pharmacy
Cardiff University
Cardiff, United Kingdom

Torsten Steinmetzer
Curacyte Chemistry GmbH
Jena, Germany

Jure Stojan
Institute of Biochemistry
University of Ljubljana
Ljubljana, Slovenia

Claudiu T. Supuran
Department of Chemistry
University of Florence
Florence, Italy

L.W. Lawrence Woo
Department of Pharmacy and
  Pharmacology
University of Bath
Bath, United Kingdom

Abbreviations

AADC amino acid decarboxylase
ACE angiotensin 1 converting enzyme
Adiol androstenediol
ADP adenosine diphosphate
AFM atomic force microscope
AG aminoglutethimide
AhR Ah receptor
AMP adenosine monophosphate
ARNT Ah receptor nuclear transporter
ASA/ASB aryl sulfatase A/B
ATCase aspartate transcarbamylase
ATP adenosine triphosphate
BOC *tert*-butoxycabonyl
BPH benign prostatic hyperplasia
CA carbonic anhydrase
CAR constitutive androstane receptor
CD circular dichroism
ChC *Clostridium hydrolyticum* collagenase
CoA coenzyme A
CoMFA comparative molecular field analysis
COMT catechol-*O*-methyltransferase
CYP 17 17α-hydrolyase/C17-20-lyase
CYP 19 aromatase
DHEA dehydroepiandrosterone
DHPS dihydropteroate synthase
DHT dihydrotestosterone
DIQ decahydroisoquinoline
DOPA L-3,4-dihydroxyphenyl alanine
DPIV dipeptidyl peptidase IV
DTT dithiothreitol
E1 estrone
E2 estradiol
E1S estrone sulfate
ECM extracellular matrix
EMATE estrone-3-*O*-sulfamate
FAD flavine adenine dinucleotide
FGly formylglycine
FMN flavine mononucleotide
ΔG free energy change
GABA α-aminobutyric acid

GABA-T GABA transaminase
GnRH gonadotrophin-releasing hormone
ΔH enthalpy change
HDBC hormone-dependent breast cancer
HIV human immunodeficiency virus
17βHSD 17βhydroxysteroid dehydrogenase
HSP90 heat shock protein
I inhibitor
IEF isoelectric focusing
IR infrared
KNF Koshland–Nemethy–Filmer Model
LBHB low barrier hydrogen bond
LHRH luteinizing hormone-releasing hormone
MAO monoamine oxidase
MCF-7 human breast cancer cells
MMPs matrix metalloproteinases
MMPIs MMP inhibitors
MT-MMP membrane-type MMP
MWC Monod–Wyman–Changeux Model
NADP nicotinamide adenine dinucleotide phosphate
NMR nuclear magnetic resonance
NSAI nonsteroidal aromatase inhibitor
Ntn N-terminal nucleophile
OATP 1 organic anion transporter-1
OC oral contraceptive
ODC ornithine decarboxylase
P450$_{arom}$ aromatase
PAGE polyacrylamide gel electrophoresis
PARP poly(ADP-ribose)polymerase
PARs protease activatable receptors
PC prostate cancer
PDB protein data base
PEG polyethylene glycol
Pi inorganic phosphate
PR HIV protease
PXR pregnane X receptor
QSAR quantitative structure–activity relationship
5α-R 5α-reductase
RT reverse transcriptase
ΔS entropy change
S substrate
SAM *S*-adenosylmethionine
SAR structure–activity relationship
SDS-PAGE sodium dodecylsulfate polyacrylamide gel electrophoresis
STS steroid sulfatase
T testosterone
TPP thiamine pyrophosphate

Contents

1 Enzyme Structure and Function

Edward A. Meighen

CONTENTS

1.1 INTRODUCTION

Enzymes are proteins that catalyze chemical reactions. A protein is simply a polypeptide composed of amino acids linked by a peptide bond, and the term generally, but not always, refers to the folded conformation. To understand how an enzyme functions, including its binding and functional properties, it is necessary to know the properties of the amino acids and how the amino acids are linked together, including the torsion angles of the bonds and the space occupied, and the interactions of the atoms leading to the final conformations of the folded protein. Only in the folded state can a protein function effectively as an enzyme to bind substrates and act as a catalyst.

The structural organization of a protein is generally classified into four categories: primary, secondary, tertiary, and quaternary structure. Primary structure refers to the amino acid sequence of the polypeptide chain; secondary structure refers to the local conformations including the α-helix, β-strand, and the reverse turn; tertiary structure refers to the overall folding of the protein involving interaction of distant parts; and quaternary structure refers to the interaction of separate polypeptide chains. However, it is sometimes difficult to make clear distinctions between the different levels of structural classification, particularly between secondary and tertiary structure. The elements and properties of these structural levels are outlined in Section 1.2 through Section 1.4.

1.2 PRIMARY STRUCTURE

Only a limited number of amino acids are found in a polypeptide chain. All amino acids have a structure of NH_3^+-CH(R)-COO with the amino acid being in the L-configuration and not in the D-configuration, as shown in Figure 1.1 for alanine (Ala), which has a methyl group as its side chain (R). The L- and D-alanine can be readily rotated into the standard Fischer projection so that the amino group is in front of the plane on the left and right, respectively, with the carboxyl group on top and the side chain (CH_3) at the bottom, both pointed toward the back and behind the plane (see Section 5.5.4.1). The L- and D-configuration forms of an amino acid are enantiomers, as they are stereoisomers (i.e., having the same molecular formula) and have nonsuperimposable mirror images (as shown in Figure 1.1).

The total number of common naturally occurring amino acids incorporated into the protein during synthesis of the polypeptide chain is only 20. Some rare amino acids are also found in proteins and, with the exception of selenocysteine, are generated by posttranslational modification of the synthesized protein. Each of the 20 amino acids differs in the structure of the R side chain (Figure 1.2). The central carbon of the amino acid is designated as α whereas the first carbon atom on the

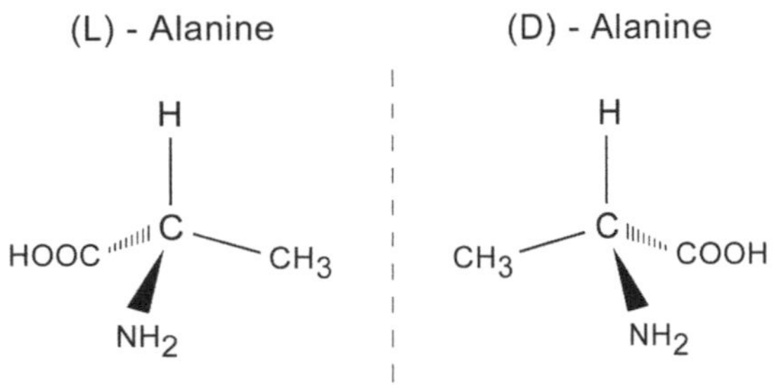

(L) - Alanine   (D) - Alanine

FIGURE 1.1 Mirror images of the two enantiomers of Ala. The COOH and NH$_2$ groups are behind and in front of the plane, respectively.

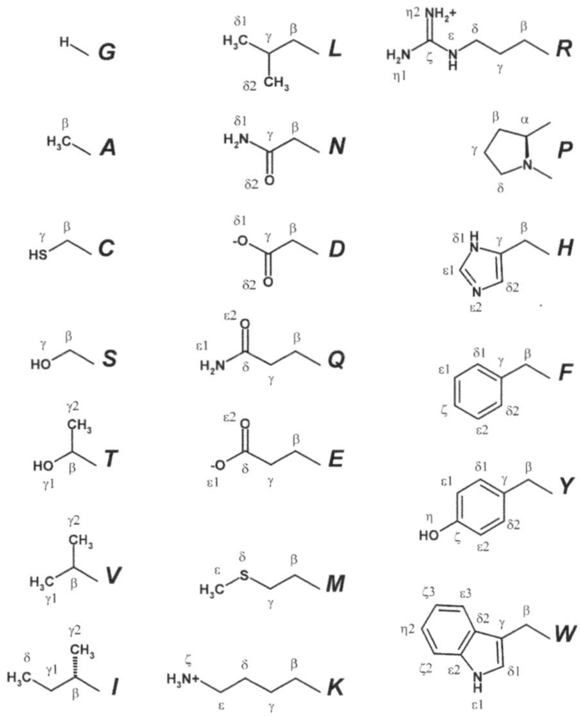

FIGURE 1.2 Structures of the side chains of the 20 common amino acids. Only the atoms of the side chain and the C$_\alpha$ of the amino acid are represented, except for Pro, which also shows the N of the backbone in the cyclic ring and the bonds to the preceding and following carbonyl groups in the peptide chain. The designations of the nonhydrogen atoms on the side chain extending from the α-carbon are also indicated.

side chain is β, and the following atoms, excluding hydrogen, are designated in order: γ, δ, ε, ζ, and η. Most amino acids have an unsubstituted β-CH$_2$ group, whereas Glycine (Gly) does not have this group and has a hydrogen on the C$_\alpha$-carbon, and Threonine (Thr), Valine (Val), and Leucine (Leu) are bifurcated at the β carbon near the polypeptide chain, which has consequences in the folding of the protein. Simi-

TABLE 1.1
Properties of Amino Acids

Amino Acids by Hydrophobicity	Codes		Percentage	pK$_a$	Area (Å^2)	Volume (Å^3)
Isoleucine	Ile	I	5.9	—	175	167
Valine	Val	V	6.7	—	155	140
Cysteine	Cys	C	1.6	8.4	135	109
Phenylalanine	Phe	F	4.1	—	210	190
Leucine	Leu	L	9.6	—	170	167
Methionine	Met	M	2.4	—	185	163
Alanine	Ala	A	7.7	—	115	89
Glycine	Gly	G	6.9	—	75	60
Tryptophan	Trp	W	1.2	—	255	228
Serine	Ser	S	7	—	115	89
Threonine	Thr	T	5.6	—	140	116
Tyrosine	Tyr	Y	3.1	10.1	230	194
Histidine	His	H	2.3	6.1	195	153
Proline	Pro	P	4.9	—	145	113
Asparagine	Asn	N	4.3	—	160	114
Glutamine	Gln	Q	3.9	—	180	144
Aspartic Acid	Asp	D	5.3	3.9	150	111
Glutamic Acid	Glu	E	6.5	4.1	190	138
Arginine	Arg	R	5.2	12.5	225	174
Lysine	Lys	K	6	10.8	200	169

Source: From Volume: A.A. Zymatin. (1972). *Progress in Biophysics*, 24, 107–123; Area: C. Chotia. (1975). *Journal of Molecular Biology*, 105, 1–14; Percentage: A. Bairoch. (2003). Amino acid scale: Amino acid composition (%) in the Swiss-Prot Protein Sequence data bank. http//ca.expasy.org/tools/pscale/A.A. Swiss-Prot.html.

larly Pro forms a cyclic ring with the δ-CH$_2$ covalently linked to the backbone nitrogen, leading to the side-chain residues being close to the polypeptide backbone and limiting the flexibility of the backbone.

Table 1.1 gives a list of these amino acids, their designations in the standard three-letter and one-letter codes, their frequencies in proteins, the pK$_a$'s of the R side chains, and some of their key properties relating to polarity and size. The average frequency of the amino acids (Table 1.1) in proteins is 5%, with Cysteine (Cys), Tryptophan (Trp), Methionine (Met), and Histidine (His) being present at relatively low frequencies (<2.4% each), whereas Leu is present at 9.6% and Ala at 7.7%, and the remaining amino acids at between 3 and 7% frequency.

About half the side chains are polar or charged, whereas the other half are nonpolar. The amino acids are listed in order in Table 1.1 based on their relative hydrophobicity (dislike of water), with the polar and charged amino acids being the least hydrophobic due to their capability of forming strong hydrogen or ionic bonds or both. Consequently, the type of side chain is critical in the formation of these bonds and even of van der Waals contacts, the primary forces that overcome the unfavorable energy required to place the polypeptide in the final active conformation required for enzymic function. These forces will determine to a major degree whether the amino acid is buried in the central part of the protein or remains on the surface exposed to solvent because many (but not all) hydrophobic groups are found in the central regions of the protein, out of contact with water, with primarily polar or charged molecules on the surface. An understanding of these forces, given in the following text, is thus important in an understanding of not only how the folded protein is stabilized but also how the enzyme interacts with other components including substrates, inhibitors, proteins, and other macromolecules.

1.2.1 VAN DER WAALS INTERACTIONS

Van der Waals interactions occur between all atoms and arise due to the increasing attraction of temporal electrical charges (induced dipoles) as atoms approach one another, offset on close contact by the strong repulsion of overlapping electronic orbitals. The maximum attraction occurs at an optimum distance equal to the sum of the atoms' van der Waals radii. Typical van der Waals radii are 1.2 Å for hydrogen, 1.4 to 1.5 Å for oxygen and nitrogen, and 2 Å for carbon. As van der Waals contacts exist between all atoms, this energy force can contribute to the folding of the protein by having highly complementary surfaces interact with the closer packing of the atoms leading to an increase in the number of van der Waals contacts and interaction energy.

1.2.2 HYDROGEN AND IONIC BONDS

The hydrogen bond arises from the sharing of an H atom between two electronegative atoms (such as O, N, and S), with the hydrogen atom being covalently attached to one of the atoms. The most common hydrogen bonds are those between the NH of the amino group and the oxygen of the carbonyl group of the peptide backbone; however, most side chains can form a hydrogen bond by accepting or donating a hydrogen atom or both, except those containing only nonpolar groups. Ionic bonds arise through interactions of charges of opposite polarity and are thus limited to Lys, Arg, Glu, and Asp, at least at pH 7, with Cys, His, and Tyr being capable of being charged in the physiological pH range in the appropriate microenvironment. Both bonding interactions cause the atoms to approach in closer contact than by the sum of their van der Waals radii. Consequently, the distance between the hydrogen atom and the electronegative atom in a hydrogen bond is only about 2 Å, whereas the sum of their van der Waals radii would be 2.6 to 2.7 Å. The strength of a hydrogen (or even an ionic) bond is quite weak in water as hydrogen bonds can readily form with water, and the highly polar solvent weakens ionic attractions. However, the relative strengths of hydrogen bonds and ionic bonds in proteins are much stronger as the protein microenvironment generally has a much lower dielectric constant (lower polarizability) than water.

1.2.3 HYDROPHOBIC INTERACTIONS

Hydrophobic bonds or attractions arise from the increase in entropy (freedom or randomness) that accompanies the release of water into the bulk solvent on interaction of two surfaces. The hydrophobic bond is not a true bond, in the sense that the atoms do not come in closer contact than the sum of the van der Waals radii. However, these contacts contribute strong binding forces to the folding of the protein (due to changes leading to an increase in the entropy of water) that extend well beyond those contributed by the van der Waals interactions. The strength of a hydrophobic bond formed by an amino acid side chain is dependent on the accessible surface area of the interacting side chains, as water in direct contact with the protein surface has lower entropy than the bulk water free in solution. As amino acids come in contact with each other, thus decreasing the accessible surface area for interaction with water, some of the water will be released from the protein surface into the bulk solution with a resultant increase in entropy of the released water. The strength of this interaction is decreased by the presence of any polar or charged groups that can interact with water or other groups by hydrogen or ionic bonds. Reagents that decrease the entropy of the bulk water, such as the denaturants of urea, guanidine hydrochloride, or sodium thiocyanate, when added in high concentrations to the protein solution, will also decrease the strength of the hydrophobic bond as the water released will not gain as much entropy. In contrast, high concentrations of phosphate and sulphate that actually increase the entropy of the bulk water will strengthen the hydrophobic attraction. Indeed, these reagents are often used in hydrophobic chromatography for purification of enzymes. Proteins that bind to hydrophobic columns can often be eluted by sodium thiocyanate as it decreases the strength of the interaction, whereas proteins that cannot bind to a hydrophobic column can often be made to bind by adding high concentrations of phosphate or sulfate to increase the strength of the hydrophobic interaction. It should be noted that as the energy derived from an increase in entropy equals $-T\Delta S$, the strength of the hydrophobic attraction increases with temperature.

A commonly used term related to the hydrophobicity of an amino acid is hydropathy, which is simply a measure of the amino acid's "feeling" (pathy) about water (hydro). Consequently, the hydrophobicity (dislike) or hydrophilicity (like) of an amino acid side chain reflects its hydropathic character, and both are similar measures starting from the opposite ends of the scale. There are many hydropathy or polarity scales in the literature reflecting the interaction of amino acid side chains with water. These scales are based on the relevant frequencies of amino acids in different microenvironments in proteins (e.g., buried or exposed) or the relative preference of amino acid analogs for liquid water compared with organic solvents or the vapor phase and, although similar, differ to some degree depending on how the hydropathic character of a given amino acid side chain is measured and weighted.

Table 1.1 gives the relative order of hydrophobicity of the amino acids based on the average of the rankings of the hydropathy of each amino acid from a number of the more popular scales. Only amino acids listed above methionine in Table 1.1 make a reasonably strong contribution to the hydrophobic interactions, at least in most hydropathy scales. In general, amino acids without polar groups are listed as having the highest hydrophobicity, with the charged amino acids at pH 7 being the most hydrophilic. The overall character of an amino acid is a measure of the ability to form hydrophobic bonds based on the accessible area of the side chain, countered by the ability of polar groups to interact with water.

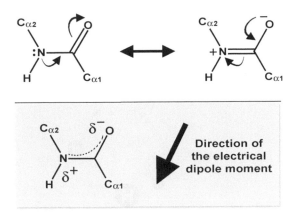

FIGURE 1.3 Resonance and charge of the planar peptide bond showing the electrical dipole moment.

1.2.4 PEPTIDE BONDS

The amino acids are linked together by a peptide bond that arises from the reaction of the amino group with the carboxyl group of another amino acid. The primary property of the peptide bond is its planar nature, which is due to the resonance of the electrons between the peptide bond and the carbonyl group, leading to a partial positive charge on the nitrogen and a partial negative charge on the oxygen and also giving the peptide bond some double-bond character as well as a small-charge dipole (Figure 1.3).

The preferred planar structure is the trans position shown in Figure 1.4, with the largest substituents (the incoming and outgoing polypeptide chains) on opposite sides of the peptide bond. Alternatively, the trans position for the peptide bond is often defined by the hydrogen on the nitrogen and the oxygen of the carbonyl being on opposite sides of the peptide bond. The other planar structure for the peptide bond is the cis configuration, with the large incoming and outgoing polypeptide chains (i.e., the α-carbons) being on the same side of the peptide bond.

Figure 1.4 shows that in the trans orientation, the R side chains are located quite far from each other in adjacent amino acids in the peptide chain, whereas the R groups are in much closer contact in the cis orientation. Due to the greater opportunity for steric overlap in the cis position compared with the trans position, the frequency of cis bonds to trans bonds is much lower (~0.3%). About 95% of cis bonds have Pro contributing the nitrogen to the peptide bond because the difference in stability favoring the trans over the cis structure is only about 20:1 for Pro. This occurs because the side chain of Pro bends back and covalently links with the nitrogen in the peptide bond, and thus the difference in potential structural overlap with the preceding R group is not as disfavored for Pro in the cis configuration compared with the trans position as that found for the other amino acids. Consequently, about 5% of Pro is present in cis bonds, whereas the other 19 amino acids are only present about 0.003% of the time in cis bonds. As crystal structures of proteins become more closely refined to the atomic level, the percentage of cis bonds

trans - peptide bond

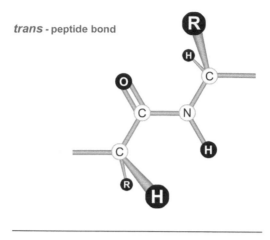

cis - peptide bond

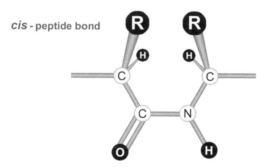

FIGURE 1.4 Trans and cis peptide bonds depicting the closer contact of the R side chains and peptide backbone in the cis configuration.

may increase to a small degree due to the tendency to assume that the much more common trans bond is present at any particular position during analyses of the electron density in the crystal structure. A point to recognize is that the direction of the polypeptide is defined from the amino terminal to the carboxyl terminal of the polypeptide and, consequently, the direction of the peptide bond is from the carbonyl to the NH group.

1.3 SECONDARY STRUCTURE

1.3.1 TORSION ANGLES

Aside from the amino acid side chains, the folding of the polypeptide is dependent upon the three torsion angles that occur for the bonds between any two adjacent backbone atoms (i.e., the carbon of the carbonyl, the α-carbon, and the nitrogen of the amino group). These three torsion or rotational angles for the backbone atoms of the polypeptide are referred to as psi (ψ), omega (ω), and phi (φ). The bond torsion angles are the angles between two planes each defined by three backbone atoms in a row, with the zero reference position being the cis configuration (0°). One plane is defined by two adjacent atoms and the previous backbone atom, whereas the second plane is defined by the same two atoms and the following backbone atom. Clockwise rotation of the second plane relative to the first plane from the cis position of the two planes leads to a positive angle, from 0 to +180°, whereas counterclockwise rotation leads to a negative angle, from 0 to −180°, with the latter angle being the same position as +180°.

The torsion angle ω for the peptide bond is quite simple to define, as one plane is given by the carbon and nitrogen in the peptide bond and the preceding α-carbon and the other by the same peptide atoms and the following α-carbon (dark triangles, Figure 1.5). When the peptide bond is in the reference cis position, the two α-carbons (on the incoming and outgoing peptide chains) are in a plane on the same side of the peptide bond. Rotation of the second plane relative to the first by 180° leads to the highly preferred trans position shown in Figure 1.5. In this representation, the dark gray shaded region containing the two triangular planes defined by the peptide bond and the preceding and following α-carbons, respectively, with a ω torsion angle of 180° leads to a common planar area extending across the gray rectangle. Note that the direction of the polypeptide is from front to back or bottom to top.

The other two torsion angles of the backbone polypeptide are defined in the same way. The ψ angle defines the rotation of the α-carbon relative to the carbon of the carbonyl group, and the φ angle defines the rotation of the nitrogen relative to the α-carbon. For the ψ angle, the two planes (triangular regions) are defined by the two carbon backbone atoms and the preceding and following nitrogen in the polypeptide backbone, whereas for the φ angle, the two planes are defined by the nitrogen and C_α backbone atoms and the preceding and following carbon of the carbonyl group. The same bond angles and relative positions of the atoms will be observed independent of the direction that one looks down the polypeptide chain. However, as the direction of observation is often defined in textbooks, this can lead to confusion due to the difficulty in visualizing the structure in three dimensions. Often the ψ and φ angles are defined by looking from the carbonyl carbon and the nitrogen, respectively, towards the α-carbon. Alternatively, and perhaps more simply, one can follow the direction of the polypeptide chain from the amino terminal towards the carboxyl terminal. In either case, the same torsion angles and relative positions of the backbone atoms would be observed.

The position of the polypeptide chain in three-dimensional space can, consequently, be defined by the two torsion angles ψ and φ for each of the amino acids and by the torsion angle ω for the peptide bonds. The value of ω for the peptide bond is almost always 180° due to its planar nature and the preference for the trans position. Both the ψ and φ angles have a much wider latitude in values, although they are restricted by the potential overlap of the steric space occupied by the backbone polypeptide and the amino acid side chains. Consideration of the energetic aspects led Ramachandran to develop a plot of the ψ angles vs. φ angles to readily reveal the more energetically favorable positions for each amino acid. Accordingly, this well-known plot, shown in Figure 1.6, was called the Ramachandran plot.

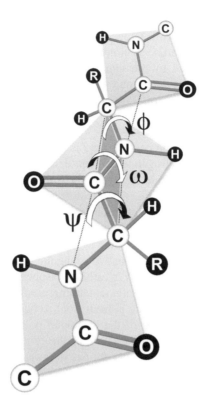

FIGURE 1.5 Torsion angles and the planar peptide bond. The atoms in the peptide bond and the preceding and following backbone carbon atoms are all in one plane (gray). The direction of the polypeptide containing amino acids in the trans configuration is from front to back. The torsion angles are labeled with the direction of positive rotation. The gray planar region arises as the two planes defined by the two atoms in the peptide bond and the preceding and following backbone carbons, respectively, indicated by the dark gray triangular areas (enclosed by dotted lines), have a ω torsion angle between them of 180° (trans) and thus are in the same plane. Rotation of 180° would give the cis configuration (0°), also putting them in the same plane. In contrast, the bond before (ψ torsion angle) and after (φ torsion angle) can have angles other than 0° or 180° as the preceding and following planes defined by the triangular areas (enclosed by dotted lines) can rotate relatively freely compared to the planar peptide bond.

1.3.2 RAMACHANDRAN PLOT

Figure 1.6 shows Ramachandran plots for the amino acids of two proteins: one protein contains a high α-helix content (a), and the second protein contains a high β-strand content (b). Most amino acids have combinations of the torsion angles in the energetically most favored positions (the darkest areas), with some amino acids having torsion angles in allowed (gray) or generously allowed (lighter gray) positions, and there are even a few amino acids with torsion angles in positions unfavored (white areas) from an energetic standpoint. Two major regions in which most amino acids are located have negative φ angles (−170 to −50°) and ψ angles in the range

(a)

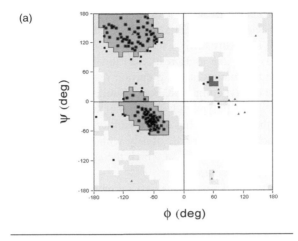

ϕ (deg)

(b)

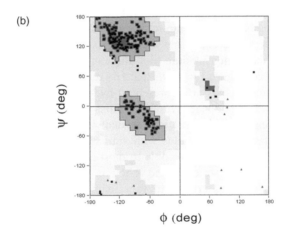

ϕ (deg)

FIGURE 1.6 Ramachandran plots showing the preferred and allowed combinations of the torsion angles (ψ, ϕ) for the positions of the amino acids of (a) the Rapamycin-associated protein (1FAP) and (b) a mutant of the green fluorescent protein (1YFP). The four-character alphanumeric character in brackets is the identifier for that protein in the PDB. Preferred regions for the torsion angles are given in dark gray, with allowed and nearly allowed regions given in light gray and very light gray, respectively, whereas nonallowed regions are given in white. The position of the combination of torsion angles for each amino acid in the protein is given by a square except for Gly residues, which are represented by triangles. Note the preponderance of Gly residues in the less preferred regions.

of -60 to $+20°$ or extending from $+100$ to about $+180°$. These two most favored combinations of angles (corresponding to minimum energy) are the central locations for amino acids in the right-handed α-helix ($\psi = -47°$, $\phi = -57°$) and in the β-strands ($\psi = -119°$, $\phi = +113°$ or $\phi = -139°$, $= +135°$ in parallel and antiparallel β-sheets, respectively) described in the following text. In this regard, it is evident that the protein at the top has a higher proportion of its amino acids in α-helixes, whereas the protein at the bottom has a greater proportion of its amino acids in β-strands.

Amino acids in proteins can fall outside this range, particularly between the preferred locations for amino acids in the α-helix and β-strand, in which the unfavorable steric interactions are still relatively low.

The amino acids with the most restricted torsion angles are Val, Ile, and Pro. For Val and Ile, the bifurcation at the β-carbon results in greater opportunities for steric overlap with the polypeptide backbone. Similarly, the cyclic ring of Pro results in closer contact with the polypeptide backbone, and the favored angle of −60° of the N-CH$_\alpha$ bond of the cyclic ring has less flexibility. It is important to note, however, that all the amino acids have some flexibility with respect to their ψ and φ angles, even those in α-helix and β-strands, in which the combination of ψ and φ angles is repeated throughout the structure.

A third region showing some preference for amino acids is located with positive ψ and φ angles in the upper-right quadrant of the Ramachandran plot. Although a number of amino acids have this combination of torsion angles, which are the angles expected for a left-handed helix (ψ and φ ranging from +50 to +60°), an extended left-handed α-helix of more than one turn has not yet been detected in proteins. As the amino acid side chains contribute to these unfavorable steric interactions, Gly, which does not have a side chain, has the least restrictions on its combinations of ψ and φ angles in the Ramachandran plot and can more readily exist in different conformations. Indeed, Figure 1.6 shows that Gly residues (represented by triangles, whereas all other amino acids are represented by squares) account for about 50% or more of amino acids outside the favorable regions and for even a higher percentage in the unfavorable regions. This result is consistent with Gly being at the most highly conserved sites in families of proteins with similar structure, due to the ability of Gly to assume configurations inaccessible to most other residues that are necessary for the enzyme to retain its structure and function. Other highly conserved sites in an enzyme are the residues critical to functioning in the active site, including nucleophilic residues taking part in the catalytic reaction.

Other amino acids, however, can still have positive angles, but their torsion angles are generally centered about the location expected for amino acids in a left-handed helix. Aside from Gly, a relatively high proportion of the few amino acids with positive values are Asp, Asn, Glu, and Gln. The presence of a nucleophilic amino acid with an unfavorable ψ and φ set of angles, and thus under a relatively unfavorable energetic strain to fold into this conformation, may at times indicate that it is involved in a key catalytic step.

An excellent way to view the relative locations of amino acids in the Ramachandran plot is to enter the protein data bank (PDB) site on the Internet and then select a specific protein for analyses. Select Geometry and then Ramachandran Plot, and then enter the Interactive Ramachandran Plot, in which it is possible to locate the positions of each type of amino acid in the plot for the protein being analyzed.

Only two structures with repeated ψ and φ angles are commonly found in proteins: the right-handed α-helix and the β-pleated sheet. As enzymes are generally relatively compact structures and α-helixes and β-strands extend in a linear fashion, it is clearly necessary that the polypeptide turn back across the protein at the ends of each α-helix and β-strand so that a compact structure can be obtained. The reverse turns were first recognized in antiparallel β-sheets and often are referred to as β-

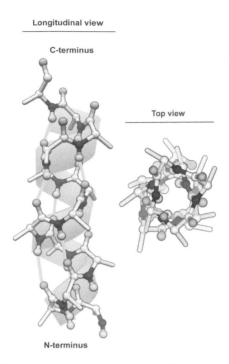

FIGURE 1.7 Longitudinal and top view of an α-helix. Dark atoms are nitrogen, and gray atoms are oxygen. Only the hydrogens on the nitrogen are indicated. Hydrogen bonds are given by the gray lines. Note that the peptide bonds are perpendicular to the helix axis and the side chains (represented by the straight bond) point away from the helix and back towards the N-terminus of the α-helix.

turns. As a rough estimate, about 25 to 30% of the residues in proteins are present in each of α-helixes, β-strands, and reverse turns or loops, with the remaining 10% being unclassified or in random coil-type configurations. Consequently, a clear understanding of the basic properties of α-helixes, and β-strands and β-sheets and reverse turns provides a solid basis for recognizing the structure of all enzymes.

1.3.3 α-Helixes

Figure 1.7 gives the side and top views of an α-helix. All α-helixes in proteins are right-handed, analogous to a right-handed screw, with torsion angles of amino acids in actual helixes in proteins varying about the highly favorable ψ and φ angles of (–57°, –47°) for an ideal α-helix. These repeated values of the torsion angles allow for the optimal formation of hydrogen bonds parallel to the helix axis, from the carbonyl of the nth amino acid to the NH of the $(n + 4)$th amino acid (as shown in Figure 1.7), running from bottom to top. The direction of the polypeptide chain is thus important in defining the position of the peptide-backbone hydrogen bonds. Because all carbonyls point towards the carboxyl terminal, and there is a partial negative charge on the carbonyl and a partial positive charge on the imide (see Figure 1.3), the sum of these small dipoles leads to a charge dipole along the helix axis with a net charge of about +0.5 e.s.u. near the amino end of each helix. This positive charge may often be influential in interactions with negatively bound substrates or cofactors when the amino end of the helix is located near the active site.

All amino acid side chains point toward the outside of the helix, as well as slightly back towards the amino terminal as depicted in Figure 1.7, in which all side chains are represented as a methyl group (i.e., as Ala). One helical turn requires 3.6 residues and, consequently, each amino acid results in a rotation about the helix of 100°. Depending upon the properties of the amino acids in the helix, the external surface of the helix could be hydrophobic, suggesting that it lies in the interior of a protein or in a membrane. Alternatively, it could be all polar, suggesting that it is exposed completely to solvent, or it could be amphipathic with one side being hydrophobic and the other side polar, suggesting that one side is buried and the other exposed. By plotting the type of amino acid (polar or hydrophobic) on a circular plot, designated as an helical or Edmundsen wheel, the hydrophobic or polar prop-erties of the sides of an α-helix can be recognized, indicating the type of environment in which the helix would reside in the protein.

The length of the helix is extended in the longitudinal direction by 1.5 Å for each amino acid, or 5.4 Å for each turn. As the width of most compact folded proteins is in the range of 30 to 40 Å, most helixes will not extend more than 20 residues (30 Å) before changing their direction; otherwise, they would extend out into solution and could not interact with other amino acid residues in the protein. It should be noted, moreover, that most helixes also have a slight twist and thus are not linear.

1.3.4 β-Sheets

Amino acids in β-strands forming part of β-sheets have repeated ψ and φ angles located in the upper-left quadrant of the Ramachandran plot at the most favorable energy. Two types of β-sheets can form: antiparallel and parallel (Figure 1.8), with idealized ψ and φ angles of (–139°, 135°) and (–119°, 113°), respectively, for the amino acids in the β-strands.

Hydrogen bonds form between the peptide NH and CO groups of amino acids on different β-strands, with their organization dependent on whether the strands are parallel (running in the same direction) or antiparallel (running in the opposite direction). In the parallel β-sheet, the NH and CO groups of one amino acid form hydrogen bonds with the corresponding CO and NH groups of two different amino acids in a parallel strand separated by one amino acid. In the antiparallel sheet, the NH and CO groups hydrogen-bond with the respective CO and NH groups of the same amino acid on an antiparallel strand. The antiparallel sheet is slightly more stable than the parallel β-sheet and, consequently, smaller β-sheets with fewer β-strands will more often be found to be antiparallel than parallel. Moreover, β-sheets, just like α-helixes, are often twisted with greater distortion for the antiparallel compared with a parallel β-sheet, as illustrated in Figure 1.8. Mixed β-sheets also occur quite often with various combinations of antiparallel and parallel strands.

The amino acid side chains extend alternately above and below the β-sheets, and the sheet is not flat but pleated, with the positions of the residues in the β-strands

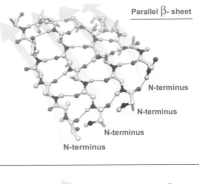

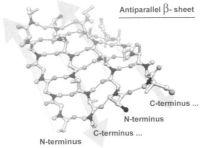

FIGURE 1.8 Parallel and antiparallel β-pleated sheets. Dark atoms are nitrogen and gray atoms are oxygen, with only hydrogens on the nitrogens being depicted. Side chains are represented by the straight bond and are found alternately on each side of the β-strands, but not exactly at 180° as the positions of the atoms are not for an idealized β-sheet but are taken from the coordinates of crystallized proteins. The hydrogen bonds between the β-strands are indicated by gray lines. Note that the atoms come in and out of the plane in the three-dimensional structure and the structural positions in the β-strand are repeated for every second amino acid. A clear twist in the β-sheet can readily be recognized in the antiparallel β-sheet.

being repeated every two residues. Consequently, this structure is often referred to as a β-pleated sheet. In Figure 1.8, the side chains are represented as methyl groups (i.e., as Ala), and the coordinates for the β-sheets have been taken directly from the structures of specific proteins and thus vary to some degree from the locations of atoms in an idealized β-sheet. Each side of the β-sheet can be analyzed for its hydrophobic or polar properties by considering the nature of alternate amino acids, analogous to analyzing an α-helix for the hydropathic properties of its amino acids on a helical wheel. Consequently, one face of a β-sheet could be primarily hydrophobic and the other could be polar, indicating that one side is buried and the other exposed to water, or both sides of the β-sheet could have similar polarity or hydrophobicity. Such β-sheets can stack one on top of the other with the primarily hydrophobic faces interacting with one another.

The length of a β-sheet, just like that of an α-helix, should not extend much more than 30 to 40 Å. As the β-sheet is extended by about 3.2 Å (3.1 Å for parallel and 3.3 Å for antiparallel β-strands) per amino acid, most β-strands will not be much longer than 10 residues. Similar to α-helixes, β-strands are often twisted away from a linear structure, sometimes with a curvature exceeding 20° or more per residue.

1.3.5 REVERSE TURNS AND LOOPS

As described above, the α-helix and β-strand generally must turn after extending 30 to 40 Å at the most. The minimum number of amino acids required for such a turn is four, unless a large energetic strain is introduced due to the amino acids in the turn assuming more unfavorable torsion angles. Of course, turns with an even greater number of amino acids exist.

The structure of reverse turns with four amino acids has been reasonably well defined, with different combinations of preferred torsion angles existing for the second and third amino acids in the turn. The common property of these four amino acid reverse turns is a hydrogen bond from the CO of the first amino acid to the NH of the fourth amino acid (i.e., from the nth to the [n + 3]rd amino acid, extending from the amino terminal to the carboxyl terminal). The three most common types of turns (I, II, and III) have ψ and φ torsion angles of (–60°, –30), (–60°, 120°), and (–60°, –30°) for the second amino acid and (–90°, 0°), (80°, 0°), and (–60°, 30°), respectively, for the third amino acid in the turn. As the torsion angles for the third amino acid of the Type II turn are highly unfavorable, a Gly residue must be at this position. In addition, Pro is often in the second position because the preferred position for its ψ and φ angles are –60° and +150°, respectively. Four other residue turns have been classified (IV, V, ...), and the mirror-image turns (with the same torsion angles as Type I, II, etc., except for multiplication by –1) also occur with a reasonable but lower frequency. In many instances Gly must be present in these mirror-image turns at the second or third amino acid or both because the torsion angles are too unfavorable to accommodate amino acids with side chains. As the torsion angles in these turns can deviate to a reasonable degree from the preferred angle, it is more difficult to classify the turns than the α-helixes or β-strands, in which the torsion angles are repeated over a number of amino acids.

1.3.6 PREDICTION OF α-HELIXES, β-SHEETS, AND REVERSE TURNS IN PEPTIDE SEQUENCES

Predictions of whether a certain sequence will form an α-helix, β-strand, or reverse turn can be made based on the frequencies of the different amino acids in the respective structures in the crystal structures of proteins. The differences in preferences of amino acids for a particular structure are generally not large and arise primarily due to their different capabilities in assuming appropriate torsion angles. In α-helixes, Glu, Leu, and Ala are found about 30 to 40% more frequently than predicted simply on the basis of amino acid composition. Similarly, Val and Ile are found about 40 to 50% more frequently in β-strands than expected, presumably due to their more restricted torsion angles arising from the bifurcation at the β-carbon of the side chain. For reverse turns, Gly and Pro are quite favored, being found with almost twice the expected frequency, whereas Ser, Asp, and Asn are found 30% more frequently than predicted by amino acid composition. By adding up the probabilities of the amino acids being present in the different structures over a short sequence range (six to ten amino acids), predictions can be made along the entire polypeptide chain about the type of structure that would be favored at any specific sequence in the folded protein.

1.3.7 PREDICTION OF THE HYDROPATHY OR POLARITY OF PEPTIDE SEQUENCES

Analogous to the prediction of the type of structure of a polypeptide, the probability of a given sequence being in a hydrophobic or hydrophilic microenvironment can be deduced from the relative hydropathy of the amino acids (see Table 1.1) in the sequence. By adding the relative hydrophobicities or hydrophilicities over short sequences (six to ten amino acids), the probable microenvironment of that sequence can be predicted. Consequently, a sequence rich in nonpolar amino acids would be in a hydrophobic environment, whereas sequences rich in polar amino acids would be in a hydrophilic environment. The relative order (Table 1.1) as well as the relative weight one gives to different amino acids is quite variable depending on what specific scale is used from the literature to generate a hydropathy plot. Alternative analyses of the hydropathy of different sides of α-helixes or β-strands are described in Section 1.3.3 and Section 1.3.4.

1.4 FOLDING OF THE PROTEIN INTO SPECIFIC CONFORMATIONS

1.4.1 TERTIARY STRUCTURE

The folding of a polypeptide into its three-dimensional structure involves balancing a number of negative and positive forces. Negative forces primarily involve the loss of entropy by the polypeptide backbone and its amino acid side chains on forming the folded protein conformation as well as on the formation of some less favorable torsion angles. Positive forces involve the formation of hydrogen bonds, hydrophobic attractions, electrostatic bonds, and van der Waals contacts. The exact contribution of each of these large forces is not well defined for any protein, but the net result clearly leads to a final conformation with a negative free-energy stability, often estimated to be in the neighborhood of −10 kcal.

The final structure is determined by the specific amino acid sequence. As an almost infinite number of torsion angle combinations is theoretically possible for even a small protein, a process testing all possible combinations of torsion angles would take too long. Thus, it is clear that the folding of any protein must follow a pathway in which only a limited number of conformational intermediates are formed during the folding process. Much research has been conducted to recognize the key and initial intermediates in the protein-folding pathway; however, relatively little progress has been made due to the extreme difficulty in detecting unstable intermediates. A folding pathway involving an initial step consisting of formation of a proximal secondary structure element (e.g., short α-helixes or β-strands or both) followed by condensation of these elements by interaction of the side chains (e.g., by forming hydrophobic bonds) has been proposed. An alternative pathway could involve interaction of specific amino acid chains (perhaps by hydrophobic bonds that, in turn, stabilize the formation) and interaction of local secondary structural elements. This unstable structure then could form a nucleus to help stabilize the formation of other secondary structural elements and, eventually, lead to the final conformations.

Although the final conformation is determined by the primary structure of the protein, other elements can influence the rate of the process and the yield of the folded protein. Three major factors that come into play in the folding process of some proteins are protein disulfide isomerase, Pro cis or trans isomerase, and the chaperones. Protein disulfide isomerases catalyze the shuffling of disulfide bridges, thus eliminating incorrect disulfide bonds, whereas Pro cis or trans isomerases increase the rate of the cis or trans isomerization of peptide bonds. Chaperones are proteins found in prokaryotic and eukaryotic cells that stabilize proteins in a partially unfolded state, preventing nonspecific aggregation and providing the opportunity for the protein to fold correctly, thus increasing the efficiency of protein folding.

1.4.2 QUATERNARY STRUCTURE

Most enzymes are polymeric rather than monomeric and thus contain multiple copies of the polypeptide subunits. Proteins containing one type of polypeptide are referred to as homopolymers, whereas those containing more than one type of polypeptide are referred to as heteropolymers. Oligomeric proteins are homopolymers that contain identical subunits, where a subunit is defined to be simply part of a larger molecule and may or may not contain more than one polypeptide. Consequently, hemoglobin with a structure of two α and two β polypeptides ($\alpha_2\beta_2$) is an oligomer as it contains two identical $\alpha\beta$ subunits. It is also correct to state that hemoglobin contains four subunits composed of two α and two β polypeptides.

The most common type of polymeric structures are dimers and tetramers. In *Escherichia coli*, dimers and tetramers account for 38 and 21%, respectively, of a set of proteins corresponding to about 10% of the proteins in this bacterium (Table 1.2). Monomers account for 19% of the proteins, whereas polymeric proteins, including multienzyme complexes, account for the remaining 81% of the structures analyzed. Of these proteins, 79% are homopolymers (including monomers), whereas 21% are heteropolymers. Because of the greater ease in analysis of simpler proteins leading to the greater availability of their structural and subunit data, it would be expected that the relative numbers of higher-order polymeric proteins would be somewhat higher for the complete set of *E. coli* proteins. Moreover, the relative percentage of heteropolymers would also be expected to be higher as different protein subunits held together by weak interactions in the cell may be dissociated upon extraction (and dilution) from the cell. It should be noted that the concentration of proteins in eukaryotic and prokaryotic cells is in the range of 100 to 150 mg/ml, whereas most proteins are extracted into relatively dilute solutions (< 5 mg/ml). Consequently, protein interactions in the cell may not be detected on analysis of the extracted proteins unless the subunit interactions are strong.

TABLE 1.2
Subunit Composition of *Escherichia coli* Proteins

Subunits	Homopolymer (%)	Heteropolymer (%)
One	19	—
Two	31	7
Three	4	1.4
Four	17	4.3
Five–Eleven (Odd)	0.6	0.8
Six	5	0.3
Eight	0.8	1.6
Ten	0.3	0
Twelve	1.2	0.6
Twelve Plus	—	5
Total	79	21

Source: Data compiled from D. S. Goodsell and A. J. Olsen. (2000). *Annual Reviews in Biophysical and Biomolecular Structure*, 29, 105–153, for 372 of the proteins listed under *E. coli* in the Swiss-Prot Protein Sequence data bank.

The forces involved in forming a polymeric enzyme are the same as those that are required in forming the secondary and tertiary structure of the folded polypeptide. Folded protein subunits, for example, may have hydrophobic patches on the surface. By interaction of the hydrophobic patches from different subunits, a more stable polymeric structure is formed, with the hydrophobic area buried in the protein at the subunit contact sites. Polar interactions also contribute to the oligomerization of proteins.

The subunits of most enzymes are arranged in a symmetrical manner as such an arrangement results in closed subunit contacts and a specifically defined structure. The most common types of symmetry are cyclic and dihedral. Cyclic structures, designated as C_N, have a single N-fold axis of rotation and include all monomers (C1), dimers (C_2), and trimers (C_3), and a few higher-order structures. Dihedral structures, designated D_N, have 2N identical units related by one N-fold rotational axis and N twofold rotational axes. Tetramers are most often in dihedral (D_2) symmetry. Protein structures with a larger number of subunits are also found with dihedral symmetry. Shown in Figure 1.9 is a representative model for the assembly and structure of *E. coli* aspartate transcarbamylase (ATCase). This enzyme contains six catalytic (C) subunits and six regulatory (R) subunits composed of two catalytic trimers and three regulatory dimers. The catalytic trimeric subunits are bound together by interactions with the three regulatory dimers, which form a bridge from a catalytic polypeptide in one trimer to a catalytic polypeptide in the other trimer. ATCase has one rotational axis of threefold symmetry and three twofold rotational axes and, thus, has D_3 dihedral symmetry (it has 2N = 6 identical subunits composed of one C and one R polypeptide). In Figure 1.9, the axis of threefold symmetry can be viewed as coming directly out of the paper for the top view of the assembled

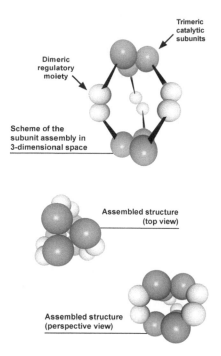

FIGURE 1.9 Subunit assembly and structure of *E. coli* aspartate transcarbamylase. The enzyme is composed of six catalytic polypeptides of 33 kDa (dark gray) and six regulatory polypeptides of 17 kDa (white). The catalytic polypeptides form trimer catalytic subunits that are bridged by three regulatory dimers. A small cavity between the two catalytic trimers in the assembled structure has been exaggerated for emphasis.

ATCase; a rotation of 120°, 240°, or 360° each gives the same structure. Similarly, there are three twofold rotational axes in which the catalytic polypeptide can be rotated 180° (i.e., from top to bottom), replacing one of the catalytic polypeptides and generating the same structure.

Other higher orders of symmetry exist, including cubic symmetries (octahedral, tetrahedral, and icosahedral) with additional rotational axes and those with rotational symmetries coupled to translational symmetries, allowing unlimited extension of the structure leading to helical and planar structures. Most of these higher-order symmetrical structures are found for storage, structural, and transport proteins and not for enzymes.

A number of reasons have been proposed for the preponderance of polymeric proteins and multienzyme complexes. Among these reasons are increased stability, reduction in contact with water as the relative surface area compared with the size of the protein decreases with increasing molecular weight, and the formation of structural elements needed in the cell. For enzymes, the creation of complexes allows substrate channeling from one subunit to another and the transfer of reactive intermediates that could be hydrolyzed in the aqueous environment. Allosteric regulation in which the binding or activity at one site affects the binding or activity at another site is clearly one major advantage of having oligomeric enzymes. Indeed, this is true for ATCase, in which the binding of the CTP inhibitor to the regulatory subunits affects the activity of the catalytic subunits.

1.5 POSTTRANSLATIONAL MODIFICATION

Although proteins are synthesized in biological organisms from condensation of only 20 amino acids (as well as selenocysteine) during the translation of mRNA, once folded into a three-dimensional structure, they can readily be modified. Formation of disulfide bridges between two Cys residues in close proximity is one simple modification that occurs in some proteins, leading to cross-linking of the polypeptide chains. In other instances, the polypeptide chain may be cleaved by proteolytic enzymes. Generation of the shorter polypeptide hormones from larger proteins or activation of proteolytic enzymes often occurs by cleavage of the polypeptide chain. The amino terminus may be modified in proteins by acylation with formyl, acetyl, or tetradecanoyl groups, by methylation, or may be removed by aminopeptidases. Modification or cleavage at the carboxyl terminus can occur but is less common.

Glycosylation of proteins resulting in the covalent incorporation of oligosaccharides is quite common, particularly for membrane and secretory proteins. The most common saccharides found in glycoproteins are glucose, mannose, fructose, and galactose as well as the N-acetyl-derivatives of glucosamine, galactosamine, and neuraminic acid. These sugar units are found as part of complex and generally branched oligosaccharides covalently linked through an N-acetylgalactosamine group to a Ser or Thr residue or via N-acetylglucosamine to an Asn residue. Moreover, the glycoprotein may be heterogeneous as different combinations of sugars can be incorporated in different molecules at the same amino acid site.

Another common modification of proteins is phosphorylation. The most common residues to be phosphorylated are Ser, Thr, and Tyr; however, other residues including Glu, Lys, His, and Arg can be phosphorylated on occasion. Methylation of Lys, Arg, His, and Asp residues and acylation on nitrogen, oxygen, or sulfur can readily occur in some proteins. Oxidation of Pro to hydroxyproline and Lys to hydroxylysine occurs for a number of Pro and Lys residues in collagen, but this modification is generally not found in other proteins. Generation of a functional enzyme may require covalent incorporation of a coenzyme into the protein, as occurs with biotin and lipoic acid on Lys residues and phosphopantetheine on Ser residues (see coenzymes below), as well as in a few instances when FMN or FAD are covalently linked to His, Tyr, or Cys residues. Even more interesting is the complete generation of new functional groups from regions of the polypeptide without the addition of exogenous groups. In histidine carboxylase, a pyruvyl group is formed by internal cleavage of the polypeptide chain, which functions in a manner equivalent to pyridoxal phosphate, in effect giving the enzyme its own coenzyme coded by the gene. In the green fluorescent protein, a conjugated chromophoric system is produced by cyclization of a set of amino acids in the polypeptide chain, allowing for the adsorption of light between 350 and 450 nm and the fluorescence of blue-green light. It should be noted, however, that most of the posttranslational modifications are limited to only a few proteins or a small family of proteins. Only glycosylation, phosphorylation, peptide chain cleavage by proteolysis, and disulfide bridge formation are found on a more common basis.

1.6 STRUCTURAL CLASSIFICATION

As the number of crystal structures of proteins is rising rapidly with tens of thousands of protein structures listed in the PDB, families of proteins with related structures are starting to be classified according to the content of the secondary structural elements (i.e., α-helix and β-sheets) as well as the arrangement of the structural elements. Three major structural classes for proteins common to all schemes are proteins having primarily α-helixes, primarily β-strands, or a mixture of both α-helixes and β-strands. The latter class is often divided into proteins with primarily alternating α-helixes and β-strands and proteins with regions of both α-helixes and β-strands. Different classification schemes also include additional classes not always directly based on these secondary structure elements (e.g., membrane or small proteins). The two major classification schemes currently being used are designated as SCOP (Structural Classification of Proteins) and CATH (Classes/Architecture/Topology/Homologous Superfamily) and can readily be accessed on the Internet.

Simple combinations of the secondary structural elements (i.e., α-helixes and β-strands) are referred to as motifs or supersecondary structure and have long been used as a characteristic trait to help assist in recognizing different families of proteins, including specific types of functions. For example, a major class of dimeric regulatory proteins controlling transcription contains a helix–turn–helix motif that allows each subunit to bind to the major groove in DNA. However, due to the relative simplicity of these motifs and thus their presence in many proteins, identification of the function of the protein or its structural family is extremely difficult. Within the major classes of proteins, structural subclasses are being systematically recognized based on how multiple secondary structural elements are related in three-dimensional space. These large structural units, folds or motifs, are the central core of domains that are relatively compact folded units having more limited contacts with the remainder of the folded protein. Proteins may consist of one or more domains even if they consist of a single polypeptide chain. Conversely, domains may also consist of more than one polypeptide although they are often formed by the folding of a contiguous stretch of one polypeptide. Recognition of structural domains consisting of combinations of multiple α-helixes and β-strands formed in an organized geometry in three-dimensional space has led to significant advances in the classification of proteins and thus in determining their potential function.

The most famous of these structural units is the Rossman fold found in most, but not all, of the very large group of dehydrogenases using NAD(P)(H) as a substrate. The Rossman fold itself consists of two linked β-α-β-α-β motifs in which the three β-strands are parallel and the α-helixes are oriented in the opposite direction. This arrangement results in the binding of the adenine nucleotide portion of

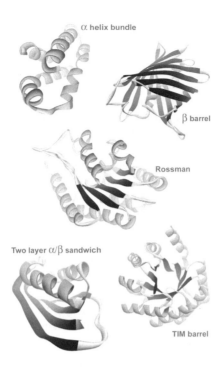

FIGURE 1.10 Structural folds for different domains. The folds after excision of other structural regions were taken from different proteins given in the PDB as follows: α-helix bundle, human FK506 binding protein Rapamycin-Associated Protein (1FAP); β-barrel, yellow version of green fluorescent protein (1YFP); Rossman fold, platelet-activating factor acetylhydrolase (1WAB); two-layer α/β sandwich, bovine testis acylphosphatase (2ACY); and TIM barrel, *E. coli* neuraminate lyase (1NAL). The β-strands are generally given in darker gray than the α-helixes.

NAD(P)(H) to the first β-α-β-α-β motif and the niacin nucleotide portion to the second β-α-β-α-β unit. The Rossman fold–containing dehydrogenases would thus be classified as proteins with alternating α-helixes and β-strands, a group that includes a large number of other enzymes with different functions. Variations in arrangement of the secondary elements in the Rossman fold, as well as the arrangement of the structural elements in the rest of the protein, then can be used to help recognize its specific function as a dehydrogenase.

The center of Figure 1.10 shows the Rossman fold found in the platelet-activating factor acetylhydrolase. Although some variation occurs in the organization, the formation, and even the number of secondary structural elements in the β-α-β-α-β motifs in different proteins, because of the twist to the β-strand, both α-helixes in any one β-α-β-α-β motif will be on the same side of the β-sheet. One of these motifs will have its α-helixes on one side of the β-strands, and the other unit will have its α-helixes on the opposite side of the β-strands. This structure has, thus, also been referred to as a β-sandwich and can clearly be recognized in the platelet-activating factor. Many other types of structures also exist that resemble sandwiches of the α-helixes and β-strands. A two-layer α/β sandwich present in the acylphosphatase from bovine testis is shown in Figure 1.10 (bottom left).

Many structures other than the Rossman fold contain alternating β-strands and α-helixes. The most common of these structures is the $(\beta\alpha)_8$ or TIM barrel, named after the enzyme triose phosphate isomerase in which the structural fold was initially identified. It has been indicated that up to 10% of the protein structures that have been currently identified have a TIM barrel. Shown in Figure 1.10 (bottom right) is the TIM barrel found in the enzyme N-acetyl neuraminate lyase. Here, β-strands alternate with α-helixes, with the β-strands forming a barrel on the inside and the α-helixes folding back over the outside. In the ideal case, all β-strands are parallel. The active site is always found at the carboxyl end of the β-barrel.

A number of proteins contain domains with primarily α-helixes or β-strands. A common structure in helix-only domains is the α-helix bundle, generally containing four α-helixes. The helixes are in a typically twisted conformation, in contact with each other often through hydrophobic interactions of the side chains. This type of fold is shown at the top left of Figure 1.10 for the FKBP–Rapamycin-associated protein (FRAP). FRAP is one of two proteins that interact with the potent immunosuppressant, rapamycin.

Structures with domains that contain primarily β-strands are also quite common. Shown in Figure 1.10 (top right) is the β-barrel taken from a variant of the green fluorescent protein in which the β-strands wrap around to form a barrel composed of an antiparallel β-sheet with a central cavity. For the green fluorescent protein, this buried region provides an ideal microenvironment for the amino acids in a short α-helix that enter the cavity to autocatalytically react to form a conjugated derivative that can absorb and then reemit light at a higher wavelength. By modifying the microenvironment and the interactions, both of which are based on the properties of the proteins, a series of sensors with different fluorescence, absorbance, and emission spectra has been developed, the great advantage of this system being that the chromophore originates from the genetic information coding for the amino acids and does not have to be supplied independently.

A list of structural motifs has been published on the Internet under the SCOP and CATH Websites. Development of a systematic and common nomenclature for the structural motifs will be very beneficial. Recognition of the enzymic function of proteins as well as different and related kinetic steps in the catalytic pathway, and relating those functional properties to the structural arrangement of the secondary structural elements (α-helixes and β-strands) is, and will be, extremely useful for application in determining the role of proteins whose functions are as yet unknown.

1.7   ENZYME CLASSIFICATION BY FUNCTION

An excellent systematic arrangement and nomenclature for enzymes has been available for many years and is being constantly updated. The nomenclature is based on the type of reaction catalyzed, with each enzyme being denoted by EC (Enzyme Commission) followed by four 1- to 3-digit numbers separated by a period (e.g., EC 1.1.1.1). The first number gives the six major classes of enzyme function: (1) Oxidoreductases (transfer hydride), (2) Transferases (transfer groups other than hydrogen), (3) Hydrolases (substrate cleaved by water), (4) Lyases (nonhydrolytic cleavage enzymes removing or adding groups to double bonds), (5) Isomerases (catalyze structural or geometric changes within one molecule), and (6) Ligases (synthetases joining two groups together, coupled with the breakdown of ATP or a similar triphosphate). The second number refers to a subclass, its meaning being dependent on the particular class (e.g., donor for EC1 and EC2, type of bond broken or formed for EC3, EC4, and EC6, and type of isomerism for EC5). The third number breaks the subclass into smaller groups (sub-subclasses), and the fourth number is the serial number of the enzyme in its sub-subclass. For example, N-acetylcholine esterase is EC 3.1.1.7, where the first number refers to hydrolysis by water, the second number to cleavage of esters, the third number for carboxyl esters, and the fourth number is the serial number for N-acetylcholine esterases in the EC 3.1.1 sub-subclass. Some of the sub-subclasses are now quite large, with EC 1.1.1 (oxido reductases [EC1] oxidizing CHOH groups [EC 3.1.1] with NAD (P)$^+$ as an acceptor [EC 3.1.1.1]) containing close to 300 different enzyme and serial number combinations.

1.8   ENZYMES AND ACTIVE SITES

Understanding the specific interactions of enzymes at the molecular level with the compounds taking part in the catalytic reaction provides the basis not only for the engineering of new enzyme functions for applications but also for the design of pharmacological inhibitors and alternative substrates that can be used for control and prevention of disease. As a large proportion of enzymes require cofactors for activity, a general knowledge of their interactions and properties is important to understand enzyme function and structure.

1.8.1   COFACTORS

Enzyme cofactors are nonprotein molecules required for optimal activity of the enzyme. These cofactors include simple inorganic molecules, in particular, cations such as Mg^{++}, Ca^{++}, Zn^{++}, Fe^{++}, and K^+, as well as more structurally complex organic molecules. This latter group of organic cofactors has been designated as coenzymes.

The function of the coenzyme is primarily to shuttle commonly used metabolic groups from one reaction or group to another. After a coenzyme accepts or donates a mobile group (e.g., hydride, acetyl group, methyl group, etc.), the original form of the coenzyme must be regenerated for it to undergo another catalytic cycle. If the coenzyme remains tightly bound to the enzyme, then the acceptance and donation of the mobile group onto the coenzyme must be catalyzed in place. In this case, the coenzyme is referred to as a prosthetic group. For enzymes that are deemed to have a prosthetic group, the enzyme form with the bound prosthetic group is referred to as the holoenzyme, whereas the corresponding unbound free enzyme is referred to as the apoenzyme. If the coenzyme readily dissociates and is released, and the original form of the coenzyme is then regenerated free in solution by another enzyme, then it would be classified as a cosubstrate. This nomenclature is actually somewhat confusing, as an enzyme could, by definition, not be a substrate, whereas a coenzyme could be a cosubstrate and, moreover, the same coenzyme could function as a prosthetic group with one enzyme and as a cosubstrate with another enzyme.

Coenzymes are derived from vitamins as well as from normal metabolic pathways. Many enzymes utilize coenzymes during the enzymic reaction and, consequently, common features of the binding sites for coenzymes can often be recognized in a diverse set of enzymes. Among the coenzymes, there is actually only a relatively limited number that are often found at the active sites of enzymes. Knowledge of the structures, common features, and properties of their binding sites is therefore of importance, as such knowledge can often be applied to understand the function and interactions at the active sites of different enzymes. Consider, for example, the large number of enzymes in the sub-subclass of EC 3.1.1.1 utilizing NAD (P)$^+$ as an acceptor and thus containing a binding site for this coenzyme.

Figure 1.11 gives the structures of some of the most common coenzymes. One group of very common coenzymes that have different functions but have some structural resemblances is that containing the adenosine moiety, including ATP itself.

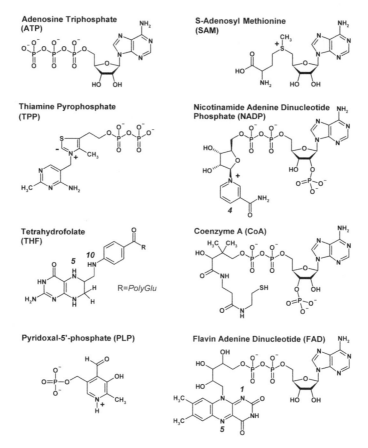

FIGURE 1.11 Structures of common coenzymes. Functional residues that are numbered on a few of the coenzymes are referred to in text.

ATP and other nucleoside triphosphates are involved in the activation of bonds by supplying an excellent leaving group (i.e., phosphoryl group), by either transferring the γ-phosphate or the AMP moiety. Four of the coenzymes listed in Figure 1.11 also contain adenosine. S-adenosyl methionine (SAM) is composed of methionine linked to adenosine and is involved in transfer of the methyl group linked to the sulfur of methionine. NAD (P)(H) has AMP linked to nicotinamide mononucleotide (NMN) in a pyrophosphate linkage and is involved in hydride acceptance and donation from the 4-position of the niacin ring. On transfer of a hydride to NAD (P)$^+$, the niacin ring is reduced and the positive charge eliminated. NADP (H) and NAD (H) only differ by the presence of a phosphate residue at the 2′ position of the ribose ring of adenosine in NADP (H). In most cases, the binding affinity of NAD (P)(H) primarily arises from interactions of the enzyme with the adenosine moiety and not with the NMN moiety. The large difference in specificity or NAD (H) and NADP (H) for different enzymes arises due to specific interactions with the 2′-phosphate moiety.

Coenzyme A is composed of phosphopantetheine (derived from vitamin B3) and AMP linked in a phosphoanhydride bond and is involved in the transfer of acyl groups linked as thioesters on the pantetheine. Phosphopantetheine is also involved in the same function as a prosthetic group linked as a phosphoester to a Ser of a small protein that transfers acyl groups (ACP, Acyl Carrier Protein). Flavine adenine dinucleotide (FAD) is composed of riboflavin phosphate (flavin mononucleotide, FMN), also linked in a phosphoanhydride bond to AMP. Both FAD (H2) and FMN (H2) are involved in hydride transfer from and to the nitrogen in the central ring (N5), along with transfer of a proton to the N1 position. Knowledge about the interaction of adenosine or AMP with enzymes and about the relevant protein structural features controlling binding is applicable to a wide range of proteins interacting with ATP, NAD (P), FAD, SAM, and Coenzyme A.

Other common coenzymes listed in Figure 1.11 are thiamine pyrophosphate (TPP) i.e., vitamin B1, tetrahydrofolate (THF), and pyridoxal phosphate (vitamin B6). TPP is involved in the transfer of two carbon units linked to the negatively charged carbanion in the thiazoline ring. Tetrahydrofolate (THF) and other folate derivatives consist of a substituted pteridine ring linked to p-amino benzoic acid which, in turn, forms an amide linkage with polyglutamate. The folates are involved in the transfer, oxidation, and reduction of the bonding of one carbon unit to the N5 or N10 position. Pyridoxal phosphate (PLP) is involved in the transfer of groups to and from the amino acids, with its carbonyl group able to readily form Schiff bases with different amino acids as substrates. PLP is often found as a Schiff base bound to a Lys residue, which can readily be displaced by amino acid substrates. Formation of the Schiff base allows for activation of bonds in the amino acid, the PLP serving as an electron sink leading to decarboxylation, or transamination or desaturation of the amino acid, depending on the particular enzyme. Coenzymes not shown in Figure 1.11 are various phosphorylated compounds including GTP, CTP, UTP, and UDP-galactose, as well as biotin, cobalamin derivatives (vitamin B12), and lipoic acid, and the less common coenzymes derived from the lipid vitamins A, D, E, K, and Q.

1.8.2   Enzyme Interactions with Substrates and Cofactors

The binding of ligands to the active sites of enzymes consists of multiple noncovalent interactions of the amino acid side chains and peptide backbone of the protein with the ligand and involves hydrophobic interactions, ionic and hydrogen bonds, and the formation of complementary surfaces to enhance van der Waals contacts. These contacts arise from many diverse locations in the primary structure that are brought together in three-dimensional space by the organization of the secondary structural elements of the protein to form the tertiary and quaternary structure. Given below are some examples of those interactions, illustrating only some of the many direct contacts between the enzyme surface and the ligand.

1.8.3   Tyrosyl tRNA Synthetase

Figure 1.12 shows the structure of the tyrosyl-tRNA synthetase with the tyrosyl-adenylate intermediate bound in the active site. The upper representation is a space-filling structure, with the negatively (red) and positively (blue) charged regions indicated on the surface, and represents most closely the structure of the enzyme. Evident in this structure are the crevices on the surface, with the partially buried intermediate (turquoise) bound at the active site. A more common representation showing the secondary structural elements with the α-helixes (red), the β-strands (blue), the reverse turns (green), and the loops (silver) is given in the center of Figure 1.12. This representation shows a twisted parallel β-sheet (in blue) lying under the substrate, allowing clear recognition of the structural motif. The β-strands are connected by α-helixes that turn back across the protein, forming a compact structure. This enzyme is part of the class of protein structures containing alternate α-helixes and β-strands. The central β-sheet clearly distinguishes it from many proteins in this class, including the large group of proteins containing a $(\beta/\alpha)_8$ barrel with the β-strands forming a barrel enclosed by alternate α-helixes. This type of representation is particularly useful in recognizing the secondary structural elements and structural motifs as shown in Figure 1.10; however, as the amino acid side chains are not given and as the space occupied by the backbone (based on their respective van derWaals radii) is not directly represented, it mistakenly gives the impression that there are large vacant areas in protein structures and does not show the direct molecular interactions of the enzyme and substrate and the close packing of the side chains in the molecule.

The lower picture in Figure 1.12 shows a close-up image of the active site of tyrosyl-tRNA synthase with the bound tyrosyl-AMP intermediate. Tyrosyl-tRNA synthetase catalyzes the reaction of tyrosine and ATP in the first step of the reaction, resulting in activation of the amino acid to form tyrosyl-AMP with the release of PPi. Without the acceptor, tRNA, for the final step of the reaction, tyrosyl-AMP remains very tightly bound to the active site, allowing crystallization of the enzyme-tyrosyl-AMP intermediate and determination of its exact position in the active site. The space-filling representation for both the tyrosyl-AMP intermediate and the enzyme allows us to readily see the substrate cavity and how close the contacts are between the enzyme and the bound intermediate. The space occupied in the protein

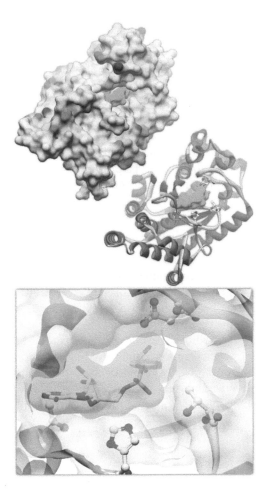

FIGURE 1.12 (See color insert following page 176.) Structure and active site of tyrosyl-tRNA synthetase containing bound tyrosyl-AMP. Top: space-filling structure of tyrosyl-tRNA synthetase with bound tyrosyl-AMP (turquoise) and the relative charge density on the surface indicated in blue (positive) and red (negative). Structural coordinates are from 3TS1 in the PDB. Middle: same structure represented by secondary structure elements; α-helix (red); β-strand (blue); and reverse turns (green). Bottom: close-up of active site using a space-filling model for enzyme and intermediate coupled with secondary structural elements and molecular structures for tyrosyl-AMP and side chains of His40, Thr45, Thr51, and Asp76 (nitrogen, blue; oxygen, red). The same side chains are also given in the center representation.

by adenosine, ribose, phosphate, and tyrosine can be readily followed in the structure of the intermediate from left to right, thus identifying locations for the interaction of specific atoms with the protein. In addition, the atomic structures of some key amino acids implicated in the binding of the substrates are also superimposed on the space-filling model for the protein. One Asp residue (Asp176) in the upper-right corner is in close contact with and forms a hydrogen bond with the hydroxyl moiety of tyrosine in the intermediate, whereas a Thr residue (Thr51) in the lower-left corner interacts with the oxygen in the ribose ring. Other residues, not shown, also help bind the substrates or intermediate or both. In addition two other residues, His40 and Thr45, are also depicted at the bottom-front of Figure 1.12 and assist in the formation of the transition state upon reaction of tyrosine and ATP. These residues are not in close contact with the tyrosyl-AMP but are at the end of the cavity that extends up from the phosphate. The separation of these residues from the intermediate can, perhaps, be more readily seen in the central plate in the color insert. These two residues are believed to form strong polar contacts with the γ-phosphate of ATP to stabilize the transition state and increase the rate of nucleophilic attack of the tyrosine carboxyl group on the α-phosphate of ATP, as mutation of these residues does not affect K_m or the binding of the substrates but increases the turnover rate of the enzyme.

1.8.4 HUMAN ALDOSE REDUCTASE

Aldose reductase preferentially catalyzes the NADPH-dependent reduction of hydrophobic and aromatic carbonyl compounds and also has the ability to reduce the more polar hexose sugars. As high levels of glucose are produced from hyperglycemia in diabetics, this enzyme has been implicated in complications arising from this disease.

Human aldolase reductase is a small monomeric protein and, in sharp contrast to the vast majority of dehydrogenases, the main structural fold is composed of a $(β/α)_8$ or TIM barrel rather than a Rossman fold (see Figure 1.10). The NADPH binding site is located at the carboxyl end of the β-barrel in accordance with the active site location of other enzymes with TIM barrels. The NADPH extends from its niacin moiety on the left to its adenine moiety on the right in Figure 1.13. The

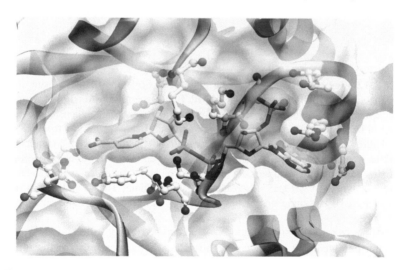

FIGURE 1.13 Space-filling structure of the active site of human aldose reductase containing bound NADP. Taken from the PDB (1ADS). Atomic structures for NADP and a number of key residues in the active site, as well as the secondary structural elements, have been overlaid on the space-filling model. Nitrogens and oxygens in key residues and NADP are dark gray to black.

tight packing of NADPH at the end of the β-barrel, as well as the molecular interactions of a number of residues with the coenzyme, are shown in Figure 1.13. Near the bottom left, Tyr209 stacks immediately below the niacin ring; Ser159 and Asn160 on the left side of the active site hydrogen-bond to the amide group of niacin, holding it tightly in place. The negatively charged pyrophosphate linking the two nucleotides is tightly bound by a number of residues, including the positive-charged lysine residues (Lys21 and Lys262) located just above the center of the coenzyme. Both the backbone NH and the side chain OH of Ser210 and Ser214 (at the bottom) also hydrogen-bond to the pyrophosphate on the opposite side to the two Lys residues. As the pyrophosphate is basically locked in the active site, a conformational change would be required for release of the NADPH by the motion of a loop between residues 213 and 217, which serves as a lid on the active site to bind the NADPH tightly. Arg268 forms an ionic bond with Glu271, both on the same side of the α-helix located immediately to the right of and in close contact with the adenine ring. The loop extending to the start of the same α-helix at the top right contains the residues giving this enzyme its specificity for NADPH over NADH, by binding the 2′-phosphate to Thr265 on top of this loop. The hydride is transferred from the C_4 position of the niacin ring on the left side of the active site, where there is a potential substrate (e.g., glucose) binding pocket that is highly hydrophobic but also contains polar residues that could assist in catalyzing the hydride transfer.

1.8.5 DIHYDROPTEROATE SYNTHASE

The active site of dihydropteroate synthase (DHPS) from *Staphylococcus aureus* is illustrated in the top of Figure 1.14. This enzyme catalyzes the reaction of *p*-aminobenzoic acid with 7,8-dihydro-6-hydroxymethylpterin pyrophosphate to form 7,8-dihydropteroate and pyrophosphate. The 7,8-dihydropteroate is the basic component of the folate coenzyme (see Figure 1.11). DHPS is the target of the sulfona-mide and sulfone derivatives that function as antibiotics, which are analogs of *p*-aminobenzoic acid and which can act as alternative substrates depleting the folate precursors, thus leading to the death of the bacteria. As resistance to sulfonamides is rising and as *Staphylococcus aureus* causes serious infections, analysis of the DHPS–substrate interactions is of interest in providing the basis for the development of new and better substrate analogs.

Figure 1.14 (top) shows a space-filling structure of the active site of DHPS bound to a close substrate analog of the hydroxymethylpterin pyrophosphate sub-strate (in yellow). The DHPS enzyme is a dimer with each subunit composed of a single domain and containing one active site. This enzyme is part of the alternating α/β structural class and has the topology of the classical $(\beta/\alpha)_8$ barrel. The active site of this enzyme is located at the carboxyl terminal of the central β barrel (see Figure 1.10), as found in other enzymes, and inspection of the right side of Figure 1.14 (top) shows the ends of the β-strands (in blue) forming the central β-barrel. The pterin ring is buried in the cavity; the pyrophosphate can be recognized (dark yellow) at the top of the cavity. A Mn^{++} ion (in bright pink) is bound through electrostatic interactions to the β-phosphate as well as to an Asn residue (Asn11) whose structure is depicted on the upper right. The β-phosphate interacts with

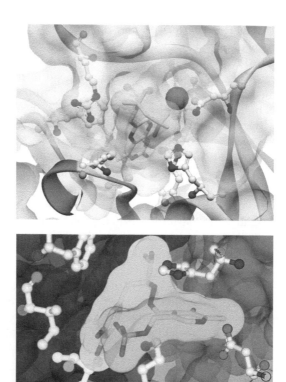

FIGURE 1.14 (See color insert following page 176.) Top: space-filling structure of the active site of dihydropteroate synthase bound with a close structural analog of the hydrome- thylpterin pyrophosphate substrate (yellow). Structural coordinates were taken from the PDB (1AD4). The molecular structure of the substrate analog is shown in the space-filling structure along with some of the amino acid side chains (Phe171, Arg52, Lys203, Arg239, His241, Asn11, Asp84, and Asp167) implicated in the active site (pyrophosphate, dark yellow; nitro- gen, blue; oxygen, pink). A bound Mn^{++} is shown as a bright pink sphere. Secondary structural elements are α-helix (red); β-strand (light blue); and reverse turns (green). Bottom: space- filling model of the active site of DOPA decarboxylase with bound pyridoxal phosphate and a DOPA analog, carbidopa. Structural coordinates are from 1JS3 in the PDB. Space occupied by the bound ligands (turquoise) contains pyridoxal phosphate in a Schiff base with the hydrazine of carbidopa. Parts of two identical subunits are shown for this dimeric enzyme, one in red and the other in purple. Molecular structures of Phe101 and Ile103 on one subunit and Thr82, Asp271, and Lys303 on the other subunit are shown (nitrogen, blue; oxygen, red; phosphorus, yellow).

Asn11, as well as with His241 (lower right) and Arg239 (immediately below His241), with the guanidinium group of Arg239 stacked over the pterin ring. The α-phosphate is bound in an ionic bond to Arg52 (top left). The pterin ring is also held in the cavity by polar and hydrophobic interactions. Three such polar interac- tions with the pterin ring are shown in Figure 1.14: Lys203 (bottom left), Asp167 (immediately right of Lys203 and below the surface), and Asp84 (below the surface and just above and to the right of Mn^{++}), whereas Phe172 (below Arg52) forms a strong hydrophobic contact. The multiple interactions involving hydrogen, hydrophobic, and ionic bonds provide a complementary interface for the tight binding of the substrate to the active site.

1.8.6 DOPA DECARBOXYLASE

DOPA decarboxylase catalyzes the conversion of L-3,4-dihydroxyphenylalanine (DOPA) into dopamine and L-5-hydroxytryptophan into serotonin using pyridoxal phosphate as a cofactor. Its role in synthesizing these key neurotransmitters has implicated DOPA decarboxylase in a number of physiological disorders including hypertension and Parkinson's disease, the latter thought to arise due to lack of dopamine. A primary treatment in Parkinson's disease is to provide DOPA into the bloodstream and block the DOPA decarboxylase with inhibitors so that DOPA will cross the blood–brain barrier before being converted to dopamine; direct treatment with dopamine is not possible as this compound will not pass the blood–brain barrier.

Figure 1.14 (bottom) shows the active site of DOPA decarboxylase containing a space-filling structure (turquoise) representing the size and shape of the bound pyridoxal phosphate and the dopamine analog carbidopa. DOPA decarboxylase is a homodimer with each of the subunits containing three domains. The major domain contains an α/β fold consisting of seven antiparallel and parallel β-strands in a mixed β-sheet surrounded by eight α-helices and encompassing the pyridoxal phosphate binding region. The active sites are located between the subunits, as shown in Figure 1.14 (bottom), with one subunit colored red and the second subunit colored purple. The turquoise region is labeled with the locations of the atoms of pyridoxal phosphate (front) and the carbidopa inhibitor (left and to the back). As this inhibitor differs from DOPA by having one extra methyl and a hydrazine rather than an amino group, the carbonyl of pyridoxal phosphate forms a Schiff base with the hydrazine rather than with the Lys303 located at the bottom of the active site.

Binding of the inhibitor and coenzyme arises through the cooperative forces of hydrophobic, ionic, and hydrogen bonds, coupled with the complementary structure of these molecules to the protein that allows optimization of the number of van der Waals contacts. Some of the key residues involved in binding the coenzyme and substrate are depicted. A His residue (H192, top right) stacks over the pyridoxal ring, providing a potentially strong binding interaction as well as hydrogen bonds to the carboxyl group of carbidopa. Asp271 (lower right) forms a strong ionic interaction with the protonated nitrogen on the pyridoxine ring of pyridoxal phosphate. On the lower left and just behind the active site, Thr82 hydrogen-bonds to the 3'-hydroxyl on the benzene ring of carbidopa. The phosphate is stabilized by a number of hydrogen bonds as well as by the positive charge at the amino end of an α-helix arising from its charge dipole. In this enzyme, the subunit interactions of the enzyme are closely connected to the binding of the substrates in the active site. Two hydrophobic residues, Ile101 and Phe103, are in van der Waals contact distance with the benzene ring of the carbidopa and are contributed by a different subunit (in purple) than the other residues shown in the active site.

1.9 MEASUREMENT OF ENZYME LIGAND
INTERACTIONS

1.9.1. INDEPENDENT BINDING SITES

The binding of a substrate (or other ligand) to an enzyme site, represented by the simple interaction of enzyme and substrate, $E + S \rightleftharpoons ES$, is given by the following equation, assuming that all the sites bind the substrate independently with a dissociation constant equal to K_d:

$$K_d = (E_f)(S_f)/ES = (n - r)E(S_f)/rE = (n\ r)(S_f)/r \qquad (1.1)$$

The terms E_f and S_f refer to the concentrations of unbound binding sites (not free E) and free ligand, respectively. For an enzyme containing n identical binding sites, as found in many oligomeric proteins, we can substitute $(n - r)$ E for E_f and rE for ES, in which r is the average number of bound ligands per enzyme molecule at any particular substrate concentration and E is the enzyme concentration.

Rearrangement of the equation leads to the following Scatchard equation:

$$r/(S_f) = n/K_d - r/K_d \qquad (1.2)$$

A plot of $r/(S_f)$ vs. r gives a straight line, as shown in the Scatchard plot (Figure 1.15), with the slope being equal to $-1/K_d$ and the intercept on the abscissa being equal to n, the number of binding sites for the substrate or ligand. The Scatchard plot provides an excellent method for analyzing the binding interactions of any two components, provided that all the sites act independently and have the same dissociation constant.

Binding analyses using the Scatchard plot require determination of the enzyme concentration and the amount of bound and free ligand. This can be relatively difficult as both the concentration of free and bound ligand in the same solution must be determined under conditions in which a thermodynamic equilibrium is maintained. One method involves equilibrium dialysis, in which the smaller binding partner (substrate) transfers freely through a dialysis membrane with the larger component (enzyme) remaining on one side of the membrane. The amount of ligand can now readily be measured on both sides of the membrane by using radiolabeled substrate or by deducing the absorption, fluorescence, or some other property of the ligand that is directly related to its concentration. The ligand concentration in the presence of the enzyme is equal to the sum of the concentrations of the free and bound ligand, whereas on the other side of the membrane, the ligand concentration corresponds only to that for the free ligand. Provided that the concentrations on the two sides of the membrane are significantly different, the concentration of the bound ligand can be determined and r calculated by dividing by the enzyme concentration. Measurements are then made at different substrate concentrations, and the values of $r/(S_f)$ vs. r plotted over the appropriate range. Other methods exist, including gel filtration and centrifugation, which also partially separate free and bound ligands but still allow thermodynamic equilibrium to be maintained between the enzyme and sub-

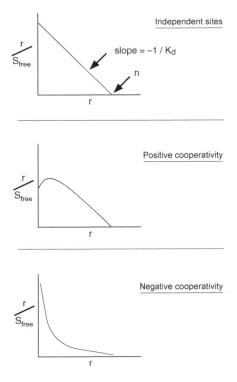

FIGURE 1.15 Scatchard plot for enzymes with identical binding sites for a ligand, in which the sites are independent or exhibit negative or positive cooperativity. The moles of a ligand bound per enzyme molecule (r) divided by the free ligand concentration (S_{free}) is plotted vs. r.

strate. These latter methods are also particularly advantageous for studying the interactions of large molecules (e.g., for enzyme–enzyme interaction), as the Scatchard equation can be applied to study the interaction of any two molecules.

There are also convenient methods to measure the fraction ($y = r/n$) of bound ligands. One of the simplest methods is to follow any perturbation of the spectral (absorption, fluorescence) or other properties of the enzyme or ligand on interaction. In most instances in which a spectral property is altered on interaction of the two binding molecules, a plot of the signal change can be used to obtain the dissociation constant. The signal change (Δx) will be proportional to the amount of bound ligand (ES), with the maximum change (Δx_{max}) corresponding to the total concentration of enzyme binding sites (nE). The term $\Delta x/\Delta x_{max}$ only gives the fraction of protein with bound ligand, ($y = r/n$), and thus it is not possible to determine n without independent data. Substitution of the bound and total sites by Δx and Δx_{max}, respectively, then gives a plot of Δx vs. $\Delta x/(S_f)$ with a slope equal to $-K_d$. If the bound amount is much lower than the free ligand concentration at all points, as occurs in many experimental sets, then the total substrate concentration (S) can be used instead of S_f.

$$K_d = (E_f)(S_f)/ES = (nE - ES)(S_f)/ES$$

$$ES = nE - K_d(ES)/(S_f)$$

$$y = ES/nE = r/n = S_f/(K_d + S_f) = \Delta x/\Delta x_{max}$$

$$\Delta x = \Delta x_{max} - K_d(\Delta x)/(S_f) \tag{1.3}$$

1.9.2 ALLOSTERIC BEHAVIOR — HOMOTROPIC INTERACTIONS

If binding at one site on the oligomeric enzyme affects the binding at another site, then the Scatchard plot will not be linear (Figure 1.15) and a plot of the fraction of bound ligand (y) vs. S will not give a hyperbolic plot (as shown in Figure 1.16) for enzymes with independent sites. These interactions between ligand binding sites may either increase or decrease the apparent binding of a ligand to a site, leading to positive or negative cooperativity, respectively.

This type of interaction has been called allosteric behavior to distinguish this effect from the interactions of two ligands at the same site. The interactions of identical ligands binding at two identical sites are called homotropic interactions, whereas heterotropic interactions refer to the interactions of different ligands binding at two different sites. For the allosteric enzyme ATCase in Figure 1.9, heterotropic interactions occur between the regulatory sites binding the CTP inhibitor and the catalytic sites binding the substrates, whereas homotropic interactions occur between the substrate-binding sites on the catalytic subunits.

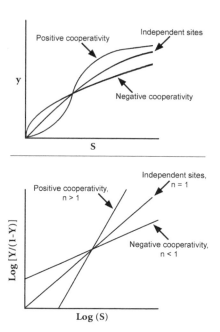

FIGURE 1.16 Plot of fractional saturation (y) vs. S (top) and the Hill plot (bottom) for enzymes with multiple sites for a ligand, in which the sites are independent or exhibit positive or negative cooperativity.

Positive cooperativity arises when the binding of the first molecule to an enzyme enhances the interaction with the second, with subsequent molecules effectively causing an apparent decrease in the dissociation constant for that ligand. Figure 1.16 shows that a plot of y vs. S for positive cooperative binding gives a sigmoid-shaped curve and that a plot of the bound vs. the free ligand in the Scatchard plot in Figure 1.15 will initially rise to a maximum and then decrease as the sites become saturated. There are two main models to account for positive cooperativity. The Monod–Wyman–Changeux (MWC) model of allosteric behavior proposed that there are two states of the oligomeric enzyme with different affinities for the ligand. As the initial molecules will primarily bind to the sites in the high-affinity state, the equilibrium between the low-affinity and high-affinity states will be shifted to the state with high affinity, in effect creating more high-affinity sites even though some of the sites will be filled. This increase in high-affinity sites results in an increase in the apparent affinity of the ligand for the enzyme, resulting in positive cooperative behavior. An alternative model, often referred to as the sequential or Koshland–Nem-ethy–Filmer (KNF) model, proposes that after the binding of the first ligand, the actual affinity at the second and subsequent sites increases, presumably due to a conformational change. The MWC model is both simple and elegant, and the number of parameters used to fit the data is relatively few. The KNF model uses different dissociation constants for each site on the enzyme and may give a closer fit of the data to the model simply because of the higher number of variable parameters.

The KNF model, however, is advantageous for understanding negative cooper-ativity, as the MWC model does not provide an explanation for this effect. Figure 1.16 (top) shows that for a negative cooperative enzyme, in a plot of y vs. S, the relative binding decreases at higher ligand concentrations compared with the hyper-bolic binding curve expected for an enzyme with independent binding sites. Simi-larly, the Scatchard plot for an enzyme exhibiting negative cooperative behavior between binding sites (Figure 1.15) will curve, with the slope decreasing as the enzyme becomes saturated. It is important to note, however, that this type of behavior is not only consistent with negative cooperativity but also with simple heterogeneity in enzyme preparations for ligand binding arising either from heterogeneous enzyme molecules or from heterogeneous sites within the same enzyme molecule or both. Indeed, this latter cause may be the reason for some of the examples of enzymes with negative cooperative behavior in the literature.

The relative degree of cooperativity between sites can be obtained by using the Hill equation. The value n in the Hill equation corresponds to a hypothetical number of bound molecules needed to be bound to the enzyme to give the degree of cooperativity for full occupation of sites in any molecule ($E + nS \rightleftharpoons ES_n$). Conse-quently, K_d would be given by the equation.

$$K_d = (E_f)(S)^n/(ES_n) \qquad (1.4)$$

Rearrangement and replacement of the dissociation constant by the affinity constant ($K_a = 1/K_d$) and substitution by y, the fraction of bound enzyme ($ES_n = yE$; $E_f = (1 - y)E$), gives the ratio of the bound enzyme to the free enzyme.

$$ES_n/E_f = y/(1 - y) = (S)^n/K_d = (S)^n(K_a)$$

$$\log (y/(1 - y)) = \log K_a + n \log S \tag{1.5}$$

A plot of the log of bound/free enzyme [log $(y/(1 - y))$ vs. log S] gives a slope corresponding to n. The substrate concentration given in the equation is generally the total substrate concentration, as most if not all analyses are conducted under conditions in which the amount of bound substrate is very low relative to the total substrate concentration. As binding is not completely cooperative, the plot generally curves. However, a relatively linear region can be obtained using data from the central region of the binding curve (from $y = 0.1$ to 0.9). Values of n will be above or below unity depending on whether there are positive or negative cooperative interactions between ligand binding sites, as shown in Figure 1.16 (bottom).

1.9.3 ALLOSTERIC INTERACTIONS BETWEEN TWO DIFFERENT LIGANDS — HETEROTROPIC INTERACTIONS

The interaction of one type of ligand at one site of an enzyme can affect the binding of a different ligand at another site. These interactions are referred to as heterotropic interactions, in contrast to homotropic interactions involving binding at only one type of site. Consider the most common cases (ordered and random binding), in which the enzyme interacts with two different ligands, A and B, but a reaction does not take place and an equilibrium can therefore be established. The apparent dissociation constant for A, K_a^*, is equal to the concentration of A times the sum of the concentration of all species without bound A divided by the concentration of all species with bound A, as indicated in the following equations for each set of pathways:

$$\begin{array}{ccc} & K_a & K_b' \\ E & \rightleftharpoons EA & \rightleftharpoons EAB \\ & A & B \end{array} \tag{1.6}$$

$$K_a^* = (E_f)(A)/(EA + EAB) = (E_f)(A)/(EA)(1 + B/K_b') = K_a/(1 + B/K_b')$$

$$\begin{array}{ccc} & K_b & K_a' \\ E & \rightleftharpoons EB & \rightleftharpoons EAB \\ & B & A \end{array} \tag{1.7}$$

$$K_a^* = (E_f + EB)(A)/(EAB) = EB(1 + K_b/B)(A)/(EB)(A)/K_a' = K_a'(1 + K_b/B)$$

$$\begin{array}{ccccc} & & EA & & \\ K_a & \nearrow & & \searrow & K_b' \\ & E & & EAB & \\ K_b & \searrow & & \nearrow & K_a' \\ & & EB & & \end{array} \tag{1.8}$$

$$K_a^* = (E_f + EB)(A)/(EA + EAB) = (E_f)(A)(1 + B/K_b)/EA(1 + B/K_b')$$
$$= K_a(1 + B/K_b)/(1 + B/K_b')$$
$$K_a^* = K_a (K_b'/K_b)(K_b + B)/(K_b' + B)$$

In all cases, the presence of the second ligand, B, affects the interaction with A. If A binds before B, then K_a^* will decrease from K_a to zero as B increases in concentration because this ligand simply pulls the equilibrium toward EAB, causing an apparent increase in affinity for A. If B binds before A, then K_a^* will decrease to K_a' as B is increased. For random binding of the two ligands, as B is increased, K_a^* will change from K_a to $K_a (K_b'/K_b) = K_a'$, either decreasing if B has a higher affinity for EA than E or increasing if B binds tighter to E than EA. The heterotropic interactions described here for the binding of two substrates to the enzyme also represent the effects of other ligands including metal ions, anions, cations, and even protons that bind at one site and affect the binding of substrate at another site. The effects of these heterotropic interactions on binding will also be more complex if there are homotropic interactions for either of the different types of binding sites.

1.10 SPECIFICITY, PROTEIN ENGINEERING, AND DRUG DESIGN

The specificity of an enzyme for different substrates is most often given by the term k_{cat}/K_m (see Chapter 3), reflecting the rate of turnover of the enzyme substrate complex (k_{cat}), and the term K_m, an apparent dissociation constant equal to the concentration of substrate needed to saturate 50% of the enzyme sites. Of these two terms, the major focus in drug design and protein engineering is on changes to enzyme ligand interactions reflected in the term K_m for substrates or the dissociation constant, K_d, for inhibitors and other ligands.

The interactions of ligands with enzymes result in multiple contacts with the protein, so that very little space is left between the protein and the ligand and only part of the ligand is exposed to the solvent. The close packing of enzyme and ligands is illustrated for the active sites of the tyrosyl-tRNA synthetase, dihydropteroate synthase, and DOPA decarboxylase enzymes in Figure 1.12 and Figure 1.14 as well as for aldose reductase in Figure 1.13. In fact, the packing is even closer than illustrated in these figures as the space-filling structures were generated taking into account only the van der Waals radii of atoms other than hydrogen. This close packing leads to strong bonding arising from the large number of van der Waals contacts due to the complementary surfaces, the formation of hydrogen and ionic bonds, and the hydrophobic interactions.

Due to the large number of bonding interactions, the specificity of an enzyme for different ligands can be extremely discriminating, thus distinguishing small differences in the structure of the ligand and leading to a range of K_ms or K_ds for ligand–enzyme interaction. In general, the active sites of most, but not all, enzymes can discriminate between very small changes in ligand structure. Isoleucine tRNA-synthetase, for example, binds Val about 100 times more weakly, indicating that the interaction of the enzyme with the δ-CH_3 of Ile creates the extra binding affinity.

Clearly, small changes in the ligands or the enzymes or both can cause dramatic changes in the interactions. Protein engineering involves mutagenesis of the amino acids, and rational design of the active site would thus often involve replacement of amino acids with amino acids of somewhat similar properties in order not to disturb protein–protein interactions or cause large conformational changes. Many of the properties to be considered during this process are given in Table 1.1 and Figure 1.2. Among the key properties is the actual architecture of the amino acid side chain as reflected in the volume and shape and the polarity of the side chain. Consequently, exchange of a Val side chain by Thr would have relatively little effect on spatial considerations but would change the polarity. Substitution of Val by Ile may also prove to be a reasonable mutation as polarity now would not be affected, and as Ile, like Val, has a bifurcated β-carbon; the extra methyl group might be accommodated in the active site at the end of the location of either of the β-methyl residues of the Val residue. However, substitution of Val by Leu may prove to be a drastic change, even though it has similar polarity, as it does not have a bifurcated β-carbon but a bifurcated γ-carbon and thus differs in shape as well as in size. Although substitution of larger amino acids by smaller amino acids avoids any spatial restrictions, it can also clearly cause loss of ligand–enzyme interactions. By changing the properties of the enzyme and its interactions with substrates using protein engineering, the active site can be redesigned to affect the specificity and lead to the preferential accommodation of different substrates.

Drug design is the complementary face of protein engineering as the goal is often to design inhibitors that can interact strongly at the active site rather than to redesign the protein. However, even in this case, many of the same considerations that go into protein engineering must be taken into account with respect to specific interactions with the enzyme concerning the architecture (volume and shape) and the properties of the drug. In short, rational design of the active site to generate changes in enzyme–ligand interaction, be it protein engineering or drug design, requires understanding and balancing of the basic properties of the amino acid side chains and polypeptide backbone of the protein and the bound ligand.

ACKNOWLEDGMENTS

The scientific, computer, and design talents of Leo Yen-Cheng Lin and his generous contribution of time in the preparation of the figures are very much appreciated. Thanks also go to Marlene Gilhooly for preparation of the manuscript. Financial support from the Canadian Institutes of Health and Research and McGill University is gratefully acknowledged.

BIBLIOGRAPHY

Books

Fersht, A. (1999). *Structure and Mechanism in Protein Science: A Guide to Enzyme Catalysis and Protein Folding*, New York: W.H. Freeman.

Horton, H.R. (2002). *Principles of Biochemistry*, 3rd ed., Upper Saddle River, NJ: Prentice-Hall.

Internet

CATH Protein Structure Classification Database (produced by Pearl, F.M.G., Stillitoe, I., Dibley, M., Thorn, J. and Orengo, C.A.), http://www.biochem.ucl.ac.uk/bsm/cath_new/mdex.html.

Enzyme Nomenclature, Recommendations of the Nomenclature Committee of the International Union of Biochemistry and Molecular Biology on the Nomenclature and Classification of Enzyme-Catalysed Reactions (prepared by G.P. Moss), http://www.chem.qmul.ac.uk/iubmb/enzyme/.

ExPASy Molecular Biology Server, http://us.expasy.org/.

National Center for Biotechnology Information, http://www.ncbi.nlm.nih.gov.

PDB Protein Data Bank, http://rcsb.org/pdb/.

SCOP (Structural Classification of Proteins), http://scop.berkeley.edu/.

Journals

Brick, P., Bhat, T.N., and Blow, D.M. (1988). Structure of tyrosyl-tRNA synthetase refined at 2.3 Å resolution: Interaction of the enzyme with the tyrosyl adenylate intermediate. *Journal of Molecular Biology*, 208, 83–98.

Burkhard, P., Dominici, P., Borri-Voltattorni, C., Jansonius, J.N., and Malashkevich, V.N. (2001). Structural insight into Parkinson's disease treatment from drug-inhibited DOPA decarboxylase. *Nature Structural Biology*, 8, 963–967.

Hampele, I.C., D'Arcy, A., Dale, G.E., Kostrewa, D., Nielsen, J., Oefner, C., Page, M.G.P., Schonfeld, H-J., Stuber, D., and Then, R.L. (1997). Structure and function of the dihydropteroate synthase from *Staphylococcus aureus*. *Journal of Molecular Biology*, 268, 21–30.

Wilson, D.K., Bohren, K.M., Gabbay, K.H., and Quiocho, F.A. (1992). An unlikely sugar substrate site in the 1.65 Å structure of the human aldose reductase holoenzyme implicated in diabetic complications. *Science*, 257, 81–84.

2 Mechanisms

Bruce A. Palfey

CONTENTS

0-415-33402-0/05/$0.00+$1.50

2.1 INTRODUCTION

Enzymes use a number of strategies to increase reaction rates by many orders of magnitude over those of uncatalyzed reactions. These strategies will be described in this section. It is worth noting that although enzymatic catalysis has been studied for more than a century, the basis for the catalytic power of enzymes remains an unsettled question, still generating controversy and new research. In spite of the lingering (and new) controversies, significant progress towards understanding enzymatic catalysis has been made in the past century. This section will attempt to introduce the salient features of what is known and to provide a description of the newly emerging explanations for the catalytic power of enzymes.

Enzymes obey the standard laws of chemistry. That is, explanations of catalysis by enzymes conform to the chemical principles derived from the study of simpler molecules. However, because enzymes are large, complex, and highly organized molecules, it may be anticipated that they can apply normal chemical forces selectively to substrate molecules in ways not possible in the random and fluctuating environment of nonenzymatic solution-phase reactions. In a sense, then, an enzyme active site may be regarded as a solvent that is preorganized with the proper arrangement of reactive groups. We will examine the basis for enzymatic catalysis in terms of the ideas of physical organic chemistry. Therefore, we will first examine a few basic concepts that govern all chemical reactions.

2.2 BASIC CONCEPTS

2.2.1 THERMODYNAMICS

The basis for enzyme catalysis lies in controlling the energetics of the reaction. Because discussions of reaction energetics require the language of chemical thermodynamics, some key concepts are presented here. For an in-depth treatment, the reader should consult any introductory physical chemistry text. Chemical thermodynamics relates the equilibrium composition of reacting systems to energy changes. Energy, defined by physicists as the capacity to do work, takes many forms. In chemistry, the most important energetic quantity is free energy. Changes in free energy, abbreviated ΔG, control the extent and direction of a chemical reaction. The extent of a chemical reaction is quantitatively described by the equilibrium constant K, which is calculated by multiplying the equilibrium concentrations of all the products raised to the power of their stoichiometric coefficients, divided by the equilibrium concentrations of all the reactants raised to their stoichiometric coefficients. This is related to the molar free energy change for a reaction by

$$\Delta G^0 = -RT \ln K \qquad (2.1)$$

where R is the gas constant (1.987 cal mol^{-1} K^{-1}), and T is the absolute temperature. In a thermodynamically favorable reaction, most of the reactants will convert to products, and the equilibrium constant will be a large number. From Equation 2.1 it is evident that a high value of K corresponds to a negative value of ΔG; this is termed an *exergonic* reaction. If the value of $K < 1$, $\Delta G > 0$, and the *endergonic* reaction does not produce a high yield of products, it is unfavorable. ΔG itself is composed of two other thermodynamic quantities — enthalpy (ΔH) and entropy (ΔS). The enthalpy of a reaction in most biochemical applications is the heat given off or taken up. When the reaction is *endothermic*, heat is absorbed, and $\Delta H > 0$; when the reaction is *exothermic*, heat is liberated, and $\Delta H < 0$. At the molecular level (an interpretation that is not required by classical thermodynamics), the enthalpy of a reaction reflects the potential energy changes caused by breaking the bonds of the reactants and making the bonds of the products; most chemists think about reaction energetics in terms of enthalpy. However, the entropy changes in a reaction can be just as important in determining whether a reaction is favored.

Entropy is commonly described as a measure of disorder. At a molecular level, this means that as the number of energy states accessible to a molecule increases, the distribution of the population spreads throughout these states, resulting in more heterogeneity or disorder and a greater entropy. The fundamental thermodynamic quantities ΔG, ΔH, and ΔS are related by

$$\Delta G^0 = \Delta H^0 - T\Delta S^0 \tag{2.2}$$

It can be seen from Equation 2.2 that when $\Delta H < 0$ and $\Delta S > 0$, both favor a reaction (where $\Delta G < 0$), and that the influence of ΔH and ΔS on ΔG can be counteracting.

An important property of ΔG, ΔH, and ΔS is that they are *state functions*; that is, the differences in these thermodynamic parameters obtained for converting a reactant to a product do not depend on the route taken in the transformation. Therefore, the free energy change for converting a reactant to a product is independent of the reaction mechanism. A reaction may involve one intermediate, five intermediates, or none; the total free energy change, obtained by adding the value for each step, will be the same. This is equivalent to saying that the equilibrium constant of the overall reaction is unchanged regardless of the reaction pathway. The important result to enzymology is that enzymes cannot alter the equilibrium constant of a reaction; they can only accelerate the attainment of equilibrium. This result can also be arrived at by considering how a catalyst such as an enzyme appears in a chemical equation. Catalysts are not consumed in a reaction and therefore will appear on both sides of a balanced chemical equation and cancel from the expression for the equilibrium constant. While enzymes cannot influence the equilibrium composition of a reaction mixture, the reactions of molecules bound to enzymes may have very different equilibrium constants than their corresponding values in solution because the enzyme–substrate complex is a different molecule than the free substrate.

2.2.2 TRANSITION-STATE THEORY

Catalysis by enzymes (or other entities) is most often discussed in terms of *transition-state theory* (also known as *activated complex theory*). Transition-state theory formulates the rate constant for a reaction in terms of thermodynamic quantities pertaining to the reactants and the *transition state* — the configuration of atoms at the maximum in free energy as bonds of the reactants are broken and bonds of the product are formed. The methods of statistical thermodynamics and quantum mechanics can, in principle, relate the structures of the molecules to their corresponding thermodynamic properties. Therefore, transition-state theory offers the hope that rate constants may be predicted from first principles. Reasonable success in matching theory to experiment has been obtained for gas-phase reactions, in which simple molecules and a limited number of interatomic interactions allow the accurate application of quantum and statistical thermodynamic methods. Reactions in solution are significantly more complex because of the numerous interactions between solvent

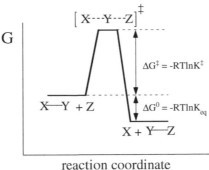

$$\left[\, \text{X---Y---Z} \,\right]^{\ddagger}$$

G

$$\Delta G^{\ddagger} = -RT\ln K^{\ddagger}$$

$$\text{X—Y} + \text{Z}$$

$$\Delta G^{0} = -RT\ln K_{eq}$$

$$\text{X} + \text{Y—Z}$$

reaction coordinate

FIGURE 2.1 Reaction coordinate diagram. The diagram represents the free energies (G) of reactants (X-Y and Z), products (X and Y-Z), and the transition state for their interconversion (X- - - Y - - - Z). The reaction rate is controlled by the height of the free-energy barrier between the reactants and the transition state ($\Delta G^{\ddagger}$), whereas the equilibrium constant depends on the difference between the free energies of the products and reactants (ΔG^{0}).

molecules and the reactants, making exact solutions to the mathematical problems impossible. Nonetheless, the concepts embodied in transition-state theory for gas-phase reactions are widely applied to solution-phase reactions in at least a qualitative way and serve as a useful framework for discussion.

Transition-state theory assumes that reactants are distorted along coordinates that will lead to the breakage of some bonds and/or the formation of new bonds. As bonds distort, the total energy of the system changes. Because the reaction starts with quasi-stable molecules, distorting the reactants requires energy. Eventually, the old bonds will be completely broken and new bonds will be formed, creating new (quasi) stable molecules — the products. Distortions in molecular geometry *en route* to the stable products will eventually liberate energy as the stable products are formed. Therefore, there must be some configuration of atoms between the reactant and product states in which the energy reaches a maximum. This state is the transition state (Figure 2.1). Because the transition state is at an energy maximum, it is inherently unstable and has a lifetime on the order of the time needed for a bond vibration, ~10^{-13} sec. This short lifetime precludes any hope of isolating a transition state or even observing it directly by classical methods. However, the transition states of some gas-phase reactions have been probed by femtosecond spectroscopy.

Transition-state theory makes a critical simplifying assumption — that the transition state may be treated as if all modes of motion are at thermal equilibrium. This allows the methods of statistical thermodynamics to be used to calculate the equilibrium constant between the reactants and the transition state. It is further assumed that the transition state decomposes to products with a characteristic frequency, making the rate of reaction equal to the product of this frequency (a very large number) multiplied by the concentration of transition states (a very small number). In deriving the rate expression, the reaction coordinate is treated as a harmonic oscillator with a frequency equal to the decomposition of the transition state and, after judicious approximations, this frequency cancels, leaving the equation

$$k = \frac{k_B T}{h} K^{\ddagger} = \frac{k_B T}{h} \exp\left(-\frac{\Delta G^{\ddagger}}{RT}\right) \tag{2.3}$$

where k is the rate constant of the reaction, k_B is the Boltzmann constant, T is the absolute temperature, h is Planck's constant, and R is the gas constant. $K^{\ddagger}$ is the pseudo-equilibrium constant between the reactants and the transition state, and $\Delta G^{\ddagger}$ is the free energy of activation (the free energy difference between the reactants and the transition state), which is calculated according to the standard relation between an equilibrium constant and a free-energy change. Equation 2.3 shows that as $\Delta G^{\ddagger}$ increases, the rate constant of the reaction decreases. This effect can be quite dramatic. For instance, a 1-kcal mol^{-1} change in the free energy of activation (smaller than the energy liberated by forming a hydrogen bond) causes a 5.4-fold change in the rate constant at room temperature. It should also be noted that the rate constant is highly sensitive to temperature. Temperature dependence is dominated by the exponential term, and it is rare that experimental data are accurate enough for the linear factor of T outside the exponential to be discernible. Finally, it should not be forgotten that this widely used expression rests on some curious (some would say dubious) assumptions, chief of which is the treatment of the transition state as a species at equilibrium with the reactants. At best, this cannot be considered a normal chemical equilibrium; at worst, some workers would argue that it is not valid. Nonetheless, Equation 2.3 has been successful in many applications and guides the thinking of most enzymologists who are trying to understand the origin of catalysis.

A thermodynamic interpretation of $\Delta G^{\ddagger}$ allows it to be split into its entropic and enthalpic components

$$\Delta G^{\ddagger} = \Delta H^{\ddagger} - T\Delta S^{\ddagger} \tag{2.4}$$

allowing the transformation of reactants to the transition state to be discussed using standard thermodynamic language. The enthalpy of activation ($\Delta H^{\ddagger}$) is usually positive (endothermic) because the quasi-stable bonds of the reactants are being broken as the transition state is attained, requiring energy, whereas energy is yet to be liberated from the formation of the bonds of the product. The free energy of activation is also influenced by the entropy of activation $\Delta S^{\ddagger}$. This quantity is generally interpreted in terms of the change in disorder between the reactants and the transition state, with $\Delta S^{\ddagger} > 0$ decreasing the value of $\Delta G^{\ddagger}$ and, therefore, increasing the reaction rate. For simple unimolecular reactions in the gas phase, the entropy of activation will be positive if the structure of the transition state is "looser" than that of the reactant — that is, certain vibrational or rotational potential energy wells become shallower in the transition state, causing the spacing of the quantized energy levels to decrease, thereby allowing more quantum states to become occupied. The entropy of activation is highly sensitive to the molecularity of the reaction, with higher-order reactions being disfavored. This is because there is a loss of translational and rotational freedom, i.e., an ordering of the system, when two or more molecules are brought together in a random collision to form a transition state.

The situation becomes more complex in solution because the solvent functions as more than a container for the reactants. When water, a very polar molecule, is the solvent, solutes are generally "coated" with a shell of water molecules oriented to favor their polar and hydrogen-bonding interactions. Thus, in unimolecular reactions in aqueous solution, the solvation shell generally reorganizes in order to allow the transition state to form; sometimes the rate of solvent reorganization is what actually controls the rate of reaction. For bimolecular (or higher-order reactions), the reactants are generally not able to react until they can make direct contact, which requires the desolvation of the two (or more) reactants. In this case, the reactants form an encounter complex that is in a solvent "cage." Within the cage, the reactants are able to collide frequently and eventually react. The involvement of solvent in solution reactions has effects on both $\Delta H^{\ddagger}$ and $\Delta S^{\ddagger}$, and these can be difficult to predict. Solvent reorganization may require the breaking of solute–solvent interactions, increasing the enthalpy of activation, but it may also free a reactive group from solvent shielding and allow it to interact more effectively with another functional group, decreasing $\Delta H^{\ddagger}$. The contribution of solvent reorganization to $\Delta S^{\ddagger}$ is equally difficult to predict. For instance, if a charge is generated in the transition state, the solvent surrounding the charge would become more oriented due to dipole interactions, lowering $\Delta S^{\ddagger}$; conversely, the dissipation of the charge in the transition state would allow a polar solvent to become less oriented, increasing $\Delta S^{\ddagger}$ and increasing the reaction rate.

The thermodynamic parameters of activation (or the parameters describing competing theories) are determined by measuring the temperature dependence of the rate constant. Usually, the logarithm of the rate constant decreases linearly with an increase in the reciprocal of the absolute temperature. From a classical Arrhenius analysis (actually introduced by van't Hoff), two parameters are obtained — the activation energy, E_a, and the Arrhenius prefactor or frequency factor, A. These empirical descriptions of the temperature dependence of a rate constant can be related to the thermodynamic interpretation of transition-state theory, giving

$$\Delta H^{\ddagger} = E_a - RT \qquad (2.5)$$

and

$$\Delta S^{\ddagger} = R \ln A - R \ln\left(\frac{k_B T}{h}\right) \qquad (2.6)$$

Note that the enthalpy of activation is not identical to the Arrhenius activation energy, although the distinction is frequently ignored in common usage. At room temperature, the difference is about 0.5 kcal mol^{-1}, which is close to experimental accuracy.

With this general background in mind, we will now examine some of the strategies and forces employed by enzymes to speed up reaction rates. Enzymes frequently reorganize a reaction by spreading the transformation over several steps, thus keeping the activation barriers of each step low. Enzymes may also speed up particular steps by bringing various forces (e.g., electrostatic and van der Waals interactions) to bear in a very organized way. Enzymes probably bring these forces to bear by stabilizing transition states, but may also promote quantum mechanical tunneling (see Section 2.4.10). Enzymes might also make covalent intermediates with prosthetic groups, forming intermediates that cannot otherwise be formed. Although much attention is deservedly given to the means by which enzymes accelerate reactions, it is also important to note that a critical function of enzymes is sometimes to slow or prevent side reactions that would lead to unwanted products if left unchecked. An enzyme's ability to suppress particular reactions is sometimes referred to as *negative catalysis* and is important for a wide variety of enzymes such as aspartate aminotransferase, aminoacyl-tRNA synthetases, triose phosphate isomerase, flavoprotein hydroxylases, and cytochrome P450s, to name just a few examples.

2.3 CATALYTIC STRATEGIES

2.3.1 CHANGE OF MECHANISM

Enzymes speed reactions compared to their rates in solution. Because the rate of a reaction depends on its mechanism, enzymes must alter the mechanism or its energetics during catalysis. Changes made to reaction mechanisms by enzymes may be classified into three groups:

1. Enzymes change the molecularity of many reactions. Although the uncatalyzed reaction of two substrates would be a bimolecular reaction, after the substrates bind to the enzyme, their reaction becomes unimolecular.
2. Forces within the active sites of the enzymes cause the energy barrier for a particular reaction to be lower than that in solution. A number of strategies that accomplish this have been identified that take advantage of the prepositioning of important protein residues.
3. Enzymes often divide a reaction into steps with several intermediates. Doing less chemistry at once keeps the activation barriers lower. Each of the reactions in a sequence is likely to be accelerated by organized functional groups in the active site, mentioned in group 2. It appears that there are limits to the amount of catalytic machinery that can be packed into the space of one active site. Thus, for complicated reactions (and even some surprisingly simple ones), conformational changes in the protein, the substrate, or a prosthetic group are used to move catalytic components to and from the site of chemistry when needed.

When the chemical transformation is beyond the capability of the natural amino acid side chains, prosthetic groups having the appropriate reactivity are used.

2.3.2 PROPINQUITY

The ability of an enzyme to form a complex with its substrate has long been recognized to be critical for rate acceleration. This quality has been called by many names, including propinquity and approximation. In order for two substrates to react, or for a catalytic group to accelerate a reaction, the partners must first come together in the proper orientation. The entropy decrease required for the association of these reactants contributes unfavorably to the total entropy of activation, which also reflects the entropy changes involved in the bond-making and breaking events. However, if two substrate molecules first bind to an enzyme, then the bond-breaking or bond-making step is no longer a part of the bimolecular reaction. Thus, splitting the overall process into two reactions with two transition states decouples the physical association of the reactants from the chemistry; consequently, the entropy of activation for the overall process is split across the two transition states, and the overall rate is enhanced. This effect has been demonstrated by comparing a number of unimolecular model reactions and their bimolecular analogs, giving rate enhancements as high as 10^6-fold higher for preorganized unimolecular systems.

2.3.3 TRANSITION-STATE BINDING BY ENZYMES

Examination of Equation 2.3 shows that as $\Delta G^{\ddagger}$ decreases (the activation barrier decreases), the rate constant for a reaction increases. In the context of transition-state theory, enzymes speed reactions by lowering the value of $\Delta G^{\ddagger}$. It is generally envisaged that enzymes lower the energy of the transition state by increasing the strength of the interactions between the enzyme and the transition state compared to the interactions between the enzyme and the reactants, consequently lowering the barrier shown in Figure 2.1. This is often referred to as transition-state binding, though, with a lifetime of only $\sim 10^{-13}$ sec, it is not an ordinary ligand. It is preferential binding or the stabilization of the transition state that causes the reaction to accelerate. A transition state could be stabilized by an enzyme through the strengthening of interactions to the enzyme as the geometry of the reactant distorts to that of the transition state. Geometric factors — distances and angles — determine the energetics for the noncovalent intermolecular forces that are important in enzymology, such as charge–charge, charge–dipole, dipole–dipole, and van der Waals interactions (described in Section 2.4.2). Thus enzymes can catalyze reactions by prepositioning groups to interact strongly in the transition state but only weakly in the reactant state. This underscores an idea that is not always appreciated — although enzymes bind substrates, there is no catalytic advantage in binding a substrate tightly. Rather, catalytic advantage is obtained when full binding energy is realized in the transition state. This concept was first enunciated in modern terms by Pauling, but its essence was stated by J.B.S. Haldane before transition-state theory and its attendant language had been invented. The importance of transition-state binding remains controversial and is currently being debated vigorously from both experimental and theoretical viewpoints.

A quantitative assessment of the catalytic ability of an enzyme (and therefore the tightness of transition-state binding) is obtained by comparing the rate of the catalyzed reaction to that of the uncatalyzed reaction. This has been done in only a handful of cases because of the challenge in determining the rates of uncatalyzed biochemical reactions, many of which are extremely slow and subject to competing side reactions. However, some data are available, obtained by measuring rates at high temperatures and extrapolating to biochemically relevant temperatures. From these measurements, rate accelerations of up to ~10^{19}-fold have been estimated, suggesting that some enzymes can stabilize transition states by as much as ~26 kcal mol^{-1}.

If an active site has such a large potential for binding a molecule, then stable mimics of transition states should be possible that take advantage of this and bind very tightly. Transition-state analogs have in fact been designed and synthesized with varying degrees of success, and sometimes show promise for use as drugs (see Chapter 5). Perhaps the most thorough and elegant demonstration of transition-state binding comes from work on the immucillins — inhibitors of purine nucleoside phosphorylase. The transition state for the phosphorolysis of guanosine was determined through a combination of kinetic isotope effect studies, which compared the rates of reaction of substrates containing a light or a heavy isotope at particular positions in the substrate (^{1}H vs. ^{3}H, ^{12}C vs. ^{14}C, and ^{14}N vs. ^{15}N) and quantum mechanical calculations. Upon determining a transition state that was consistent with the experimental data, molecules were designed and synthesized that mimicked the shape and charge distribution of the transition state. These molecules were potent inhibitors of the enzyme, with dissociation constants of the order of 10^{-11} M. Structures of the enzyme with substrate, product, or transition-state analogs were obtained by x-ray crystallography. Comparisons of these structures showed that a large number of hydrogen bonds become shorter in the transition-state structure, providing, in aggregate, considerable stabilization upon forming the transition-state complex.

2.4 CATALYTIC TACTICS AND INTERMOLECULAR FORCES

We will now consider in a more detailed way some of the molecular tactics and intermolecular forces that enzymes use to implement their catalytic strategies.

2.4.1 STEREOELECTRONIC CONTROLS

The conformation of a group in a molecule may alter its reactivity. For example, many eliminations are more favorable from an *anti* conformation. (Groups are anti when they have rotated and are diametrically on the opposite sides of the C–C bond, i.e., their dihedral angle is 180°.) This enhancement of reactivity in certain conformations is known as stereoelectronic control. It happens because some conformations of a molecule are better suited for the orbital interactions that stabilize the nascent product. Substrate-binding sites of enzymes can be very selective for conformation of a substrate. Therefore, it should not be surprising that stereoelectronic effects appear to be important in some enzyme reactions. Enzymes that utilize pyridoxal phosphate provide excellent examples. The pyridoxal phosphate prosthetic group in many enzymes forms covalent adducts with amino acid substrates and provides a planar electron-withdrawing π-system that stabilizes carbanion formation on the substrate, which is then involved in subsequent reactions. The carbanion is often formed either by the removal of the α-proton by an active-site base, or by decarboxylation (see Chapter 5). Selectivity is achieved by positioning the proper site of

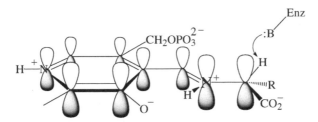

FIGURE 2.2 Stereoelectronic selectivity in pyridoxal enzymes. Pyridoxal-amino acid Schiff's base adducts may react by deprotonation of the α-carbon, decarboxylation, or sometimes by reactions on the amino acid side chain. One means enzymes use to control the reactivity is to orient the substituents of the tetrahedral α-carbon so that the reactive group is held perpendicular to the planar, conjugated pyridoxal π-system. When the electrophilic leaving group (H^+ or CO_2) departs, the orbital containing the lone electron pair that is left behind is in the proper orientation to be delocalized into the π-system, lowering the energy barrier of the proper reaction.

reactivity perpendicular to the plane of the prosthetic group, enabling the nascent negative charge to be delocalized by overlapping with the π-system (Figure 2.2).

2.4.2 VAN DER WAALS INTERACTIONS

Molecules are clouds of electrons in which nuclei are embedded. At any instant in time, there is a dipole moment, which averages to zero in a nonpolar molecule as electrons explore the space available to them (defined by the wave function). When two molecules approach each other, these transient dipole moments become correlated, resulting in attractive electrostatic interaction. This dipolar interaction is a very short-range phenomenon, weakening with the sixth power of the distance. However, if two molecules approach each other too closely, electrostatic repulsion dominates (this increases with the twelfth power of the approaching distance). The two opposing effects create a preferred separation distance characterizing each atom — the van der Waals radius — which can be thought of as defining the surface of the molecule. The energy in van der Waals interactions is relatively low, but can accumulate to become quite significant. For instance, the van der Waals attractive energy for two methylene groups is estimated to be 2 kcal mol^{-1}; in enzyme–ligand complexes, where there is a large degree of surface contact, this can add up to tens of kcal mol^{-1}.

2.4.3 HYDROGEN BONDS

"Classical" hydrogen bonds occur when a hydrogen atom that is covalently bonded to an electronegative atom (oxygen or nitrogen) is only a short distance from an electronegative atom that has a lone electron pair (again, usually oxygen or nitrogen). Experimentally determined distances between heavy atoms involved in a hydrogen bond show that they are slightly closer than the distance expected from their van der Waals radii, suggesting that the proton pulls the two together. The typical hydrogen bond is mostly ionic in character, resulting from the partial positive charge that resides on a hydrogen atom in the polar covalent bond to the electronegative *donor* atom and the partial negative charge that resides on the lone pair of the electronegative *acceptor* atom. The ionic character of hydrogen bonds makes their strength distance dependent but less sensitive (though not insensitive) to direction.

Estimates for the free energy of formation of typical hydrogen bonds in aqueous systems vary widely — in the range of ~2 to 5 kcal mol^{-1}. The free energy for forming a hydrogen-bonded pair in an aqueous system is relatively small because it also involves a loss of hydrogen bonds — those bonded to the solvent. In order to form a hydrogen bond between a donor molecule and an acceptor molecule, the hydrogen bonds between the donor and water, and the acceptor and water, must be broken. Pairing the donor and acceptor will form a new hydrogen bond, and the water liberated from solvation will also form new hydrogen bonds. Thus, the net change in enthalpy is usually close to zero. However, there is usually a favorable gain in entropy because water no longer orders itself around the separated donor and acceptor. Due to the importance of electrostatic interactions in hydrogen bonds, the energy of their formation is sensitive to the dielectric constant of the immediate environment. Surroundings with high dielectric constants shield the charges of the participants, reducing the attractive force. However, the interior of enzymes can provide a solvent-free, low-dielectric (i.e., hydrophobic) environment, drastically increasing the energy of hydrogen-bond formation.

The position of the proton in a hydrogen bond is determined by a double-well potential energy profile (Figure 2.3). The potential energy of the proton in a normal hydrogen bond is lowest at its covalent bond distance from the donor atom. However, when the hydrogen bond is strengthened by (1) moving the donor and acceptor atoms closer together, (2) matching the pK_a values of the donor and acceptor atoms, and (3) lowering the dielectric of the surrounding medium, the potential energy profile becomes symmetric, and the barrier separating the two wells shrinks. Such a hydrogen bond is called a *low-barrier hydrogen bond* (LBHB). In the extreme case, the barrier disappears and the potential energy for the movement of the proton between the heavy atoms becomes a single well. LBHBs are thought to be very favorable energetically, with estimates as high as 15 kcal mol^{-1}. The importance of LBHBs in enzymology is being vigorously debated. LBHBs have been detected in several enzymes by kinetic isotope effects and nuclear magnetic resonance spectroscopy, but their contribution to rate enhancement remains controversial. If estimates of their energies of formation are correct, it is possible that they can be key factors in stabilizing transition states.

2.4.4 Acid/Base Catalysis

Acid/base catalysis — the participation of an enzyme-bound functional group in transferring protons — is a nearly ubiquitous feature of enzyme reactions. Every ionizable protein functional group, as well as groups provided from substrates or prosthetic groups, has been invoked as acid or basic catalysts in enzymology. Two categories of proton transfer reactions may be considered: those to or from noncarbon atoms (heteroatoms) and those to or from carbon atoms. The transfer of protons to or from heteroatoms is generally facile, as evidenced by the high rates of proton

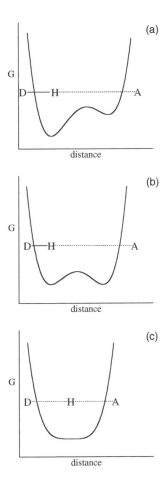

FIGURE 2.3 Potential energy wells for hydrogen bonds. The free energy of hypothetical hydrogen-bonding systems is plotted as a function of the distance between the hydrogen bond donor atom D, which is fixed at the left axis, and the hydrogen bond acceptor atom A, which is fixed some distance away. (a) The pK_a value of the donor is higher than that of the acceptor, causing the equilibrium position of the proton to be on the donor. (b) The pK_a value of the donor and acceptor are the same, resulting in a symmetrical potential energy well. The proton is equally likely to reside near D or A, though only the former is shown. (c) When the pK_a values of D and A are matched and the distance between the two is small enough, the barrier to proton transfer disappears, resulting in a very strong hydrogen bond that finds the proton bound tightly between both heavy atoms.

transfer in aqueous solution. Therefore, the proton transfer reactions themselves do not generally contribute a major energetic barrier to the overall reaction being catalyzed, but may be viewed as being secondary to the primary bond-breaking and -making reactions of the heavy atoms. In this case, protonations or deprotonations accelerate reactions by neutralizing the excess charge that develops in the transition state as bonds to heavy atoms (i.e., nonhydrogen atoms) are formed and/or broken. Two models for catalytic proton transfers between heteroatoms have been considered. In one model, the heteroatom–proton bond does not break in the transition state; that is, the proton remains at a minimum potential energy even at the transition state, and actual proton transfer occurs afterwards across the low-potential energy barrier typical for reactions in solution. Although the proton remains in a potential energy well in the transition state, it becomes more strongly hydrogen-bonded, providing transition-state stabilization. As an example, the addition of a nucleophile to a carbonyl group may be catalyzed by an enzymic acid. As the nucleophile forms a bond to the carbonyl carbon, the partial negative charge that develops on the oxygen would cause it to become a better hydrogen-bonding partner to the catalytic acid, which would protonate the oxygen only after the transition state for nucleophilic addition. This is in contrast with the alternative model for acid/base catalysis, in which the proton is "in flight" at the transition state and is therefore no longer in a potential energy well. This situation would occur when the changes in bonding of the heteroatoms involved in the proton transfer have occurred to such an extent that the pK_a of the heteroatom has changed to make proton transfer energetically favorable. For instance, when a nucleophile attacks a carbonyl group, the carbonyl oxygen ($pK_a \sim -1$) is transformed into an alkoxide group ($pK_a \sim 15$). If the transition state occurs midway between these extremes, the pK_a of the oxygen atom will be ~ 7, a value that is accessible to enzymic acids, and proton transfer will occur so that the negative charge then resides on the enzymic conjugate base, which then becomes the more stable locus as the reaction leaves the transition state. While both models for acid/base catalysis appear feasible, most experimental data, including solvent kinetic isotope effects and substituent effects, suggest that the former is operative and protons are not "in flight" in the transition state. However, some likely exceptions, characterized by unusually large solvent kinetic isotope effects (the ratio of rates when the reaction is performed in H_2O and D_2O), have been noted in which proton transfer is coupled to electron transfer.

Being catalysts, enzymes must eventually regenerate their reactive forms, including the correct ionization states of acid/basic catalytic groups, in order to undergo the next catalytic cycle. Curiously, the regeneration of the proper protonation state can sometimes be the rate-determining step in catalysis even though proton transfers between heteroatoms are usually rapid in aqueous solution. Such transfers can be much slower between the active site of an enzyme and bulk aqueous solution when the acid/base group is at least partially shielded by the protein, or when the group is involved in strong hydrogen bonds that must be broken prior to reaction with solution. Rate-determining regeneration of the correct ionization state of the catalyst leads to an *iso* steady-state kinetic mechanism and has been observed for several enzymes such as fumarase and pepsin.

Hydrogen bonds may also be a precursor to proton transfer reactions, especially in so-called proton-relay networks or "proton wires." In such situations, a network of hydrogen bonds connects a solvent-inaccessible reaction site to the solvent, allowing protons to be transferred when needed, yet, retaining the advantages of an active site where bulk solvent has been excluded (see Section 2.4.6; Figure 2.4). Proton transfer along such networks can be rapid compared to the reactions they are coupled to because of the relatively low energy barrier to proton transfer between hydrogen-bonded partners. Examples of enzymes that use proton-transferring hydro-

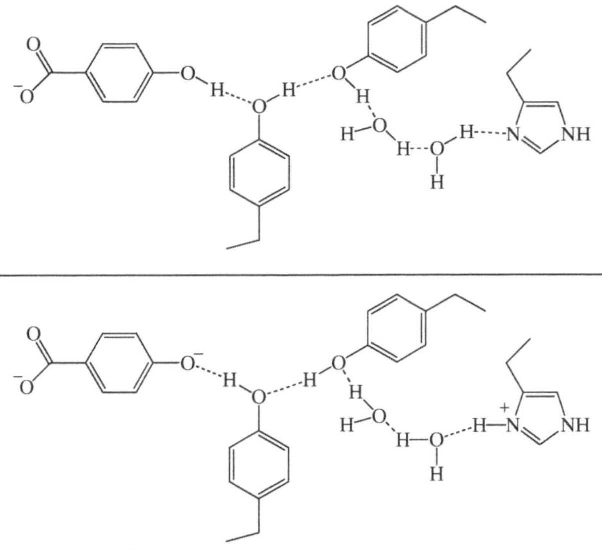

FIGURE 2.4 A proton transfer network in the enzyme *p*-hydroxybenzoate hydroxylase. This proton transfer network consists of hydrogen bonds in the core of the protein between the phenolic oxygen of the substrate *p*-hydroxybenzoate (on the left side), two tyrosines, two water molecules trapped within the protein, and finally terminates after ~12 Å at a histidine that contacts the surface (Palfey, B.A., Entsch, B., Massey, V., and Ballou, D.P. (1994). *Biochemistry*, 33, 1545–1554). This network allows the substrate to be activated by transient deprotonation without exposing the active site to bulk solvent, which would cause the rapid decay of a reactive flavin hydroperoxide intermediate (not shown). The top panel shows the network when the substrate is protonated. The bottom panel shows the network after the substrate has been deprotonated.

gen-bond networks include alcohol dehydrogenase, *p*-hydroxybenzoate hydroxylase, and the cytochromes P450.

We will now briefly consider the catalysis of proton transfers to and from carbon. These reactions occur frequently in biochemistry when enolates are reaction intermediates. From organic chemistry we know that protons on carbon are not very acidic unless the negative charge formed by deprotonation is stabilized by a neighboring electron-withdrawing group; the pK_a of an unactivated proton is estimated to be ~40, well beyond the biochemically accessible range. Therefore, deprotonation of carbon acids in biochemistry requires one or more activating groups, which are generally carbonyls. Even when a proton on a carbon is highly activated and therefore thermodynamically acidic, it is the experience of organic chemists that deprotonations of carbon acids (and the reverse protonations) are very slow. Nonetheless, enzymes deprotonate (and protonate) carbon compounds rapidly. The basis for this difference in behavior lies in the contrast between the random nature of the solvation shell found in solution and the highly organized active sites of enzymes. In solution, deprotonation of a carbonyl compound to form an enolate requires the solvent to reorganize to accommodate the developing negative charge on the oxygen atom so that strong hydrogen bonds form. The necessary large reorganization of the solvent is slow and limits the rate of reaction. In the active sites of enzymes that catalyze enolizations, protein groups that can form strong hydrogen bonds or otherwise stabilize a negative charge are found in crystal structures to be properly positioned near the oxygen atoms that will bear the charge. Thus, preorganization allows rapid deprotonation of carbon. Enzymes maximize the rate enhancement by matching the pK_a of the enzymic hydrogen-bond donor with the pK_a of the nascent enolate, resulting in a low-barrier hydrogen bond upon deprotonation.

2.4.5 ELECTROSTATICS

The interaction of charges can have a strong influence on reaction rates; some argue that this is the primary origin of the catalytic power of enzymes. Hydrogen bonds and acid/base catalysis are electrostatic effects that are important enough to be treated in their own sections (see preceding sections). However, beyond the catalytic effects of specifically positioned protons, constellations of charged amino acid side chains can create regions of positive or negative electrostatic potential. Similar effects can be produced by the amide groups of asparagine and glutamine side chains or the dipoles of the peptide groups in the protein backbone. The combined effects of peptide dipoles can be concentrated by an α-helix, though little is gained by increasing the length of the helix beyond ~2 turns. Electrostatic fields due either to side-chain charges or dipole moments from the protein structure can interact with bound molecules, causing changes in pK_a values and stabilizing developing charges in transition states and intermediates. These effects are sensitive to the dielectric constant that intervenes between the charge/dipole and the substrate/transition state/intermediate, and can be amplified by excluding water (see the following section).

2.4.6 DESOLVATION

Excluding solvent from the reaction site can be critically important in two ways: (1) it modifies the reactivity of substrates and protein side chains, and (2) prevents side reactions that could intercept reactive intermediates. In the former case, the exclusion of water decreases the local dielectric constant, favoring neutral over charged protonation states, thereby altering the pK_a values of groups and allowing proton transfers to occur that would not be possible in solution. Many unstable intermediates must also be protected from reaction with water by preventing access to it. There are many enzymes that generate highly basic intermediates that could be protonated by water, or intermediates with excellent leaving groups whose elimination could be accelerated by specific acid or base catalysis, or highly electrophilic intermediates that could be trapped by water. In all cases, barring the access of the solvent to the active site prevents unwanted reactions and maintains the integrity of the catalytic cycle.

2.4.7 NUCLEOPHILIC CATALYSIS

Enzymes sometimes use nucleophilic side chains or nucleophilic prosthetic groups to form an adduct with the substrate or a substrate fragment. The side chains of serine, threonine, cysteine, histidine, aspartate, glutamate, and tyrosine have been identified as nucleophiles in a variety of enzymes, including proteases, esterases, kinases, and lyases, to name just a few. Nucleophilic catalysis is frequently involved in group transfer reactions, in which a group on one substrate is transferred to an acceptor site on another substrate. Rather than catalyzing the direct reaction between the two substrates, the enzyme breaks the reaction into partial reactions in which the group is transferred first to the enzymatic nucleophile and then to the ultimate acceptor. Breaking the reaction into several steps may lower the activation barriers, affording energetic advantages. Also, the division of a reaction into discrete steps involving different substrates often allows modularity in the structure of the enzyme, so that separate domains perform separate tasks. Such modularity would greatly facilitate the evolution of new enzyme activities by allowing the recombination of genes encoding for protein domains tailored to bind specific substrates or catalyze particular reactions. Enzymes can increase the nucleophilicity of a group by excluding water from the active site. Additionally, specific interactions with protein residues can activate a nucleophile. For instance, serine acts as a nucleophile in a number of proteases and is activated by a hydrogen bond to a histidine which, in turn, is hydrogen-bonded to a buried aspartate. This "catalytic triad" is thought to generate the alkoxide form of the serine side chain for attack on the amide carbonyl of a protein substrate.

2.4.8 ELECTROPHILIC CATALYSIS

The 20 genetically encoded amino acids provide an abundance of nucleophilic groups but no electrophilic groups (with the exception of a proton donated by an acid, but this is in a category of its own). Nonetheless, many biochemical reactions require catalysis by an electrophile, necessitating the recruitment of electrophilic prosthetic groups. These can be either small organic molecules, such as pyridoxal phosphate, or metal ions. Electrophilic catalysts speed reactions by forming an adduct with a substrate and activating them for the loss of an electrophilic-leaving group such as CO_2 or H^+. When an electron-deficient group leaves a molecule — for instance, in a deprotonation reaction — it will leave behind an electron pair. It is the job of the electrophilic catalyst to stabilize this new lone pair, usually by delocalization in the case of organic prosthetic groups, or by direct charge neutralization in the case of metals.

2.4.9 REDOX CATALYSIS

A large number of biochemical reactions are net redox reactions. In addition to these, there are many reactions that do not involve a net redox change but do require a reactive intermediate to be generated by redox chemistry. Although some redox reactions occur directly between two substrates, in a large number of cases a substrate oxidizes or reduces the enzyme directly, which subsequently reacts with the next substrate. Consequently, a large number of enzymes do redox chemistry. Proteins themselves are capable of some redox reactions, but there are also many prosthetic groups that catalyze redox reactions, as described in the following text.

Redox reactions may be broadly classified into two groups: those in which one electron is transferred, and those in which two electrons are transferred. The former group necessarily involves unpaired electrons. Because radicals often react readily with molecular oxygen, these reactions have been studied to a lesser extent than those involving formal two-electron transfers. Nonetheless, both types of reactions are important in enzyme mechanisms. The usual considerations of transition-state stabilization through electrostatic interactions, hydrogen bonding, etc., are applicable in redox reactions, as well as in quantum mechanical tunneling, to be discussed in the next section. However, the energetics of redox reactions are usually discussed in terms of redox potentials, which are related to free energy differences by

$$\Delta G^0 = -nF\Delta E \qquad (2.7)$$

in which ΔE is the difference in redox potentials of two reactants (in volts), n is the number of electrons transferred, and F is Faraday's constant (23061 cal $V^{-1}mol^{-1}$). Usually, electrons are transferred in the thermodynamically favored direction ($\Delta E > 0$) in redox steps in catalytic cycles. The redox potentials of enzyme-bound reactants are very frequently different from those of the reactants in solution. If a redox potential of an enzyme-bound molecule or group differs from that in solution, then a thermodynamic cycle may be written showing that the protein must preferentially bind one redox state over the other (Figure 2.5).

Such preferential binding of redox states is very common and offers one means for the protein to control reactivity. Controlling the reactivity of redox centers is very important because they can be highly reactive and are conceivably vulnerable to side reactions, many of which could lead to toxic products. Remarkably, side reactions do not frequently happen with enzymes, underscoring the importance in

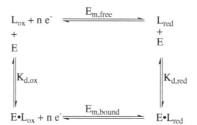

FIGURE 2.5 Thermodynamic "box" showing the effect of protein binding on the redox potential of a ligand L. L can be in an oxidized form or reduced by n-electrons (generally n = 1 or 2). The redox potential of L in solution is given by $E_{m,free}$. Its strength of binding to the enzyme E is given by the dissociation constant K_d, which is the equilibrium constant for the *dissociation* reaction. (The *lower* the value of K_d, the *tighter* the binding.) Both the redox potentials and the dissociation constants are thermodynamic functions directly related to free-energy changes. The free-energy change in going from the upper-left corner of the box to the lower-right corner must be identical for the two possible paths. If $K_{d,ox}$ is greater than $K_{d,red}$ (the protein binds reduced L tighter than oxidized L), then the E•L$_{ox}$ will accept electrons more avidly than free L$_{ox}$ — that is, the $E_{m,bound} > E_{m,free}$.

FIGURE 2.6 Thiol–disulfide interchange reactions. Thiols become very nucleophilic upon deprotonation. The thiolate R_3-S^- is shown attacking an electrophilic disulfide bond to form a new disulfide and liberating a new thiolate. Such thiol–disulfide interchange reactions can be extremely fast.

enzymology not just of accelerating the desired reaction but also of inhibiting competing side reactions. Preferential binding of redox states is just one means that proteins use to control the reactivity of redox-active groups. Additionally, proteins can restrict the access to the redox center, thereby permitting the reaction only of the proper substrate. However, some molecules are not easily excluded. For instance, it has been demonstrated that molecular oxygen can quench the fluorescence of buried tryptophan residues, suggesting that such a small nonpolar molecule is capable of penetrating the fluctuating protein matrix (more on dynamics in the following text). In principle then, molecular oxygen cannot be excluded from redox centers with which, based on thermodynamics, it could react. Yet the reactions of many reduced enzymes with molecular oxygen are quite slow. One hypothesis explaining the suppression of oxygen reactivity recognizes that in the reduction of O_2, the superoxide radical anion, O_2^-, is an obligate intermediate due to spin conservation. The protein environment can suppress the formation of the anionic intermediate by being hydrophobic or even by being negatively charged.

Perhaps the most notable redox functionality available to proteins is the thiol group of cysteine, which can be oxidized in pairs to form a disulfide and can react with other thiols or disulfides (Figure 2.6). Thiol–disulfide interchanges are facile reactions that are critical components of the catalytic cycles of many enzymes, especially those involved in energy conservation, detoxification, and protein folding. Occasionally, enzymes are discovered containing an atom of selenium in place of the expected sulfur in cysteine. Selenocysteine, a naturally occurring amino acid, is substituted for cysteine during protein translation in several enzymes. Selenium is in the same group as sulfur in the periodic table, and therefore has similar chemical properties. However, its larger size and lower electronegativity make it more reactive as a nucleophile but also more susceptible to oxidation by molecular oxygen, which can be problematic in the isolation of active selenoenzymes.

The reactions of cysteine (or selenocysteine) constitute the only two-electron redox chemistry available to the natural amino acid side chains. However, in recent years it has become clear that the side chains of tyrosine or tryptophan may be modified to form redox-active quinone cofactors. Quinones are highly electrophilic and capable of being reduced by one or two electrons. The best-studied quinoproteins also contain bound copper and are involved in the oxidation of amines by molecular oxygen.

As radical reactions have been discovered and studied in detail, it has been found that some enzymes form protein radicals during catalysis. Cysteine is oxidized to cystinyl radical, tyrosine is oxidized to the tyrosinyl radical, and tryptophan is oxidized to the tryptophanyl radical. The protein backbone itself can also participate in redox chemistry, especially at the α-carbon of glycine, in the form of a catalytically relevant radical. Peptide backbone radicals are stabilized by the *captodative* effect, in which the radical is stabilized by orbital overlap with an electron-donating group (the amide nitrogen of the glycine) and an electron-withdrawing group (the carbonyl of the glycine). Protein radicals are usually strong oxidizing agents, involved in reactions with substrates that are difficult to oxidize.

Redox prosthetic groups, comprised of organic molecules, metals, or metal complexes, provide further reactivity for enzyme reactions. These prosthetic groups are adapted for a variety of chemical tasks, including the oxidation or reduction of organic and inorganic molecules, electron transfer reactions, and photochemistry. A detailed account of the fascinating chemistry of the large number of redox prosthetic groups is beyond the scope of this account, but some general features are worth mentioning. Many redox prosthetic groups are limited to single-electron transfers (e.g., heme and iron–sulfur clusters), some are limited to two-electron transfer reactions (e.g., NADH), whereas some can react either by one or two electrons (e.g., flavins). The protein environment generally promotes one mode of reactivity of a redox prosthetic group over the often large number of possibilities and will adjust the redox potential to a value suited for catalysis. Whereas the protein environment often retards side reactions to a very large extent, it is usually not possible to eliminate them completely. Thus, protein-bound redox prosthetic groups often react with artificial reductants or oxidants, providing a convenient experimental tool. In normal catalysis, many redox-active prosthetic groups are reduced or oxidized by substrates in single-electron transfers (sometimes at a large distance), in two-electron steps by a hydride transfer, or by forming covalent adducts as intermediates that decompose with a change in oxidation state.

2.4.10 QUANTUM MECHANICAL TUNNELING AND PROTEIN DYNAMICS

Until now, we have considered enzymes to be catalysts that obey transition-state theory. According to this view, the reactants cannot become products unless they have enough energy to distort their structures to that of the transition state, thereby crossing the potential energy barrier separating reactants from products. However, quantum mechanical tunneling allows for many reactions to proceed even if enough energy is not available to distort the reactant to the transition-state structure. According to quantum mechanics, matter has both particle and wave properties. The quantum mechanical wave function describes the distribution of a particle over space in terms of its kinetic and potential energy. Classically, a particle cannot be found in a region of space where its kinetic energy is lower than its potential energy. Quantum mechanics allows particles to penetrate these classically forbidden zones as long as the potential energy in the zone is not infinite; this penetration is known as quantum mechanical tunneling. Thus, if an atom is being transferred in a chemical reaction from one site to another across a finite activation barrier, there is a finite probability that the particle will appear as the reaction product on the other side of the energy barrier due to quantum mechanical tunneling, even though the particle did not have enough kinetic energy to reach the transition state. The actual importance of tunneling depends on the distance over which the particle must move and the shape (width and height) of the potential energy barrier separating reactant from product. The particle/wave duality of quantum mechanics allows a wavelength λ to be calculated for a particle of mass m with a given kinetic energy E according to the de Broglie relationship

$$\lambda = \frac{h}{\sqrt{2mE}} \tag{2.8}$$

where h is Planck's constant. Quantum mechanical tunneling will only contribute significantly to chemical reactions when the wavelength of the transferring particle is similar to the distance over which it is transferred. This limits the importance of tunneling effects to reactions involving the transfer of light particles — electrons or hydrogens. If one assumes a chemically reasonable kinetic energy of 2.4 kcal mol^{-1}, then an electron has a wavelength of ~18 Å (roughly the diameter of a small protein), whereas the much heavier hydrogen atom has a wavelength of ~0.5 Å, similar to the distance it would move in a reaction between two molecules in van der Waals contact.

The long wavelength of the electron makes long-range biological electron transfer reactions feasible, and these are frequently observed, both in intramolecular electron transfers between redox prosthetic groups within electron-transport proteins and in intermolecular electron transfer reactions between different proteins. The factors that control the rates of electron transfer are currently the subject of a large amount of theoretical and experimental research. It is clear that one important factor is the distance between the reacting redox centers; as the distance increases, the probability of tunneling, and consequently the rate of electron transfer, decreases. However, there is a growing consensus that the intervening medium between reacting redox centers is also important, and that a long-range electron transfer can actually be composed of a series of shorter electron jumps through favorably disposed orbitals on the atoms of the protein itself. A great deal of effort is being directed towards understanding the factors that favor a particular electron-transfer pathway.

The tunneling distance for hydrogen is much shorter than for electron tunneling, so that the hydrogen donor and acceptor must be in van der Waals contact for tunneling to occur. Direct contact of the reactants also allows for hydrogen to be transferred semiclassically "over the barrier," as described by transition-state theory, so that in principle both mechanisms may be operational. Because the de Broglie wavelength of a particle is dependent on the mass, substituting a proton with deuterium often causes anomalously large decreases in reaction rates when tunneling is the important means of hydrogen transfer, although large kinetic isotope effects are not a required symptom of tunneling. A consensus is emerging that tunneling is likely to be a contributor in most, if not all, hydrogen transfer reactions.

Protein dynamics — the thermally driven periodic oscillations of the atoms and groups of the protein — appears to play an important role in promoting both electron and hydrogen tunneling. The structures of enzymes are not static but are always in motion. These motions include the very rapid vibrations of individual atoms, the somewhat slower vibrations and rotations of larger groups such as amino acid side chains, and the yet slower movements of large elements of secondary structure and domains (sometimes referred to as protein "breathing"). Because the precise arrangement of the atoms of the active site determines the height and width of the activation energy barrier for the reaction of a substrate, the shape of this barrier fluctuates with the protein structure. Therefore, protein dynamics creates a family of reaction coordinates, each characterized by its own reaction rate. The width of the energy barrier is critical in determining the probability of hydrogen tunneling because the de Broglie wavelength of the hydrogen atom is similar to the distance it must move during a reaction. If a protein vibration were to squeeze the hydrogen donor and acceptor together, the width of the potential energy barrier separating reactants from product would decrease, thereby promoting hydrogen transfer by tunneling. In another instant, the distance between a hydrogen donor and acceptor might increase as the structure of the enzyme changes. The longer distance through the potential energy barrier would then decrease the probability of hydrogen tunneling, thereby decreasing the rate of reaction by tunneling in that instantaneous configuration of the enzyme-reactant complex.

Protein dynamics is also a critical determinant in the rate of electron transfer reactions. Because the de Broglie wavelength of an electron is so large, the rate of reaction is less sensitive to the small fluctuations caused by protein dynamics in the width of the potential energy barrier separating the donor and the acceptor. However, the efficiency of electron tunneling increases as the orbitals of the donor and acceptor overlap, with the most overlap occurring when the energies of the donor and acceptor orbitals are matched. This is brought about by dynamic fluctuations in structure. Thus electron transfer reactions occur when a high-energy structure is transiently attained by the dynamic motions of the protein, as described by the Marcus theory of electron transfer rates. The importance of protein dynamics in controlling the rates of electron transfer has been evident for some time, whereas its importance in hydrogen transfer reactions has become apparent only recently. These concepts lead to two intriguing questions that are yet to be fully answered: Have enzymes evolved in a way to enhance vibrational modes that promote catalysis, and can protein dynamics increase reaction rates even when tunneling is not possible (i.e., when only heavy atoms are transferred)?

FURTHER READING

Acid/Base Catalysis

Gerlt, J.A. and Gassman, P.G. (1993). Understanding the rates of certain enzyme-catalyzed reactions: proton abstraction from carbon acids, acyl-transfer reactions, and displacement reactions of phosphodiesters. *Biochemistry*, 32, 11943–11952.

Harris, T.K. and Turner, G.J. (2002). Structural basis of perturbed pK_a values of catalytic groups in enzyme active sites. *IUBMB Life*, 53, 85–98.

Hille, R. (1991). Electron transfer within xanthine oxidase: a solvent kinetic isotope study. *Biochemistry*, 30, 8522–8529.

Rebholz, K.L. and Northrop, D.B. (1995). Kinetics of isomechanisms. *Methods in Enzymology*, 249, 211–240.

Schowen, K.B., Limbach, H.-H., Denisov, G.S., and Schowen, R.L. (2000). Hydrogen bonds and proton transfer in general-catalytic, transition-state stabilization in enzyme catalysis. *Biochimica et Biophysica Acta*, 1458, 43–62.

Electrostatics

Warshel, A. (1998). Electrostatic origin of the catalytic power of enzymes and the role of preorganized active sites. *Journal of Biological Chemistry*, 273, 27035–27038.

General Enzyme Mechanisms

Benkovic, S.J. and Hammes-Schiffer, S. (2003). A perspective on enzyme catalysis. *Science*, 301, 1196–1202.

Blow, D. (2000). So do we understand how enzymes work? *Structure*, 8, R77–R81.

Bruice, T.C. and Benkovic, S.J. (2000). Chemical basis for enzymatic catalysis. *Biochemistry*, 39, 6267–6274.

Bruice, T.C. (2002). A view at the millennium: the efficiency of enzymatic catalysis. *Accounts of Chemical Research*, 35, 139–148.

Fersht, A. (1999). *Structure and Mechanism in Protein Science. A Guide to Enzyme Catalysis and Protein Folding*. San Francisco: W.H. Freeman.

Hammes, G.G. (2002). Multiple conformational changes in enzyme catalysis. *Biochemistry*, 41, 8221–8228.

Jenks, W.P. (1987). *Catalysis in Chemistry and Enzymology*. New York: Dover.

Edited by Sinnott, M. (1998). *Comprehensive Biological Catalysis, Vol. IV/Lexicon of Terms and Concepts in Mechanistic Enzymology*. San Diego: Academic Press.

Hydrogen Bonding

Cleland, W.W., Frey, P.A., and Gerlt, J.A. (1998). The low-barrier hydrogen bond in enzymatic catalysis. *Journal of Biological Chemistry*, 273, 25529–25532.

Hibbert, F. and Emsley, J. (1990). Hydrogen bonding and chemical reactivity. *Advances in Physical and Organic Chemistry*, 26, 255–279.

Jeffrey, G.A. (1997). *An Introduction to Hydrogen Bonding*. Oxford, U.K.: Oxford University Press.

Mildvan, A.S., Massiah, S.S., Harris, T.K., Marks, G.T., Harrison, D.H.T., Viragh, C., Reddy, P.M., and Kovach, I.M. (2002). Short, strong hydrogen bonds on enzymes: NMR and mechanistic studies. *Journal of Molecular Structure*, 615, 163–175.

Palfey, B.A., Entsch, B., Massey, V., and Ballou, D.P. (1994). Changes in the catalytic properties of *para*-hydroxybenzoate hydroxylase caused by the mutation Asn300Asp. *Biochemistry*, 33, 1545–1554.

Propinquity

Bruice, T.C. and Pandit, U.K. (1960). *Journal of the American Chemical Society*, 82, 5858–5865.

Redox Catalysis

Edited by Sinnott, M. (1998). *Comprehensive Biological Catalysis, Vol. III/Radical Reactions and Oxidation/Reduction.* San Diego: Academic Press.

Transition State Binding

Kicska, G.A., Tyler, P.C., Evans, G.B., Furneaux, R.H., Shi, W., Federov, A., Lewandowicz, A., Cahill, S.M., Almo, S.C., and Schramm, V.L. (2002). Atomic dissection of the hydrogen-bond network for transition-state analogue binding to purine nucleoside phosphorylase. *Biochemistry*, 41, 14489–14498.

Schramm, V.L. (2002). Development of transition-state analogues of purine nucleoside phosphorylase as anti-T-cell agents. *Biochimica et Biophysica Acta*, 1587, 107–117.

Wolfenden, R. and Snider, M.J. (2001). The depth of chemical time and the power of enzymes as catalysts. *Accounts of Chemical Research*, 34, 938–945.

Transition State Theory

Albery, J.W. (1993). Transition-state theory revisited. *Advances in Physical and Organic Chemistry*, 28, 139–170.

Connors, K.A. (1990). *Chemical Kinetics. The Study of Reaction Rates in Solution*, pp. 187–243. New York: VCH.

Tunneling and Dynamics

Kohen, A. and Klinman, J. (1998). Enzyme catalysis: beyond classical paradigms. *Accounts of Chemical Research*, 31, 397–404.

McCammon, J.A. and Harvey, S.C. (1987). *Dynamics of Proteins and Nucleic Acids.* Cambridge, U.K.: Cambridge University Press.

Warshel, A. (2002). Molecular dynamics simulations of biological reactions. *Accounts of Chemical Research*, 35, 385–395.

2.5 CYTOCHROME P450 EXAMPLE

W. Edward Lindup and Neil R. Kitteringham

2.5.1 INTRODUCTION

The cytochrome P450 enzymes are one of the largest extended families of proteins known to man, with over 400 members identified throughout the animal and plant kingdoms. P450s have been the focus of intense research efforts for over 40 years and not only has this provided a detailed understanding of the function of heme-based biological catalysts, it has also yielded insights into gene regulation, cell signaling, and, most recently, the generation of nitric oxide. We shall not be able to cover these aspects here. The purpose of this review is to focus specifically on the role of P450 in pharmacology, with particular emphasis on the clinical importance of this enzyme system, and to highlight its significance using therapeutic agents as examples.

To describe the pharmacology of P450 we must consider its action on drugs, its physiological role in aiding excretion, its toxicological role in generating harmful metabolites, and also the action of drugs on the P450 system. This latter aspect becomes more important as our understanding grows regarding enzyme function and its pharmacological modulation, both through gene regulation and inhibition.

2.5.2 BACKGROUND AND HISTORICAL CONTEXT

The discovery of P450 can be traced back to 1958 when two independent research workers (Klingenberg and Garfinkel) published papers that identified a pigment in rat liver that produced an unusual spectrum upon reduction with carbon monoxide. The absorption maximum at 450 nm gave the name to cytochrome P450. Prior to this, the vast oxidative capacity of the liver with respect to foreign compounds was already well established, mainly through the pioneering work of R.T. Williams. However, it was not until the isolation of P450 by the Japanese researchers Omura and Sato in 1964 that the precise nature of P450 was characterized as a heme-containing protein. The identification of a second P450 enzyme with a slightly different binding spectrum (cytochrome P448) in 1967 indicated the possibility of multiple forms of the enzyme. This was the subject of scientific debate for some years until the advent of molecular biology led to the discovery of the many hundred isoforms known today. An additional point of interest was that the administration of chemical carcinogens to experimental animals increased the activity of cytochrome P450 and, in particular, increased the amount of P448. Such an increase in enzyme activity is usually called enzyme induction, and the intriguing link between this process and chemical carcinogenesis became the focus of intense research.

Unfortunately, this wealth of knowledge resulted in a confusing period for scientists attempting to keep up with the development of P450 identification because classification at that time was based, as with other enzyme systems, upon substrate specificity. As it began to emerge that one of the distinguishing features of P450 enzymes was the *lack* of substrate specificity, it became apparent that the same P450 isoform was referred to by different names in different laboratories. A classification based on gene sequence homology was therefore developed. Thus, 74 major families have now been designated (labeled CYP1, CYP2, CYP3, etc.), each of which contains genes of at least 40% sequence homology. Within each family, several subfamilies (CYP1A, CYP1B, etc.) can reside, provided that the genes within these subfamilies show at least 55% homology. The final Arabic numeral in the classification (CYP1A1, CYP1A2, etc.) designates a unique isozyme (theoretically a 100% homology, but with a 3% allowance for natural mutations). This system, which is updated periodically, has provided a robust and serviceable framework for the growth of knowledge in P450 research. With respect to man, it is only necessary to focus on the four main families as shown in Table 2.1.

Each family contains a relatively small number of subfamilies with a cluster of individual isozymes (or isoforms) in each. The relative proportions of these major isoforms in human liver are indicated in the pie chart (Figure 2.7) along with their contribution to overall drug metabolism. The distribution, substrate preference, and polymorphic expression for each isozyme have been defined by a combination of selective inhibitors, specific antibodies, and heterologous expression systems which has provided a solid framework for defining the precise role and importance of P450 enzymes in clinical pharmacology.

TABLE 2.1
Cytochrome P450 Isozymes Involved in the Metabolism of Drugs and Xenobiotics in Man

Subfamily	Isozyme	Features
CYP1A	CYP1A1	Extrahepatic; inducible by cigarette smoke
	CYP1A2	Hepatic
CYP2A	CYP2A6	Some evidence of polymorphism
CYP2B	CYP2B6	Induced by phenobarbitone
CYP2C	CYP2C8	Induced by rifampicin
	CYP2C9	Induced by rifampicin
	CYP2C18	Induced by rifampicin; polymorphic (20%)
	CYP2C19	Induced by rifampicin
CYP2D	CYP2D6	Uninducible; polymorphic (5–10%)
CYP2E	CYP2E1	Inducible by ethanol; polymorphic (<1%)
CYP2F	CYP2F1	Found only in lung
CYP3A	CYP3A3	Found only in lung
	CYP3A4	Major human isozyme; inducible by rifampicin, phenytoin
	CYP3A5	Polymorphic (present in ~20%)
	CYP3A7	Fetal form
CYP4A	CYP4A1	
CYP4B	CYP4B1	Inducible by clofibrate

2.5.3 CHEMISTRY, BIOCHEMISTRY, AND MOLECULAR BIOLOGY OF CYTOCHROME P450

2.5.3.1 Mechanism of Action

In order to appreciate the clinical relevance of cytochrome P450, it is necessary to understand the fundamental aspects of its mode of action, how this affects the clearance of drugs from the body, and how this, in turn, may be modulated by other drugs or chemicals. The P450s responsible for foreign-compound metabolism evolved about 400 to 500 million years ago in order to allow animals to detoxify chemicals from plants eaten in the diet. The liver is the main site of expression, but other tissues also contain lower levels of P450 enzymes. Indeed, there can be more than 30 P450 isoforms expressed in a particular tissue at any one time.

The characteristics that each individual enzyme possesses include a powerful oxidizing capacity, low and frequently overlapping substrate-specificity, and often a relatively low substrate affinity. Catalysis is driven by activation of oxygen rather than binding of substrate, which enables the enzyme system as a whole to metabolize an apparently limitless array of low-molecular-weight, lipid-soluble organic molecules. In order to pass through lipid-rich cell membranes, starting with the cells lining the gastrointestinal tract and extending to cells at other sites of action such as the brain, drugs have to be designed with a reasonable degree of lipid solubility (lipophilicity). This in turn means that such drugs cannot be excreted without being first metabolized to less lipid-soluble but more water-soluble metabolites.

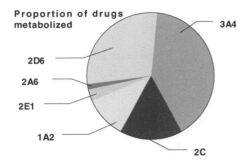

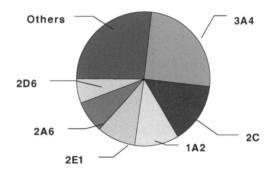

FIGURE 2.7 Relative abundance of the major hepatic isoforms of cytochrome P450 in human liver.

The substrates are partly preselected for metabolism by a number of physiological processes, which are dependent upon the general physicochemical properties of the drug molecule. These include absorption from the gastrointestinal tract (if given orally) or other sites of administration, presentation to and extraction by the liver, and further accumulation in the endoplasmic reticulum of hepatocytes. All of these serve to bring the concentration of the substrate in the vicinity of the cytochrome P450 enzymes to that of the affinity constant (K_m) for one or more of the individual enzymes.

Highly specific processing mechanisms are responsible for the insertion of cytochrome P450 enzymes into the membrane of the endoplasmic reticulum. The heme-containing part of the enzyme is exposed to the cytosolic side of the membrane (Figure 2.8) and is orientated in such a way that the substrate-binding site faces the membrane, there being two substrate access channels to the active site, thus allowing access for both lipophilic and hydrophilic substrates. The catalytic site of the enzymes, comprising the heme-binding site and the oxygen-binding site, are strongly conserved, which explains the similar requirements of the P450 enzymes to perform their catalytic function: binding of molecular oxygen, cleavage of the dioxygen bond, and the interaction with electron-transport coenzymes.

Many drugs undergo extensive metabolism by cytochrome P450 enzymes, commonly by several types of biotransformation, which may occur in parallel or in

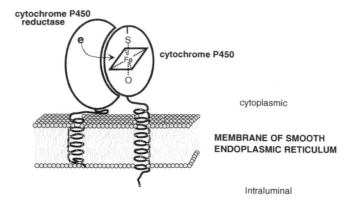

FIGURE 2.8 Schematic representation of cytochrome P450 located in the extraluminal surface of the endoplasmic reticulum membrane, indicating the relationship between P450 and its electron-donating coenzyme, cytochrome P450 reductase.

series, and each biotransformation can be catalyzed by more than one enzyme. All the enzymes are able to carry out aliphatic hydroxylation, aromatic hydroxylation, demethylation, N-hydroxylation, and other biotransformations. All of these can be rationalized from a chemical viewpoint in terms of a mechanism that involves insertion of an atom of oxygen through reaction of carbon–carbon, carbon–hydrogen, and heteroatom–hydrogen bonds. The reactive form of oxygen is thought to be bound principally to heme $[FeO]^{3+}$, although a mechanism involving $[FeO]^{2+}$ has also been postulated (see Section 5.6.1). Although the P450s are able to carry out a bewildering array of different metabolic reactions, most substrates are in fact oxidized by one of three mechanistic pathways (Figure 2.9), which differ in the initial step of the enzyme–substrate interaction.

The majority of oxidative biotransformations of drugs result in the formation of products that can undergo clearance (mainly via the kidneys but also from other organs) either with or without a subsequent phase II biotransformation (Figure 2.10). In certain circumstances that are determined by the site of oxidative attack on the substrate and the overall chemistry of the product, chemically reactive metabolites can be formed that can, if not detoxified, cause biochemical lesions that may (depending on the kinetics of damage and repair mechanisms) eventually lead to overt tissue damage (see Section 2.5.6).

Thus, overall, the role of cytochrome P450 can be viewed as the first step in the rapid removal of lipophilic compounds from the body, partly by increasing their water solubility but mainly by structurally modifying them to render them more amenable as substrates for the phase II conjugation enzymes, which produce glucuronide and sulfate conjugates.

2.5.3.2 Biochemistry of Cytochrome P450 Enzymes

P450s are distributed throughout the body although they are concentrated predominantly in the liver. Despite the promiscuity in P450s' choice of substrates, certain generalizations can be made with respect to both the distribution and substrate

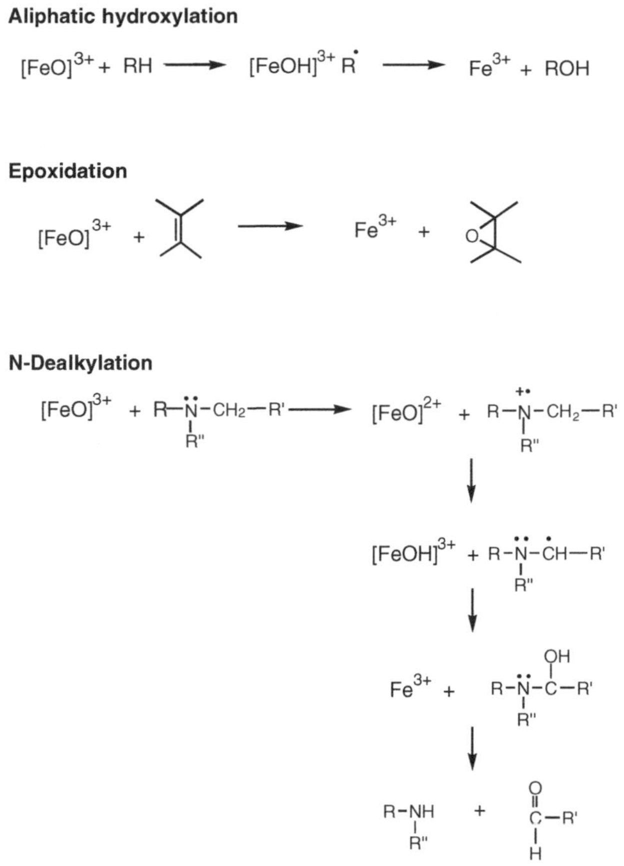

Aliphatic hydroxylation

$$[FeO]^{3+} + RH \longrightarrow [FeOH]^{3+} R^{\bullet} \longrightarrow Fe^{3+} + ROH$$

Epoxidation

$$[FeO]^{3+} + \quad \longrightarrow \quad Fe^{3+} +$$

N-Dealkylation

$$[FeO]^{3+} + R{-}\overset{..}{N}{-}CH_2{-}R' \longrightarrow [FeO]^{2+} + R{-}\overset{+\bullet}{N}{-}CH_2{-}R'$$
$$\qquad\quad\ \ \underset{R''}{|} \qquad\qquad\qquad\qquad\qquad \underset{R''}{|}$$

$$[FeOH]^{3+} + R{-}\overset{..}{N}{-}\overset{\bullet}{C}H{-}R'$$
$$\qquad\qquad\qquad\qquad \underset{R''}{|}$$

$$Fe^{3+} + \quad R{-}\overset{..}{N}{-}\overset{\overset{OH}{|}}{C}{-}R'$$
$$\qquad\qquad\qquad\quad \underset{R''}{|}$$

$$R{-}NH \quad + \quad \overset{O}{\underset{H}{\overset{\|}{C}{-}R'}}$$
$$\underset{R''}{|}$$

FIGURE 2.9 The three basic chemical mechanisms of xenobiotic oxidation reactions carried out by cytochrome P450 enzymes.

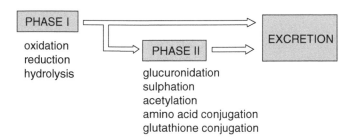

FIGURE 2.10 Classification of drug metabolism into phase I and phase II.

preferences of the different families. Some of the human P450 isoforms are covered in more detail in the following text.

2.5.3.2.1 CYP1A

The CYP1A family of P450s comprises two members, CYP1A1 and CYP1A2. CYP1A1 is mainly present in tissues other than the liver (i.e., extrahepatic) and is generally expressed in low or negligible amounts unless an individual is exposed to certain environmental chemicals, notably those in cigarette smoke, that cause enzyme induction. CYP1A1 is also expressed in white blood cells, which may provide a useful surrogate index for expression in other tissues. CYP1A2, in contrast, appears to be exclusively expressed in the liver, in which it is found constitutively. Both CYP1A1 and CYP1A2 are able to metabolize procarcinogens such as polycyclic aromatic hydrocarbons. For this reason, the CYP1A family has attracted enormous research attention as a carcinogen activator because several procarcinogens fall into this chemical class, particularly those present in cigarette smoke. Drugs such as theophylline (for asthma) and clozapine (for schizophrenia) are also metabolized by CYP1A2; this isoform is inducible by cigarette smoking, which accounts for the higher doses required to treat smokers with these drugs.

2.5.3.2.2 CYP2A

CYP2A6 is the only member of the subfamily expressed to any extent in human liver, but represents less than 1% of the total. It is also inducible by rifampicin and aromatic anticonvulsants. An inactive variant of the enzyme is present in about 1% of the population. CYP2A6, at least partly, is involved in the metabolism of coumarin (an anticoagulant) and the general anesthetics methoxyflurane and halothane (reductive pathway only; see Section 2.5.6).

2.5.3.2.3 CYP2C

The CYP2C subfamily contains at least five genes, the two most important isoforms being CYP2C9 and CYP2C19. CYP2C9 contributes to the metabolism of the anticoagulant warfarin, the antiepileptic drug phenytoin, and nonsteroidal antiinflammatory drugs such as diclofenac, whereas CYP2C19 metabolizes the tranquillizer diazepam and omeprazole (an inhibitor of gastric acid secretion used to treat ulcers). CYP2C19 (previously termed mephenytoin hydroxylase) is polymorphically expressed and is absent in about 3% of the Caucasian population. Polymorphisms have also been described in CYP2C9; these result in alteration of activity toward certain substrates only. For example, the CYP2C9*2 polymorphism may partly determine the dose of warfarin required. The CYP2C subfamily is inducible by rifampicin (an antituberculosis drug) and anticonvulsants.

The crystal structure of human CYP2C9 has recently been determined for the first time by Williams and colleagues (Figure 2.11), and it has been studied both in the absence and presence of one of its substrates, the anticoagulant warfarin. CYP2C9 catalyzes the 6- and 7-hydroxylation of S-warfarin, the active enantiomer of the drug. S-warfarin occupies a predominantly hydrophobic pocket in the enzyme's structure, and the warfarin molecule is positioned so that the likely site of hydroxylation is about 10 Å (1 nm) from the heme iron. The active site of CYP2C9 is large (about 470 Å^3) and it has been speculated that a second molecule of warfarin

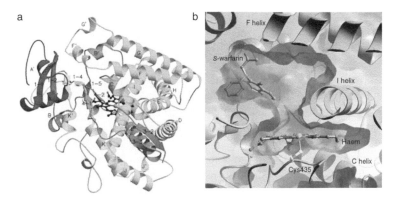

FIGURE 2.11 (See color insert following page 176.) Crystal structure of human CYP2C9, determined recently by Williams and colleagues (2003). Panel (a) shows the structure of CYP2C9 colored from blue at the N-terminus to red at the C-terminus. The heme group is shown as a ball-and-stick model in the center of the molecule. Panel (b) depicts the active site of CYP2C9, showing a substrate, *S*-warfarin, occupying the substrate-binding site. This active site is large and could be occupied either by substrates or inhibitors larger than *S*-warfarin, or multiple compounds of comparable size to *S*-warfarin. With *S*-warfarin bound in this location, the heme group remains available to metabolize additional substrate molecules. (We gratefully acknowledge copyright permission from *Astex Technology* and *Nature Publishing Group* for these diagrams).

or another drug could bind simultaneously. This would help to explain the complex enzyme kinetics displayed by CYP450 proteins and also how the metabolism of warfarin could be inhibited by another drug.

2.5.3.2.4 CYP2D

CYP2D6, the only significant human member of the 2D subfamily, has been one of the most rigorously studied P450 enzymes because of the discovery in the 1970s that 5 to 10% of the Caucasian population was deficient in the enzyme. This was the first major polymorphism to be identified within the P450 system (see the following text). The substrate-binding site of CYP2D6 has a relatively rigid structural requirement but, despite this, over 30 different drugs have been shown to be substrates for CYP2D6. These include antiarrhythmics, antipsychotics, antidepressants (including some of the newer selective reuptake inhibitors), and β-blockers. Analgesics such as codeine, oxycodone, and tramadol are also partly metabolized by CYP2D6. It is interesting to note that many of the drugs that are substrates for CYP2D6 are also neurotransmitter antagonists, which raises the possibility that CYP2D6 may have an endogenous role in neurotransmission. Indeed, it has been suggested that CYP2D6 may be functionally related to the dopamine transporter and that absence of the transporter protein may compromise neuronal response to transmitters.

2.5.3.2.5 CYP2E

This is the most important form of P450 from the perspective of the general anesthetics that are given by inhalation because CYP2E1 is largely responsible for the metabolism of many of them. It also metabolizes, and is inducible by, alcohol.

CYP2E1 is present not only in the liver but also in extrahepatic tissues such as lung and kidney. The toxicity associated with the general anesthetic halothane has been linked to its oxidation by CYP2E1 (see Section 2.5.6). Halothane can cause rare but serious liver damage, particularly if the patient receives it for successive surgical operations. Fortunately, some of the risk factors for this, such as being obese and being female, have been identified. CYP2E1 is also inhibited strongly by some compounds with anesthetic properties. In a recent study, a series of alcohols of increasing molecular size were tested for their ability to inhibit CYP2E1, and this was related to their potency as anesthetics. Curiously, a strong correlation was observed with the cutoff point at which the compounds were too large to inhibit CYP2E1, coinciding precisely with their loss of activity as anesthetics. This has allowed researchers to utilize P450 as a model for the elusive general anesthetic "receptor."

CYP2E1 is also responsible for the toxicity associated with the analgesic paracetamol. When taken accidentally or intentionally in an overdose (> 10 g), paracetamol causes severe liver damage. The effect is due to a toxic metabolite generated by CYP2E1 and can result in a long and excruciatingly painful death for the unfortunate victim.

2.5.3.2.6   CYP3A

Enzymes in this subfamily are the most abundant in human liver and account for up to 60% of the total P450, the principal form being CYP3A4. Thus, more than any other P450, the members of CYP3A determine the overall capacity of an individual's liver to oxidize drugs. They have the broadest substrate preference and are involved in the metabolism of the greatest number of different drugs. The numbers of drugs metabolized are too numerous to list here, but examples include lidocaine, alfentanil, nifedipine, midazolam, ciclosporin, many steroids including the contraceptive ethinyl estradiol, the natural steroid cortisol, and the analgesic methadone. CYP3A4 is inducible by glucocorticoids, rifampicin, and anticonvulsants such as carbamazepine and phenytoin.

2.5.3.3   Molecular Aspects of the Cytochrome P450 Supergene Family

The genes encoding the cytochrome P450 enzymes trace back almost as far as the development of life. The common ancestral gene probably appeared about 3 billion years ago and has since then diversified throughout the animal and plant kingdoms. Humans possess a large superfamily of P450 genes, present on a number of different chromosomes, which differ not only from those of other mammals but often from other individuals of identical racial origin. Genes that have adapted to fulfill a specific role are apt to undergo mutation and possibly even deletion from the gene pool when that role is no longer required. Over the last decade or so, several common defects in the expression of P450s have been discovered that, although inconsequential in physiological terms, may have marked pharmacological significance when the individual receives drug treatment.

The first of these defects to be discovered was with the enzyme CYP2D6, which is either defective or absent in at least 5 to 10% of the Caucasian population.

Allele	Mutation		Frequency
CYP2D6*3	Frameshift mutation	CYP2D6 gene ↑ base pair deletion	5%
CYP2D6*4A	Splice site mutation	CYP2D6 gene ↓ G-A transition	75%
CYP2D6*5	Gene deletion	CYP2D6 gene ←→ deletion	11%
	Rare and unknown mutations	CYP2D6 gene ?	5%

FIGURE 2.12 Polymorphisms in the isoforms of CYP2D6.

Three common polymorphisms are responsible for over 90% of the slow metabolizer phenotypes (Figure 2.12). The poor metabolizer phenotype has been linked to a few adverse drug reactions usually due to accumulation of the parent drug through reduced clearance (see Section 2.5.6). Structural polymorphisms have also been discovered in CYP2C19, CYP2C9, and CYP2A6. The pharmacological and toxicological impact, in quantitative terms, of the absence of a particular P450 isoform (either genetic or due to inhibition by concurrently administered drugs; see the following text) depends critically on (1) the fraction of the dose normally metabolized by that pathway and (2) the therapeutic index of the drug. The therapeutic index of a drug can be defined in terms of the maximum nontoxic dose tolerated, divided by the minimum effective dose. Hence, the higher this value, the better is the safety margin of the drug. Thus, drugs such as warfarin and phenytoin, which have low or narrow therapeutic indices and are largely metabolized by one P450 isoform, are more likely to cause toxicity than a drug such as propranolol that has a wider therapeutic index and is metabolized by several P450 isoforms.

Apart from structural polymorphisms, nucleotide changes have also been described in the regulatory regions of genes for P450 enzymes such as CYP1A1 and CYP2E1. These polymorphisms affect the binding of transcription factors and hence affect the level of transcription of the gene rather than the nature of the protein product. Such polymorphisms have been associated with chemical-induced diseases (see Section 2.5.6).

2.5.4 REGULATION OF THE CYTOCHROME P450 ENZYME SYSTEM

2.5.4.1 Induction

Several drugs are well-known inducers of the P450 enzymes (Table 2.2). In general, the mechanisms for the induction include:

- Increased gene transcription
- Increased mRNA stability and/or translation
- Decreased protein breakdown

Easily the most important of these is increased transcription. The induction of CYP1A1 has been the best characterized of all the P450s; induction involves the activation of a cytosolic transcription factor, the Ah receptor (AhR), which is normally inhibited by association with a heat shock protein (HSP90). Binding of the lipophilic inducer displaces HSP90, which allows the AhR to dimerize with a second protein, the Ah receptor nuclear transporter (ARNT). In the nucleus, the AhR-ARNT dimer binds to a specific enhancer region in the 5′ upstream control region of the CYP1A1 gene that causes a deformation of the DNA and allows proteins of the polymerase-binding complex to initiate transcription of the gene.

The inducing effect of barbiturates such as phenobarbital on some forms of P450 has been known for almost as long as the enzymes themselves. The mechanism is similar to that of CYP1A induction; however, other cytosolic receptors and transcription factors are involved in the induction process. CYP2C9, for example, can also be induced by steroids such as dexamethasone through a glucocorticoid receptor-

TABLE 2.2
Cytochrome P450 Isozymes Involved in the Metabolism of Drugs and Xenobiotics in Man

P450 Isoform	Substrates	Inducers	Inhibitors
CYP1A2	Theophylline	Cigarette smoke	Furafylline
	Clozapine	Omeprazole	Ciprofloxacin
CYP2A6	Methoxyflurane	Rifampicin	Tranylcypromine
	Halothane	Phenytoin	
CYP2C9	Diclofenac	Rifampicin	Sulfaphenazole
	Warfarin	Barbiturates	
CYP2C19	Diazepam	Rifampicin	Tranylcypromine
	Citalopram		
CYP2D6	Tramadol	Not inducible	Quinidine
	Codeine		
	Nortriptyline		
CYP2E1	Halothane	Alcohol	Disulfiram
	Enflurane		
	Isoflurane		
CYP3A4	Steroids	Glucocorticoids	Ketoconazole
	Alfentanil	Barbiturates	Erythromycin
	Amiodarone	Rifampicin	
CYP4A1	Testosterone	Clofibrate	—

Note: Only a few substrates, inhibitors, and inducers have been mentioned for each P450 isoform. More comprehensive lists can be found in the references given.

TABLE 2.3
Examples of Drug–Drug Interactions Produced by Either
P450 Enzyme Induction or Inhibition of P450

Drug (Inducer/Inhibitor)	Interacted Drug	Consequence
Enzyme Induction		
Phenytoin	Oral contraceptive	Unwanted pregnancy
Carbamazepine	Warfarin	Decreased blood clotting
Rifampicin	Ciclosporin	Decreased immunosuppression
Enzyme Inhibition		
Cimetidine	Warfarin	Hemorrhage
Ciprofloxacin	Theophylline	Cardiac arrhythmias
Erythromycin	Terfenadine	Serious effects on heart rhythm

responsive element in the nucleus, whereas phenobarbital and other drugs can induce via a constitutive androstane receptor (CAR) and the pregnane X receptor (PXR).

The clinical impact of enzyme induction depends on the number of different P450 isoforms affected, the magnitude of the inductive response within an individual, and also on the therapeutic indices of the affected substrates. The interaction may therefore lead to decreased efficacy through the accelerated elimination of a drug at an otherwise effective dose (Table 2.3). This effect has been observed with rifampicin and oral contraceptive (OC) pills. Ethinyl estradiol, a major component of most OCs, is metabolized exclusively by CYP3A4. This isoform is also powerfully induced by rifampicin, and the combination of rifampicin and OCs caused several unwanted pregnancies before the basis of the interaction was understood. Alternatively, enzyme induction may also lead to enhanced toxicity because there is an increase in formation of a toxic metabolite. The best-characterized example of this is with paracetamol, which damages the liver (hepatotoxicity) when taken in overdosage. Paracetamol is converted to the toxic metabolite by the P450 isoforms CYP1A2, CYP2E1, and CYP3A4. Thus, alcoholics who have induction of CYP2E1 and patients on anticonvulsants who have induction of CYP3A4 develop hepatotoxicity at lower doses of paracetamol, which is usually more severe in the case of alcoholics. *N*-acetylcysteine is given intravenously as the antidote for paracetamol poisoning. It is recommended that treatment with *N*-acetylcysteine be started at lower plasma levels of paracetamol than would be done for a patient with normal activities of these CYP isoforms.

2.5.4.2 Enzyme Inhibition

As would be anticipated with an enzyme system of such broad-based specificity, competition between ligands for a particular isoform is rife within the P450 family, with competitive inhibition between mutual substrates being a common pharmacological phenomenon. The extent to which this causes therapeutic problems will depend upon a number of factors for any particular drug:

- Therapeutic index of the drug
- Proportion of the overall metabolism of the drug conducted by the inhibited pathway
- Affinity constant (K_i) of the inhibitor relative to the affinity constant (K_m) of the drugs
- Whether or not the drug is metabolized to a pharmacologically active or toxic metabolite by the inhibited enzyme

In practice, the number of clinically significant adverse drug interactions caused by P450 inhibition is relatively small, because, to achieve a twofold increase in the plasma concentration of the drug, the target pathway must account for at least 90% of the metabolism of the drug. Examples of some P450 inhibitors are shown in Table 2.2, and examples of important interactions are shown in Table 2.3. An issue that has attracted attention recently is the interaction between terfenadine (an antihistamine that does not cause drowsiness, which is metabolized by CYP3A4) and inhibitors of this isoform such as ketoconazole (an antifungal drug) and erythromycin (an antibiotic). Concomitant administration of terfenadine with either ketoconazole or erythromycin leads to high plasma levels of terfenadine, which in some unfortunate individuals causes adverse effects on the rhythm of the heart and sudden death. Protease inhibitors used in the treatment of HIV disease are also potent P450 inhibitors. A recent case report describes the interaction between midazolam (a CYP3A4 substrate and sedative drug used before surgery) and saquinavir (a CYP3A4 inhibitor and anti-HIV drug), which needed the use of the antagonist flumazenil to reverse the sedative effects of midazolam.

Although inhibition of CYP enzymes often leads to undesirable effects and even death as described above, the inhibition of P450 has a potential for therapeutic use that is only just starting to be utilized. The high levels of P450 both in the liver and gastrointestinal (GI) tract result in extremely low bioavailabilities of a number of orally administered drugs, known as the first-pass effect. In effect, very little unchanged (active) drug reaches the general circulation in these cases because of extensive metabolism to inactive metabolites during the first pass(age) of the drug through the wall of the GI tract and/or the liver. An interesting possibility is to turn the inhibition of P450 to a strategic advantage to improve the bioavailability of highly metabolized drugs. This approach is being used with some anti-HIV drugs that are administered together with grapefruit juice, which contains a natural inhibitor of CYP3A4. Similarly, some transplant units administer ketoconazole with ciclosporin to inhibit the metabolism of ciclosporin, a relatively expensive drug, to reduce the dose and save money.

2.5.5 P450 AND THERAPEUTIC RESPONSE

There is considerable interindividual variability in the level of expression of the different P450 isoforms, which can affect the therapeutic efficacy of certain drugs. This variability may be acquired by concomitant administration of enzyme inducers or inhibitors, as alluded to in the above text, or alternatively, it may be genetically determined. With respect to the latter, it has been shown that individuals deficient

TABLE 2.4
The Effect of CYP2D6 Status on Drug Efficacy and Drug Toxicity

Drugs	Phenotype/ Genotype Affected	Effect	Mechanism
Debrisoquine	Poor metabolizer	Enhanced pharmacological effect	Reduced first-pass metabolism
Perhexilene	Poor metabolizer	Idiosyncratic hepatotoxicity and neuropathy	Accumulation due to reduced metabolism
Phenacetin	Poor metabolizer	Methemoglobinemia	Rerouting of metabolism
Codeine	Extensive metabolizer	Therapeutic failure	Active metabolite not formed
Encainide	Extensive metabolizer	Altered pharmacodynamic effect	Formation of metabolite with properties different from those of parent drug
Antidepressants	Ultrarapid metabolizer (CYP2D6L)	Therapeutic failure with conventional doses	Increased clearance due to extensive metabolism

in CYP2D6 cannot convert codeine, a prodrug, to the active component morphine, and thus do not derive any analgesic benefit from codeine. Conversely, it has recently been shown that some individuals have more than one copy of CYP2D6 as a result of gene amplification (one family has been reported to have 12 copies of the CYP2D6 gene), which can lead to therapeutic failure with drugs such as tricyclic antidepressants unless the patients are treated with doses that are higher than normally recommended.

2.5.6 P450 AND DRUG TOXICITY

Genetic deficiency of a particular enzyme may impair the metabolism of certain drugs and thus result in their accumulation with consequent toxicity. With some drugs, this is responsible for determining individual susceptibility to toxicity. Toxicity caused by many drugs has been associated with a deficiency of CYP2D6, and an individual with this genetic makeup is referred to as having the poor metabolizer phenotype (Table 2.4).

Most of these toxicities are predictable and arise from high plasma concentrations of the parent drug because of a reduced rate of elimination. The antianginal drug perhexilene has been associated with idiosyncratic hepatotoxicity and neuropathy in poor metabolizers deficient in CYP2D6. Perhexilene is a positively charged (cationic), amphiphilic compound with a lipophilic moiety and an ionizable nitrogen that becomes trapped in lysosomes and reacts with phospholipids leading to their accumulation within the lysosomes. This accumulation is believed to be a cause of the toxicity, which is dose-related with perhexilene. Thus, poor metabolizers are at higher risk because deficient metabolism of the drug leads to more extensive accumulation within the lysosomes.

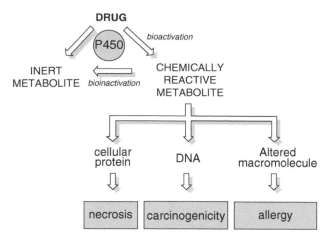

FIGURE 2.13 Metabolism by cytochrome P450 can produce chemically reactive metabolites that react with cellular macromolecules to produce different types of drug toxicity.

2.5.6.1 Liver Damage

The P450 enzymes, apart from converting drugs to stable metabolites, can also lead to the formation of toxic, chemically reactive metabolites that, if not detoxified, can combine with cellular macromolecules and lead to various forms of toxicity including cellular necrosis, hypersensitivity, and carcinogenicity (Figure 2.13). An important example is the hepatitis caused by the general anesthetic halothane. An extensive investigation called the National Halothane Study showed that two forms of hepatic injury occur with halothane, the first being a mild form that occurred in up to 20% of patients. This has received little attention in terms of mechanism but is thought to be some form of direct toxicity. The second form, "halothane hepatitis," is much more severe and is characterized by massive cell necrosis, but occurs in only 1 in 35,000 patients on primary exposure and in 1 in 3,700 patients on secondary exposure.

P450-mediated metabolism of halothane is thought to play an important role in the development of halothane hepatitis (Figure 2.14). In man, halothane undergoes 20% hepatic metabolism principally by cytochrome P450, with trifluoroacetic acid, chloride, and bromide being the main metabolites. Experiments with human liver cells (hepatocytes) have shown that halothane undergoes biotransformation to chemically reactive metabolites that combine with proteins, which are thus altered, in the liver to form several neoantigens (42 to 100 kDa). These neoantigens are able to stimulate an immune response and are similar to those identified in livers of animals treated with halothane *in vivo*. The neoantigens were localized mainly in the endoplasmic reticulum and in small amounts on the plasma membrane of parenchymal cells in intact hepatocytes *in vivo*, which would make them accessible to immune effector mechanisms. This would provoke an immune attack on the liver cells that are perceived as "foreign" because of the altered proteins on the cell surface. The neoantigens are generated as a result of oxidative biotransformation of the anesthetic

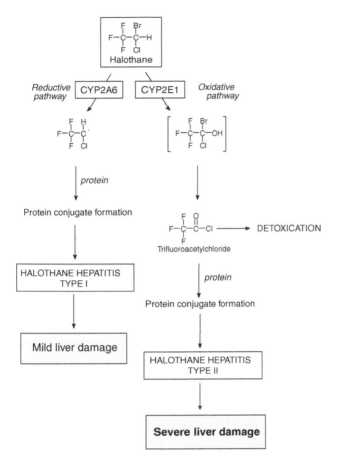

FIGURE 2.14 Halothane hepatitis. The reductive pathway of metabolism by CYP2A6 leads to a milder form of liver damage, whereas oxidative metabolism by CYP2E1 leads to rarer but much more severe damage.

that is catalyzed mainly by CYP2E1. This P450 isoform is present in substantial amounts in the centrilobular hepatocytes of both normal liver and liver in which it has been induced.

It appears that the severe idiosyncratic hepatotoxicity associated with halothane is initiated by a P450 biotransformation that occurs extensively in all individuals. Accordingly, it is likely that all experimental animals and patients exposed to halothane can generate haptenic conjugates within the liver. So why does halothane hepatitis not occur in everyone? It seems, therefore, that immune responsiveness is the susceptibility factor in halothane hepatitis. Given the role of CYP2E1 in the bioactivation of halothane, it has been suggested that pretreatment with disulfiram, a CYP2E1 inhibitor, would prevent halothane hepatitis, but this is yet to be tested clinically. Fortunately, there are several newer anesthetics that are similar to halothane but which undergo less metabolism that promise to be less damaging to the kidney.

2.5.6.2 Kidney Damage

In addition to the liver, the kidney can also be the target organ for toxicity caused by general anesthetics related to halothane. Kidney damage (nephrotoxicity) caused by methoxyflurane (2,2-dichloro-1,1-difluoro-1-methoxyethane), for example, also seems to be related to its metabolism by CYP2E1 (predominantly) to inorganic fluoride ions. The use of methoxyflurane, which is extensively metabolized (40%), is characterized by a high fluoride serum concentration that inhibits chloride transport in the ascending limb of the loop of Henle in the kidney. The administration of either enflurane or isoflurane, which rarely cause nephrotoxicity, is accompanied by much lower serum levels of inorganic fluoride because of lower rates of metabolism (3% and 1% for enflurane and isoflurane, respectively). However, this is not the only reason because sevoflurane, which is also associated with the high serum levels of fluoride, rarely causes nephrotoxicity. The explanation for this may be the high rate of intrarenal metabolism by CYP2E1, 2A6, and 3A4 (in diminishing order of activity) of methoxyflurane but not of sevoflurane.

2.5.7 P450 AND DISEASE

Given the role of certain P450 isoforms in the activation of procarcinogens, a great deal of research has been undertaken to determine whether P450 isoforms play a role in the multistep process of carcinogenesis in man. This work has concentrated on the cancers associated with cigarette smoking, such as cancer of the lung and bladder. Figure 2.12 shows the frequency of polymorphisms in the isoforms of CYP2D6 that have arisen through mutation. Attempts to find a link between lung cancer in patients and such polymorphisms in their P450 isoforms of CYP1A1, CYP2D6, and CYP2E1 have so far produced conflicting results. Thus, the role of P450 enzymes in the pathogenesis of the different types of cancer remains unclear at present, and further studies are required.

The polymorphism of CYP2D6 has also been studied in relation to its role in determining susceptibility to Parkinson's disease. In a study of 229 patients and 720 controls, it was found that patients deficient in CYP2D6, i.e., poor metabolizers, had a 2.54-fold increased risk of Parkinson's disease. It has been suggested that CYP2D6 protects critical areas of the brain such as the substantia nigra from toxic effects of chemicals such as MPTP (1-methyl-4-phenyl-1,2,3,6-tetrahydropyridine).

Patients with autoimmune hepatitis have a circulating autoantibody termed anti-LKM1; the antigen to which this antibody binds has been identified as CYP2D6. More recent studies have also demonstrated an association of anti-LKM1 antibodies with Hepatitis C virus (HCV) infection. In both instances, it has been suggested that anti-CYP2D6 antibodies may be arising as a result of molecular mimicry, in which parts of the peptide chain of two different proteins are identical. For example, there is a sequence of 33 amino acids in the peptide of CYP2D6 identical to two segments of the HCV polyprotein.

It is apparent that investigations into the role of P450 enzymes in human disease are in their infancy and are likely to become the focus of increasing research attention in the future.

2.5.8 CONCLUSIONS

The cytochrome P450 enzyme system is the most versatile biological catalyst known to man. Its importance extends into every field of medicine. It is now routine practice in the development of a new drug to characterize the metabolic pathways and the P450 and non-P450 enzymes involved in its metabolism (biotransformation). This will help to predict potential drug–drug interactions before the drug reaches clinical practice, and thereby prevent either therapeutic failure because the drug is inactivated by metabolism too rapidly, or drug toxicity when the drug accumulates and produces toxic metabolites. There is no doubt that recent advances in molecular biology have provided valuable insights into the functioning of this ubiquitous group of enzymes and will, in the future, help to identify its role in drug-induced and non-drug-induced diseases. This will make drug therapy safer and provide new mechanistic insights into disease processes to aid diagnosis and treatment.

FURTHER READING

Drug Metabolism Reviews (2002), 34, 1–450. Two issues (1 and 2) devoted to human cytochrome P450.

Lewis, D.F.V. (2001). *Cytochromes P450. Structure, Function, and Mechanism.* London: Taylor & Francis.

Williams, P.A., Cosme, J., Ward, A., Angove, H.C., Vinkovic, M.D., and Jhoti, H. (2003). Crystal structure of human cytochrome P450 2C9 with bound warfarin. *Nature*, 424, 464–468.

2.6 CARBONIC ANHYDRASE EXAMPLE

Claudiu T. Supuran, Andrea Scozzafava, and Angela Casini

2.6.1 INTRODUCTION

The carbonic anhydrases (CAs, EC 4.2.1.1) are ubiquitous metalloenzymes, present in prokaryotes and eukaryotes, being encoded by three distinct, evolutionarily unrelated gene families: the α-CAs (present in vertebrates, bacteria, algae, and cytoplasm of green plants), the β-CAs (predominantly in bacteria, algae, and chloroplasts of both mono- as well as dicotyledons), and the γ-CAs (mainly in archaea and some bacteria), respectively.[1–4] In higher vertebrates including humans, 14 different α-CA isozymes or CA-related proteins (CARP) were described, with very different subcellular localization and tissue distribution.[2–4] Basically, there are several cytosolic forms (CA I–III, CA VII), four membrane-bound isozymes (CA IV, CA IX, CA XII, and CA XIV), one mitochondrial form (CA V), as well as a secreted CA isozyme (CA VI), together with three acatalytic forms (isozymes CARP VIII, X, and XI).[2–4]

These enzymes efficiently catalyze a very simple but fundamental physiological reaction, the interconversion between carbon dioxide and the bicarbonate ion

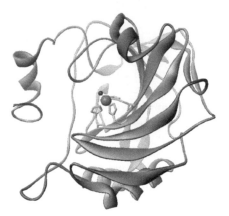

FIGURE 2.15 Human isozyme CA II (hCA II) backbone folding, zinc ion (sphere) and its ligands (three histidine residues and a water molecule).

$$O=C=O + H_2O \Leftrightarrow HCO_3^- + H^+ \tag{2.9}$$

Because CO_2 is generated in high amounts in all living organisms, CAs have a high versatility and are involved in crucial physiological processes connected with respiration and transport of CO_2/bicarbonate between metabolizing tissues and lungs, pH and CO_2 homeostasis, electrolyte secretion in a variety of tissues and organs, biosynthetic reactions (such as gluconeogenesis, lipogenesis, and ureagenesis), bone resorption, calcification, tumorigenicity, and many other physiologic or pathologic processes.[1–4]

2.6.2 MECHANISM

The general shape of this protein, containing one polypeptide chain of around 260 amino acid residues (molecular weight of 30 kDa) mainly folded as β-pleated sheets, and a catalytical metal ion are shown in Figure 2.15.

The Zn(II) ion of CAs is essential for catalysis.[4–6] X-ray crystallographic data showed that the metal ion is situated at the bottom of a 15 Å-deep active-site cleft (Figure 2.15 and Figure 2.16), being coordinated by three histidine residues (His 94, His 96, and His 119) and a water molecule/hydroxide ion.[7]

The zinc-bound water is also engaged in hydrogen bond interactions with the hydroxyl moiety of Thr 199, which in turn is bridged to the carboxylate moiety of Glu 106; these interactions enhance the nucleophilicity of the zinc-bound water molecule and orientate the substrate (CO_2) in a favorable location for the nucleophilic attack[4–7] (Figure 2.17). The active form of the enzyme is the basic one, with hydroxide bound to Zn(II) (Figure 2.17a). This strong nucleophile attacks the CO_2 molecule bound in a hydrophobic pocket in its neighborhood (the well-hidden substrate-binding site comprises residues Val 121, Val 143, and Leu 198 in the case of the human isozyme CA II[5] (Figure 2.17b), leading to the formation of bicarbonate coordinated to Zn(II) (Figure 2.17c).

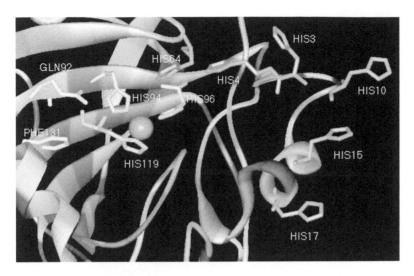

FIGURE 2.16 Details of hCA II active site. The Zn(II) ion (central sphere) and its three histidine ligands (His 94, His 96, His 119) are shown. The histidine residues (His 64, His 4, His 3, His 17, His 15, and His 10) involved in proton transfer reactions between the active site and the reaction medium are also seen. The figure was generated from the x-ray coordinates reported by Briganti et al. (1997)[9] (PDB entry 4TST). (From Scozzafava, A. and Supuran, C.T. (2002). *Bioorganic Medical Chemistry Letters,* 12, 1177–1180. With permission. Copyright Elsevier Science, 2002.)[9]

The bicarbonate ion is then displaced by a water molecule and liberated into solution, leading to the acid form of the enzyme, with water coordinated to Zn(II) (Figure 2.17d), which is catalytically inactive.[4–6,8] In order to regenerate the basic form A, a proton transfer reaction from the active site to the environment takes place, which may be assisted either by active-site residues (such as His 64 — the proton shuttle in isozymes I, II, IV, VII, and IX among others; see Figure 2.16 for isozyme II) or by buffers present in the medium. The process may be schematically represented by Equation 2.10 and Equation 2.11.

$$EZn^{2+}\!\!-\!\!OH^- + CO_2 \Leftrightarrow EZn^{2+}\!\!-\!\!HCO_3^- \overset{H_2O}{\Leftrightarrow} EZn^{2+}\!\!-\!\!OH_2 + HCO_3^- \quad (2.10)$$

$$EZn^{2+}\!\!-\!\!OH_2 \Leftrightarrow EZn^{2+}\!\!-\!\!HO^- + H^+ \quad\quad\quad (2.11)$$

The rate-limiting step in catalysis is the second reaction, i.e., the proton transfer that regenerates the zinc-hydroxide species of the enzyme.[4–6,8] In the catalytically very active isozymes such as CA II, CA IV, CA V, CA VII, and CA IX, the process is assisted by a histidine residue placed at the entrance of the active site (His 64), as well as by a cluster of histidines (Figure 2.16) that protrudes from the rim of the active site to the surface of the enzyme, thus assuring a very efficient proton transfer

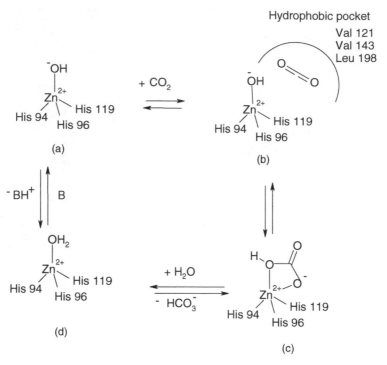

FIGURE 2.17 Schematic representation of the catalytic mechanism for the CA-catalyzed CO_2 hydration. The hypothesized hydrophobic pocket for the binding of substrates is shown schematically at step (b).

process for the most efficient CA isozyme, CA II.[10] This also explains why CA II is one of the most active enzymes known (with a $k_{cat}/K_m = 1.5 \times 10^8$ M^{-1} sec^{-1}), approaching the limit of diffusion control,[4,5,8] and also has important consequences for the design of inhibitors with clinical applications. Kinetic parameters for several CA isozymes, as well as inhibition data with acetazolamide (a clinically used inhibitor), are shown in Table 2.5.

The generally high activity of most CA isozymes as well as their abundance in target tissues in higher vertebrates is a critically important factor for obtaining inhibitors with clinical applications. Thus, these enzymes must be inhibited to the extent of 99.9% in order to achieve clinical responses, such as, for example, in the treatment of glaucoma with topically applied sulfonamide inhibitors.[2,3] The recent isolation of several CA isozymes (e.g., IX and XII) in tumor tissues also opens novel strategies for the design of antitumor therapies based on such enzyme inhibitors.[4,9,11,12]

TABLE 2.5
Kinetic Parameters for Different α-CA Isozymes,
and Their Inhibition Data with Acetazolamide
(5-Acetamido-1,3,4-thiadiazole-2-sulfonamide),
a Clinically Used Inhibitor

Isozyme	Activity Level	k_{cat} (sec^{-1})	k_{cat}/K_m (M.sec^{-1})	K_I (acetazolamide) (μM)
hCA I	Moderate	2×10^5	5×10^7	0.2
hCA II	High	1.4×10^6	1.5×10^8	0.01
hCA III	Low	1×10^4	3×10^5	300
hCA IV	High	1.1×10^6	5×10^7	0.04
mCA V	Moderate	7×10^4	3×10^7	0.06
hCA IX	High	3.8×10^5	5.5×10^7	0.025

Note: h = human; m = murine isozyme.

Source: From Supuran, C.T., Scozzafava, A., and Casini, A. (2003). *Medicinal Research Reviews,* 23, 146–189.[4]

REFERENCES

1. Smith, K.S. and Ferry, J.G. (2000). Prokaryotic carbonic anhydrases. *FEMS Microbiological Reviews,* 24, 335–366.
2. Supuran, C.T. and Scozzafava, A. (2000). Carbonic anhydrase inhibitors and their therapeutic potential. *Expert Opinion on Therapeutic Patents,* 10, 575–600.
3. Supuran, C.T. and Scozzafava, A. (2002). Applications of carbonic anhydrase inhibitors and activators in therapy. *Expert Opinion on Therapeutic Patents,* 12, 217–242.
4. Supuran, C.T., Scozzafava, A., and Casini, A. (2003). Carbonic anhydrase inhibitors. *Medicinal Research Reviews,* 23, 146–189.
5. Christianson, D.W. and Fierke, C.A. (1996). Carbonic anhydrase: evolution of the zinc-binding site by nature and by design. *Accounts of Chemical Research,* 29, 331–339.
6. Bertini, I., Luchinat, C., and Scozzafava, A. (1982). Carbonic anhydrase: an insight into the zinc binding site and into the active cavity through metal substitution. *Structure and Bonding,* 48, 45–92.
7. Stams, T. and Christianson, D.W. (2000). X-ray crystallographic studies of mammalian carbonic anhydrase isozymes. In *The Carbonic Anhydrases — New Horizons,* Chegwidden, W.R., Edwards, Y., and Carter, N., Eds., Birkhäuser Verlag, Basel, pp. 159–174.
8. Lindskog, S. and Silverman, D.W. (2000). The catalytic mechanism of mammalian carbonic anhydrases. In *The Carbonic Anhydrases — New Horizons,* Chegwidden, W.R., Edwards, Y., and Carter, N., Eds., Birkhäuser Verlag, Basel, pp. 175–196.
9. Scozzafava, A. and Supuran, C.T. (2002). Carbonic anhydrase activators: human isozyme II is strongly activated by oligopeptides incorporating the carboxytermical sequence of the bicarbonate anion exchanger AE1. *Bioorganic & Medicinal Chemistry Letters,* 12, 1177–1180.

10. Briganti, F., Mangani, S., Orioli, P., Scozzafava, A., Vernaglione, G., and Supuran, C.T. (1997). Carbonic anhydrase activators: x-ray crystallographic and spectroscopic investigations for the interaction of isozymes I and II with histamine. *Biochemistry*, 36, 10384–10392.
11. Vullo, D., Franchi, M., Gallori, E., Pastorek, J., Scozzafava, A., Pastorekova, S., and Supuran, C.T. (2003). Carbonic anhydrase inhibitors. Inhibition of the tumor-associated isozyme IX with aromatic and heterocyclic sulfonamides. *Bioorganic and Medicinal Chemistry Letters*, 13, 1005–1009.
12. Winum, J.-Y., Vullo, D., Casini, A., Montero, J.-L., Scozzafava, A., and Supuran, C.T. (2003). Carbonic anhydrase inhibitors: inhibition of cytosolic isozymes I and II and the transmembrane, tumor-associated isozyme IX with sulfamates including EMATE also acting as steroid sulfatase inhibitors. *Journal of Medicinal Chemistry*, 46, 2197–2204.

2.7 PROTEASES EXAMPLE

André J. Niestroj and Hans-Ulrich Demuth

2.7.1 INTRODUCTION

It is commonly recognized that proteases play a crucial role in living organisms, ranging from digestion of proteins to the regulation of physiological processes. Proteases fulfill specific functions in antigen presentation, processing events, apoptosis, signal transduction, extracellular matrix turnover, fertilization, and immune response. They also appear to be involved in wound healing, matrix remodeling, and differentiation. Evidence of the importance of proteases is indicated by the results on several knockout mice, which lack a gene for a specific protease; many examples in the literature show that these mice have impaired physiological functions and therefore stop thriving. As a consequence, proteases are promising targets for the development of drugs. Without knowledge of the mechanism and architecture of the active site of proteases, the development of potent drugs as therapeutically useful protease inhibitors and the prediction of possible activities of anticipated structures are not feasible. In particular, new approaches in drug design, for example, virtual high-throughput screening or computer-assisted drug design have been successful in reducing time-consuming and costly synthetic work.[1–6] Therefore, investigations of the mechanism of enzymes have been an exciting and important field for years. This chapter will focus on several examples using 3-D figures of proteases for displaying and classifying the appropriate mechanism.

2.7.1.1 Characterization of Proteases

Proteolysis is described as the hydrolytic cleavage of peptide bonds under enzymatic conditions, in which the enzymes act as catalysts for the irreversible cleavage (Scheme 2.1). The enzymes responsible for the catalysis have been named *proteases*. The synonym *peptidase* characterizes any enzyme that hydrolyzes peptide bonds. Synonymous with peptidase, it is still possible to use the term protease. Other commonly used terms in the terminology of proteolytic enzymes are *endopeptidases*

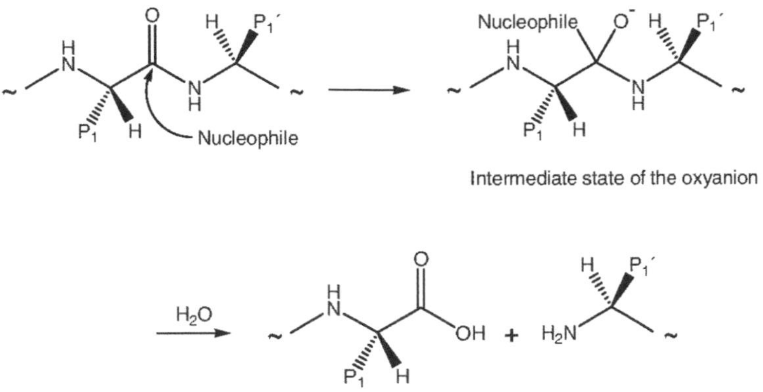

SCHEME 2.1 Hydrolysis of peptide bonds independent of the considered mechanism.

and *exopeptidases*. Endopeptidases act on the internal bonds of proteins in contrast to exopeptidases, which act only near the terminus of proteins.

The relation between the peptide bond and the favored hydrolyzed form is defined as an equilibrium state. Cleavage of a peptide bond (more precisely, an amide bond) yields hydrolyzed amino acids and a gain of a defined amount of energy. Therefore, this reaction is energetically favored. This neither means that the reaction occurs rapidly nor that spontaneous hydrolysis results. The cleavage has to pass an energetically high activation barrier. To overcome this activation barrier, very strong conditions are required to cleave peptide bonds in the laboratory (for acid hydrolysis: 6 N HCl, 24 h, 110°C). Without this activation barrier, all peptides would be immediately cleaved by water. Proteases are able to cleave peptide bonds in living cells under mild conditions by reducing the level of the activation barrier compared to the initial activation barrier (see Section 2.1).

Research on proteolytic enzymes has a long history. Investigations of proteases by affinity labeling were introduced more than 30 years ago to obtain information about the architecture of the active site and the mechanism of proteolytic enzymes. By this method, a label is used to identify or localize active-site residues of a protease. In general, an affinity label is a molecule structurally similar to a substrate or product; however, it also contains a chemically reactive group that forms a covalent bond with the protease.[7,8] Another means of identifying active-site residues is site-directed mutagenesis, involving an amino acid exchange within the putative active site. The subsequent inability to cleave a substrate highlights the importance of the mutated amino acid.[9,10] Active-site mutations have also been used to address the specific role of various other amino acids within the enzyme molecule.[11] A better understanding concerning spatial features has been gathered by x-ray diffraction and NMR spectroscopy to reveal three-dimensional structures. Both NMR spectroscopy and x-ray crystallography are powerful tools for structure determination.

Apart from the differences in localization, specificity, and function, proteases differ mainly in the ways in which they catalyze the hydrolysis of the peptide bond. Proteases are classified into four hydrolase classes according to their different catalytic mechanisms: *serine, cysteine, metallo,* and *aspartic* proteases.[12] However, it has been recently suggested that *threonine* peptidases should be added as a new class. In serine and threonine peptidases, the hydroxyl group of the active-site serine or threonine acts as a nucleophile, whereas in cysteine peptidases the nucleophile is the sulfhydryl group of the active-site cysteine. The nucleophile in metallopepti-dases and aspartic peptidases is a water molecule. Hence, one mechanistically discriminates between covalent catalyzing (serine, cysteine, and threonine) and non-covalent catalyzing (aspartate and metallo) enzymes. These mechanisms have been described elsewhere.[13] It is noteworthy that clans containing similar proteases are common; for example, clan PB contains proteases that have a cysteine residue in their active site, similar to conventional cysteine proteases contained in clans CA, CB, etc. We have followed the MEROPS peptidase database classification, which gives a better understanding of the evolutionary relationship of the proteases.

2.7.1.2 Specificity of Proteases and Spacial Features of the Active Site

Proteases act as important biochemical switches in living cells. Therefore, a strict specificity of proteolysis is necessary. The specificity of proteases is afforded by a series of factors. First of all, many proteases act at a defined pH-optimum. Further-more, the recognition of the substrate by the proteases is necessary. This comprises the optimal fitting of the substrate between the residues of the active site as well as subsite binding of the substrate. Figure 2.18 shows the superposition of the active-site residues of a cysteine protease and a serine protease. Both proteases act by a catalytic triad, and the spacial features are similar. Because of these similarities, one could also assume similarities in the cleavage behavior for substrates. The specificity is assured by the optimum binding of substrate in the positions $P_1, P_2, P_3 \ldots P_n$ to $P_1', P_2', P_3' \ldots P_n'$ with the corresponding subsites ($S_1, S_2, S_3 \ldots S_n$ to S_1', S_2', S_3'

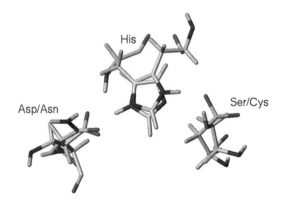

FIGURE 2.18 Superposition of the catalytic triads of the cysteine protease papain and the serine protease trypsin. The superposition was performed by overlapping several fitting points of the residues Cys 25, His 159, and Asn 175 of papain and Ser 195, His 57, and Asp 102 of trypsin. The atomic coordinates of PDB ID code 9PAP[15] and 1BRA[16] were used to construct this figure. The catalytic Cys 25 is in an oxidized form.

... S_n') and act with discrimination regarding size and polarity of substrates. For the interaction of a substrate and a proteolytic enzyme, the nomenclature of Berger and Schechter has become commonly used.[14] Based on the scissile bond, the amino acid residues of the amino terminus are numbered P_1, P_2, P_3, to P_n, etc., and the amino acid residues of the carboxy terminus are numbered P_1', P_2', P_3', to P_n', etc. The corresponding subsites of the active enzyme are numbered complementary S_1, S_2, S_3, to S_n and S_1', S_2', S_3', to S_n'. The different types of binding are hydrophobic interactions, salt bridges, and hydrogen bonds. The rate of cleavage, or at least the selectivity is influenced by the intensity of binding.

Investigations on the catalytic features of serine and cysteine proteases revealed crucial differences between the two classes of enzymes. In serine proteases, the P_1 amino acid is of primary importance for binding and specificity compared to cysteine proteases, whereas the P_2 amino acid is of greater significance for specificity.[17–19] Moreover, in serine proteases the negatively charged tetrahedral intermediate formed by nucleophilic attack is stabilized by a complex hydrogen bonding system (see Scheme 2.1). It is assumed that in cysteine proteases the electrostatic oxyanion stabilization is not so crucial for catalysis.[20–29] All these features are responsible for the differences in the recognition of substrates.

Overall, similar to these differences, proteases exhibit slight variations in the composition of the subsites for the restriction of substrates.

2.7.2 Determination of the Structure of Proteases

2.7.2.1 Initial Analytical Methods for the Characterization of Proteases

An isolated protease can be characterized by biochemical and analytical methods. Initially, the hydrolytic specificity of the cleavage is of obvious interest to biochemists. Therefore, in assays, several different substrates are examined to determine the cleavage behavior of a protease. Furthermore, analytical methods like SDS PAGE, native PAGE, and gelfiltration determine the size and oligomerization state of the protease, and isoelectric focusing (IEF) reveals the isoelectric point (iP). The primary sequence is determined either by Edman degradation or by the use of Sanger's reagent. To investigate structural features of sugar residues of proteases, lectin affinity chromatography is used. Additional characterization of a protease can be achieved by analytical chemistry methods; e.g., mass spectroscopy, which provides information about the molecular weight and has an application in the analysis of protein sequences.

2.7.2.1.1 Three-Dimensional Determination

Spectroscopy is the investigation of the interaction of electromagnetic radiation with matter. The broad electromagnetic spectrum ranges from radio waves (with the longest wavelengths and the lowest level of energy) to gamma rays (with radiation of the shortest wavelengths and the highest level of energy). Additionally, high-level energy radiation has high frequency and *vice versa*. Depending on the radiation used, different features of the investigated matter through the detected interaction become available.

The determination of the structure and conformation of proteases have been investigated by several methods, but there is insufficient space here to explain them in a detailed manner. Therefore, the most important methods used for structural determination are listed and briefly described (for additional reading, see the reference list).

IR spectroscopy provides information on functional groups present in molecules. Other important methods that reveal information on the structural features of proteases are fluorescence spectroscopy, circular dichroic spectroscopy (CD spectroscopy), and atomic force microscope (AFM). Fluorescence spectroscopy has an application in ligand-binding studies and in the quantitation of proteins. CD spectroscopy investigates the secondary structure and conformational changes of biomacromolecules.[30] The AFM observes the surface of macromolecules in subnanometer resolution. By the use of this method, the function and assembly of single molecules has been described.[31] The most important methods for the structural characterization of macromolecules are x-ray diffraction[32–35] and nuclear magnetic resonance (NMR).[36–39]

2.7.3 General Models of the Architecture of the Active Site

Early ideas on the catalytic activity of enzymes were made by E. Fischer.[40] His *key–lock theory*, also referred to as *template theory*, describes the substrate as a key and the enzyme as the corresponding lock. Analogous to a key which fits properly into a lock, the substrate is recognized and cleaved. Nevertheless, this theory fails to explain why certain analogs of substrates are excluded. To overcome this problem, Koshland elaborated on the key–lock theory and introduced the *induced fit* model, which considers aspects of the flexible properties of the substrate and the enzyme.[41] In this model, interaction of the substrate and the protease at the active site results in a transition state, and structural rearrangements lead to modified configurations in both the protease and the substrate. Lipscomb and Quiocho have published the three-dimensional structure of carboxypeptidase A, in which conformational changes were observed in the complex of protease and substrate compared to the protease alone.[42,43] In Figure 2.19, the superposition of carboxypeptidase A as mature peptidase (**I**) and in complex with the substrate glycyl-L-tyrosine (**II**) is shown. The conformational changes of the highlighted three amino acids are clearly illustrated.

2.7.4 Evolutionary Aspects

Proteases catalyze the hydrolysis of peptides and are divided into four major classes: serine, cysteine, metallo, and aspartic proteases. This classification simply characterizes the mechanism of the catalysis but gives no impression regarding evolutionary development. However, there are indications of a common origin for the evolution of proteases, starting from those with only digestive functions to those with functions in regulatory processes.[46,47] The proteases in this very complex evolutionary process have undergone either divergent or convergent evolution. Proteases of a divergent evolution have been defined to evolve from a common ancestor. They have similarities in their three-dimensional structure, active site, function, and amino acid sequence. For instance, trypsin and chymotrypsin are assumed to have evolved from

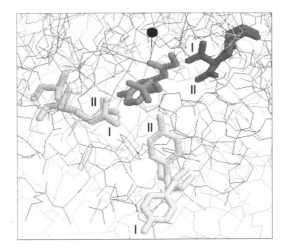

FIGURE 2.19 Superposition of carboxypeptidase A as mature peptidase and in complex with the substrate glycyl-L-tyrosine is shown. In the center, the zinc atom (black sphere) is given, surrounded by Arg 145 (green), Glu 270 (blue), Tyr 248 (yellow), and the substrate (red). The atomic coordinates of PDB ID code 3CPA[44] and 1M4L[45] were used to construct this figure.

a common origin due to homologous amino acid sequences and very similar tertiary structures. Superposition of trypsin and chymotrypsin shows structural similarity even though only about 40% of the amino acid sequences between the two proteins are identical. In contrast, proteases of convergent evolution have similarities in function and active site, but they differ in their three-dimensional structure and amino acid sequence.[47]

Using over 80 enzymes of the chymotrypsin family, Di Cera and coworkers have searched for signs of an evolutionary history of serine proteases. The authors determined that ultimately only a limited number of residues clustered around the active site characterize the organization of a phylogenetic tree of serine proteases.[48,49] For cysteine proteases, Barrett and Rawlings observed seven evolutionary origins. All cysteine proteases with a characteristic three-dimensional structure and function evolved from these origins.[50] Subsequently, an even more detailed classification concerning the evolutionary development has recently been introduced.[51] The known proteases are classified into clans, which are further subclassified into families based on structural similarities. A clan is a group of families assumed to have arisen from an evolutionary ancestor. For a clan, an evolutionary relationship is indicated by the identical order of the catalytic-site residues or similar tertiary structures. A family, in turn, consists of proteases with similar amino acid sequences. Based on these features, the MEROPS peptidase database was created. This database currently contains 2055 peptidases, divided into 42 clans and 172 families (release 6.30, June 2003).

Based on the functions, a classification of enzymes was established by the Nomenclature Committee of the International Union of Biochemistry and Molecular Biology (NC-IUBMB). This classification was published in *Enzyme Nomenclature 1992* and its supplements.[52–57]

2.7.5 EXAMPLES OF DIFFERENT ARCHITECTURES OF PROTEASES

2.7.5.1 Covalent Catalysis: Cysteine Proteases

The active sites of proteases are typically characterized by a catalytic triad or dyad, an oxyanion hole, and specificity-binding pockets. The divergence of substrate specificity is determined by the structure of the specificity pockets. Cysteine proteases are divided into the clans C-, CA, CD, CE, CF, and CH in the MEROPS peptidase database. The active-site residues mentioned here are given in the order as found in the sequence.

2.7.5.2 Catalytic Triad (Cys, His, Asn/Asp)

The largest subfamily among the class of cysteine proteases is the papain-like cysteine proteases, originating from papain as the archetype of the cysteine proteases. Papain is a plant enzyme that belongs to the clan CA and gives the family (C1) its name. The active site of proteases belonging to the clan CA consists of a histidine, an asparagine or aspartic acid, and the nucleophilic residue cysteine. In Scheme 2.2, the catalytic mechanism for cysteine proteases with a catalytic triad consisting of Cys, His, Asn/Asp is given.

In papain-like cysteine proteases, the nucleophilic cysteine residue is embedded in a highly conserved peptide sequence CGSCWAFS; active-site cysteine is underlined, whereby only a small number of proteases possess alternative residues in this region. In addition, the vicinities of the residues asparagine and histidine are highly

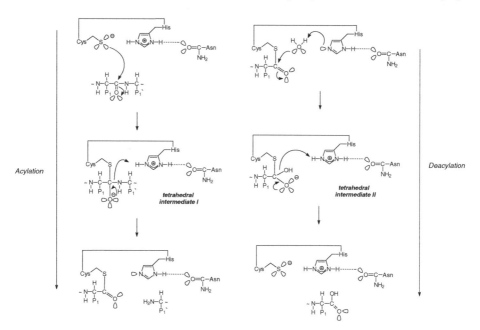

SCHEME 2.2 Catalytic mechanism for a cysteine protease with a catalytic triad consisting of Cys, His, Asn/Asp. An asparagine residue is used in the scheme.

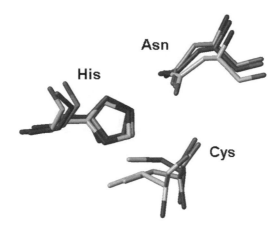

FIGURE 2.20 Superposition of the catalytic triads of papain (orange), yeast bleomycin hydrolase (gray), and procathepsin X (purple). The superposition was performed manually, overlapping the residues Cys 25, His 159, and Asn 175 of papain; Cys 73, His 369, and Asn 392 (yeast bleomycin hydrolase); and Cys 31, His 180, and Asn 200 (procathepsin X). To create this structural comparison figure, the atomic coordinates of papain (PDB ID code 1PE6),[59] yeast bleomycin hydrolase (PDB ID code 1GCB)[60] and procathepsin X (PDB ID code 1DEU)[61] were used. For better clarity, only the active site consisting of the residues cysteine, histidine, and asparagine are shown.

conserved as well as the vicinities of the active-site residue cysteine.[58] The super-position of the catalytic triads of three members of the peptidase family C1 (papain family) in Figure 2.20 gives an impression of the similarities in position and orientation of the active-site residues.

Figure 2.21 depicts the active site of papain. The endopeptidase papain was isolated from papaya and other plants. It consists of a single polypeptide chain of

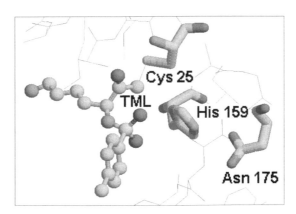

FIGURE 2.21 Representation of the active site of papain in a complex with a chloromethyl ketone substrate analog. The atomic coordinates of PDB ID code 4PAD[62] were used to construct this figure. The residues of the inhibitor are shown as ball-and-stick figures (TML = tosylmethylenyllysyl residue).

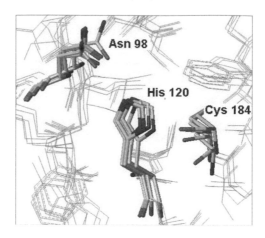

FIGURE 2.22 Representation of the active site of sortase. For better clarity, only 5 of 25 available conformers of an NMR study are shown in a capped stick representation. The atomic coordinates of PDB ID code 1IJA[67] were used to construct this figure.

212 residues with a molecular weight of approximately 21,000 Da. Papain catalyzes the hydrolytic cleavage of a broad substrate range; of relevance is the cleavage of immunoglobulins into specific fragments. In Figure 2.21, the active site of papain with the catalytic residue cysteine bound to a tosylmethylenyllysyl residue is shown. This covalent complex was formed by the attack of Cys 25 on the substrate analog chloromethyl ketone.

As an example of a structure solved by NMR spectroscopy, the active site of the transpeptidase sortase (member of the clan C-, family C60), a 206-residue polypeptide, is given in Figure 2.22. Based on more than 2100 experimental restraints, Clubb and coworkers have calculated the structure of the truncated sortase (residues 59–206), represented by 25 conformers. It is assumed that the active site consists of the residues Asn 98, His 120, and Cys 184 (numbering based on the untruncated transpeptidase). It suggests a correlation to the papain-like cysteine proteases in the catalytic mechanism, though there is no similarity in the structural features of the proteases.

Sortase is crucial in anchoring proteins to the cell wall of Gram-positive bacteria such as *Staphylococcus aureus* and *Staphylococcus suis*.[63,64] Therefore, the protease plays an important role in the pathogenesis of infections caused by Gram-positive bacteria.[65] Homologs of sortase have been indentified in several bacteria.[66] For the development of antibiotics, knowledge of the mechanism and architecture of sortase is clearly crucial, especially within the background of an increasing number of drug-resistant microbes.

2.7.5.2.1 Catalytic Triad (Glu, Cys, His and His, Glu/Asp, Cys)

In addition to cysteine proteases possessing a catalytic triad, other catalytic mechanisms for cysteine proteases are described. In clan CF, the asparagine residue of the catalytic triad is replaced by a glutamine residue (Glu, Cys, His). In clan CE, proteases acting by a catalytic triad in the order His, Glu (or Asp), Cys, are grouped.

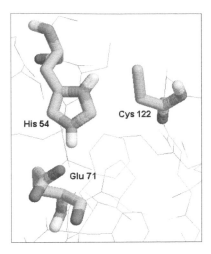

FIGURE 2.23 Representation of the active site of human adenovirus type 2. PDB ID code 1AVP[68] was used to construct this figure.

As an example, the human adenovirus proteinase 2 as a member of the clan CE (family C5) is given. Figure 2.23 shows the active site of the human adenovirus proteinase 2. It is commonly assumed that this proteinase is required for the generation of infectious viruses.[68]

2.7.5.2.2 Catalytic Dyad (Cys, His)

X-ray diffraction studies of different caspases revealed that there is no other amino acid in close proximity to the catalytic Cys and His to serve as the third member in a catalytic triad.[69] A proposed catalytic mechanism of caspase 1 describes the process of peptide bond cleavage in which the active-site cysteine attacks the carbonyl carbon atom. The cleavage is supported by the catalytic histidine. In contrast to the papain mechanism, the substrate has to fit between both active-site residues. In Figure 2.24,

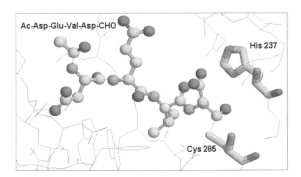

FIGURE 2.24 Representation of the active site of apopain/CPP32 in complex with the tetrapeptide Ac-Asp-Glu-Val-Asp-CHO. PDB ID code 1PAU[70] was used to construct this figure.

the active site of apopain/CPP32 is shown. Apopain/CPP32 acts as a key mediator of apoptosis and cleaves the poly(ADP-ribose)polymerase (PARP) in an early state of the apoptosis. In this crystal structure, apopain is in complex with the tetrapeptide aldehyde inhibitor, Ac-Asp-Glu-Val-Asp-CHO.

Peptidases with the catalytic dyad in the order His, Cys are grouped together in clan CD. Currently, this clan contains four families: caspases, gingipains, clostripain, and legumains. Caspases are involved in apoptosis, whereas their inhibition could regulate this important biochemical process.[71] Gingipains, isolated from the peridontopathogenic microorganism *Porphyromonas gingivalis*, are assumed to be involved in gingivitis.[72,73] Legumain is believed to exert a key role in the processing of the C-fragment of tetanus toxin for presentation by the MCH class II system.[74] Because of their importance in regulatory processes and their high specificity for the amino acid in the P_1 position, peptidases of CD clan are promising targets for the development of drugs.

2.7.5.2.3   Catalytic Dyad (Cys, Cys)

Site-directed mutagenesis on both the ER-60 and ER-72 proteases indicated strong evidence that the C-terminal cysteines from the two **CGHC** motifs serve as catalytic residues.[75,76] The proteases were formerly grouped into the family C17 in a new clan (clan CG), but this classification was deleted in release 3.1 in December 1998. To our knowledge there is no three-dimensional structure currently available.

2.7.5.2.4   Cysteine as the Nucleophilic Residue

A completely different mechanism of peptide bond cleavage is present in the autoprocessing domain of the hedgehog protein (clan CH, family C46). Hedgehog proteins are synthesized as precursors, and during intramolecular processing, a C-terminal and an N-terminal fragment with a cholesterol moiety covalently attached to its carboxyl terminus are formed.

The active-site cysteine attacks intramolecularly, as the nucleophile, the carbonyl group of the preceding amino acid residue, and a thioester linkage is generated.[77,78] The release of the N-terminal fragment is catalyzed by the nucleophilic attack of the 3β-hydroxyl group of the cholesterol moiety. Other nucleophiles like DTT (dithiothreitol) can easily replace the cholesterol. The function of the histidine residue (His 329) is possibly not just to stabilize the negative charge on the carbonyl oxygen of the preceding amino acid (Gly 257) but also the donation of a proton to the free α-amino group of the active-site cysteine and the maintenance of an appropriate orientation of reaction components. Stabilization of the reactive conformation is possibly the function of Thr 326. The nucleophilic attack of the cholesterol is probably assisted by an aspartic acid residue (Asp 303). This amino acid serves as a general base to deprotonate the 3β-hydroxyl group (Scheme 2.3).

Figure 2.25 shows the active site of the C-terminal autoprocessing domain of the hedgehog protein isolated from *Drosophila melanogaster*.

In the case of self-splicing proteins, a similar mechanism is found.[78] These proteins are examples of mechanistic class overlapping, in which the primary nucleophile was found to be Ser or Cys.

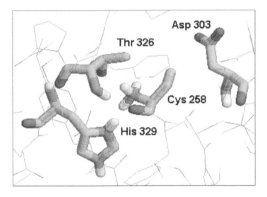

SCHEME 2.3 Catalytic mechanism for cysteine proteases with Cys as a catalytic residue.

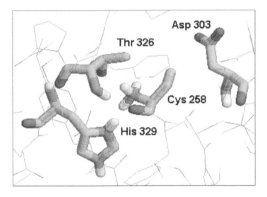

FIGURE 2.25 Representation of the active site of the C-terminal autoprocessing domain of the hedgehog protein. PDB ID code 1AT0[77] was used to construct this figure. As in previous figures, the active-site residues are shown in stick representation.

2.7.5.3 Covalent Catalysis: Serine Proteases

Serine proteases are named after their nucleophilic residue. Approximately one third of all proteases belong to the class of serine proteases. They are present in virtually all organisms and function both inside and outside the cell. Serine proteases are divided into eight clans: S-, SB, SC, SE, SF, SH, SK, and SM in the MEROPS peptidase database.

2.7.5.3.1 Catalytic Triad (Ser, His, Asp)

The general catalytic mechanism of serine peptidases based on the catalytic triad of serine, histidine, and aspartate is similar to cysteine peptidases with the replacement of cysteine by serine. The residues serine, histidine, and aspartate represent the "classical catalytic triad" of serine proteases. In addition, the nature of the oxyanion hole that has been found in the cysteine protease papain is similar to that found in the serine protease subtilisin. The mechanism of hydrolysis for serine proteases is identical in the clans PA (catalytic residues in the order of His, Asp, Ser), SB (order of catalytic residues: Asp, His, Ser), and SC (order of catalytic residues: Ser, Asp, His). In Scheme 2.4, the catalytic mechanism for serine proteases with a catalytic triad consisting of Ser, His, and Asp is given.

In Figure 2.26, Figure 2.27, and Figure 2.28, the active sites of several proteases of the previously described clans are shown. Clan PA is represented by the inhibitor complex between trypsin and the inhibitor BPO (Figure 2.26). The protease trypsin is produced by the mammalian pancreas and is involved in breaking specific peptide bonds that are formed using the carbonyl groups from Lys or Arg.

As an example of a member of clan SB, the structure of subtilisin Carlsberg is shown in Figure 2.27. Subtilisin Carlsberg, isolated from *Bacillus licheniformis*, hydrolyzes peptide bonds of proteins with a broad substrate specificity and has a preference for large, uncharged residues in P_1 position.

As an example of a member of clan SC, prolyl oligopeptidase, which is covalently bound to the inhibitor Z-Pro-prolinal, is shown in Figure 2.28. Prolyl oligopeptidase, isolated for the very first time in the human uterus as an oxytocin-degrading enzyme, is capable of cleaving peptide substrates on the C-terminal side of proline residues. It also plays an important role in the degradation of neuropeptides such as substance P and angiotensin. It is commonly assumed that these proline-containing neuropeptides are involved in learning and memory processes. Further-more, there is evidence that suggests a link to Alzheimer's.[81]

2.7.5.3.2   Catalytic Triad (His, Ser, His)

A slight variation to the classical catalytic triad is observed in clan SH, in which the aspartic acid of the catalytic triad is replaced by another histidine. The order of the catalytic-site residues is His, Ser, His in the sequence. The clan consists of only one family, the family 21 of cytomegalovirus assemblin. Superposition of the pre-sumed active-site residues of cytomegalovirus proteinase with the catalytic triads of chymotrypsin, subtilisin, and papain reveals an overlap of the position of the third member in the catalytic triad.[83] In Figure 2.29, the active site of the human herpes simplex virus type II protease is given as another example of a protease operating via a catalytic triad Ser, His, His. The human herpes simplex virus type II protease is required for viral replication, and due to the responsibility of this virus for *herpes labialis* and genital herpes, it is a promising target for intervention in these diseases.

2.7.5.3.3   Catalytic Dyad (Ser, Lys)

For years, it was assumed that all serine peptidases use a catalytic triad so that it became a dogma. Surprisingly, investigations on recently discovered enzymes brought forward proof for an existing catalytic dyad. In fact, in these enzymes, an active-site Ser was identified, but no His and Asp were found to complete the triad.

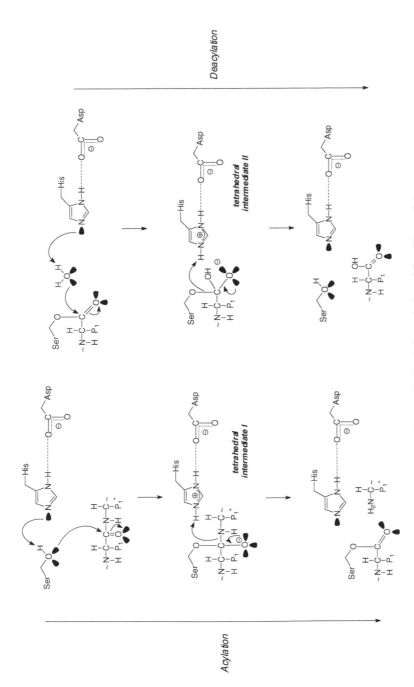

SCHEME 2.4 Catalytic mechanism for serine proteases with a catalytic triad consisting of Ser, His, and Asp.

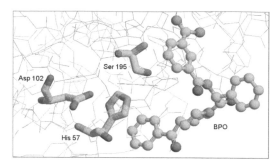

FIGURE 2.26 Representation of the active site of trypsin in complex with BPO (3-[(Z)-amino(imino)methyl]-*N*-[2-(4-benzoyl-1-piperidinyl)-2-oxo-1-phenylethyl]benzamide). PDB ID code 1EB2[79] was used to construct this figure.

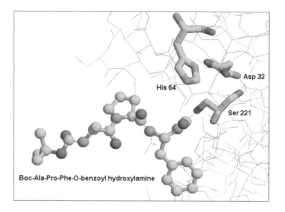

FIGURE 2.27 Representation of the active site of subtilisin Carlsberg in complex with the inhibitor, Boc-Ala-Pro-Phe-*O*-benzoyl hydroxylamine. The PDB ID code 1SCN[80] was used to construct this figure.

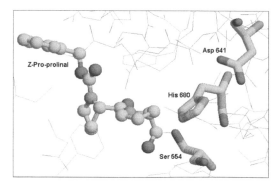

FIGURE 2.28 Representation of the active site of prolyl oligopeptidase from porcine muscle with the covalently bound inhibitor, Z-Pro-prolinal. The PDB ID code 1QFS[82] was used to construct this figure.

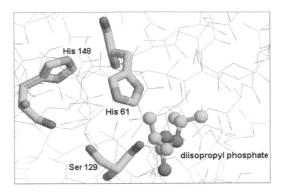

FIGURE 2.29 Representation of the active site of the herpes simplex virus type II protease with diisopropyl phosphate covalently bound to the active-site Ser 129. The PDB ID code 1AT3[84] was used to construct this figure.

So, mechanistically novel serine proteases with a Ser as the nucleophile and a Lys as the base have been discovered. Two clans of peptidases (clans SE and SF) use the Ser/Lys dyad.

Based on the determined structure of the signal peptidase I from *Escherichia coli*, also known as leader peptidase I, Paetzel et al.[85,86] proposed a mechanism for proteases cleaving with a Ser-Lys dyad. The active-site residue Lys 145 deprotonates, as a general base, the hydroxyl group of the nucleophilic Ser 90. After the nucleophilic Ser attacks the substrate scissile amide, a tetrahedral intermediate is formed. The tetrahedral intermediate has to be stabilized by an oxyanion hole that works by neutralizing the negative charge generated. Usually, oxyanion holes contain two main-chain amide hydrogens that act as hydrogen-bond donors. In the case of the apoenzyme structure of leader peptidase I, one main-chain amide hydrogen (Ser 90 NH) and one side-chain hydroxyl hydrogen (Ser 88 γOH) form the oxyanion hole.[85]

One approach for the maturation of proteins is the cleavage of large precursors. In the case of precursors consisting of the mature protein and an amino-terminal signal peptide, membrane-bound signal peptidase recognizes and cleaves this signal peptide, which results in the release of the mature protein. Examples of signal peptidases have been found and isolated from bacteria. In Figure 2.30, the active site of the bacterial signal peptidase in complex with an inhibitor, allyl (5S, 6S)-6-[(R)-acetoxyeth-2-yl]-penem-3-carboxylate, is shown.

2.7.5.4 Covalent Catalysis: Cysteine, Serine, or Threonine as Nucleophilic Residues

Clan PA comprises peptidases that have either a serine or a cysteine as the nucleophile. The catalytic mechanism is the same as classic serine peptidases. All members of the clan PA represent a chymotrypsin-like fold. The only difference between serine and cysteine peptidases in this clan is the choice of the nucleophile.

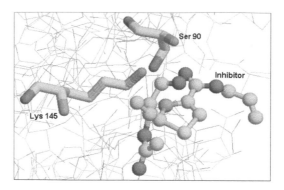

FIGURE 2.30 Representation of the active site of the bacterial signal peptidase in complex with the inhibitor, allyl (5*S*, 6*S*)-6-[(*R*)-acetoxyeth-2-yl]-penem-3-carboxylate. PDB ID code 1B12[86] was used to construct this figure.

In another clan, clan PB, enzymes with an unusual feature of a new mechanism are grouped together. They use the side chain from the amino-terminal residue that attacks as a nucleophile on the carbonyl carbon. This side chain is incorporated into a β-sheet. Brannigan has therefore suggested the name Ntn (N-terminal nucleophile) hydrolases.[87] The nucleophile is a threonine in the case of the proteasome, a cysteine in the case of glutamine PRPP amidotransferase, and a serine in the case of penicillin acylase. The crystal structures of all these aminohydrolases have been solved.[88–91] Overall, clan PB consists of two cysteine families, two serine families, and four threonine families. Because of their biochemical importance in controlled proteolysis and as promising targets for cancer chemotherapy, investigations have been focused especially on proteasomes. Investigations reveal clearly that proteasome inhibitors may prevent angiogenesis and metastasis *in vivo*, induce apoptosis in tumor cells *in vitro* and, furthermore, show some level of selective cytotoxicity to cancer cells.[92] The function of the proteasome to be a cell's shredder has a long history because it has also been found in the archaebacterium *Thermoplasma acidophilum*. The general architecture of the three-dimensional structure, which was investigated by electron microscopy studies, shows a high conservation of *Thermoplasma acidophilum* to humans.[93] The proteasome is located in the cytoplasm and nucleus, and it consists of 28 subunits overall, in which two different types of subunits, α and β, are distinguishable. Seven subunits of each type form stacked rings, the α-subunits forming the two outer rings and the β-subunits forming the two inner rings. Some of the β-subunits are responsible for the catalytic activity, whereupon these subunits are synthesized as propeptides to obstruct their biological activity. Three of the seven β-subunits have inhibitor-binding sites suggesting chymotrypsin-like and trypsin-like activity and peptidylglutaminyl peptide hydrolytic specificity, so there are different kinds of activity of the proteasome.[94] Proteins destined for degradation have to pass, in advance, the central channel in an unfolded state. Structural determinations of the large proteasome reveal a fascinating facet of the beauty and power of nature. In Figure 2.31, the active site of the proteasome in complex with the calpain-inhibitor I is shown.

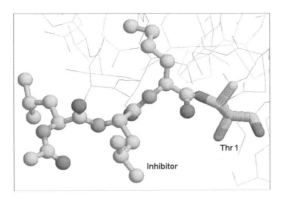

FIGURE 2.31 Representation of the active site of proteasome in complex with the calpain-inhibitor I, 2-acetylamino-4-methyl-pentanoic acid. The atomic coordinates of PDB ID code 1J2Q[91] were used to construct this figure. For better clarity, just one of the β-subunits, the subunit H, is shown.

2.7.5.5 Noncovalent Catalysis

2.7.5.5.1 Water Bound by Two Aspartic Acid Residues

Until now, six clans of aspartic peptidases are known. These are A-, AA, AB, AC, AD, and AF. In general, it is assumed that the hydrolysis of aspartic peptidases is achieved by a general acid-base catalysis mechanism. In Scheme 2.5, the catalytic mechanism for aspartic peptidases is given.

Due to the great therapeutic interest of the HIV-1 aspartic protease, the catalytic mechanism of this enzyme is the most investigated in this mechanistic class. Therefore, several x-ray structures have been solved and published with the data listed in the PDB. The HIV-1 protease acts based on an active-site consensus template, Asp 25-Gly 27. Only one of the two Asp residues of the active site in HIV-1 aspartic protease is protonated in the enzyme substrate complex.[95] One of the Asp residues acts as a general base that activates the water molecule. In Figure 2.32, the active site of the HIV-1 protease is in complex with the inhibitor VX-478, which is an *N*, *N*-disubstituted (hydroxyethyl) amino sulfonamide.

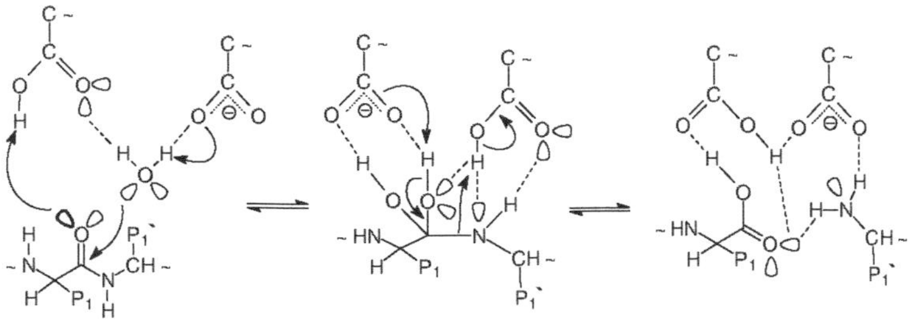

SCHEME 2.5 Catalytic mechanism for aspartic peptidases.

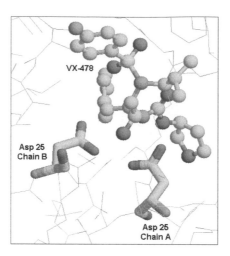

FIGURE 2.32 Representation of the active site of the HIV-1 protease in complex with the inhibitor, VX-478. The atomic coordinates of PDB ID code 1HVP[96] were used to construct this figure.

2.7.5.5.2 Water Bound by a Single Metal Ion

Only a single metal ion is involved in the catalytic mechanism of the clans M-, MA, MC, MD, and ME. All proteases of the clans use a water molecule as a nucleophile. The water is bound by a single zinc ion that is ligated to two His residues and, as the third member, either His, Asp, or Glu residues (clan MA). For clan MC, the zinc ion is ligated to His, Glu (within the motif HXXE), and His residues; or to His, Asp, and His residues (clan MD); or to two His residues (within the motif HXXEH) and a Glu residue (clan ME). In Scheme 2.6, the catalytic mechanism for monometallo-ion-dependent peptidases is given.

Several studies on carboxypeptidase A and thermolysin have suggested a mechanism for zinc peptidases.[97,98] Carboxypeptidase A is a member of the clan MC (family M14) and thermolysin is a member of clan MA (family M4) in the MEROPS peptidase database. The zinc ion is tetrahedrally coordinated by two His residues, a Glu residue, and a water molecule. The nucleophilicity of the water molecule is assisted by Glu 270, and the oxyanion is stabilized by the Zn^{2+} ion and an arginine. After the formation of the tetrahedral intermediate by the nucleophilic attack of a water molecule, the intermediate is decomposed by the transfer of the proton from Glu 270 to the leaving nitrogen. Electrostatic interactions between Glu 270 and the products may facilitate the release of the product. Recent investigations on thermolysin give a different catalytic mechanism for this protease. After the replacement of the activated water by the substrate, a histidine residue assists the nucleophilic water to attack the scissile peptide bond.

In Figure 2.33, the active site of carboxypeptidase A is shown. Carboxypeptidase A is a digestive enzyme with preference for the hydrolysis of substrates with hydrophobic N-terminal residues. The protease has been isolated from pig, dogfish, and cattle pancreas. Figure 2.33 shows the protease in complex with the inhibitor

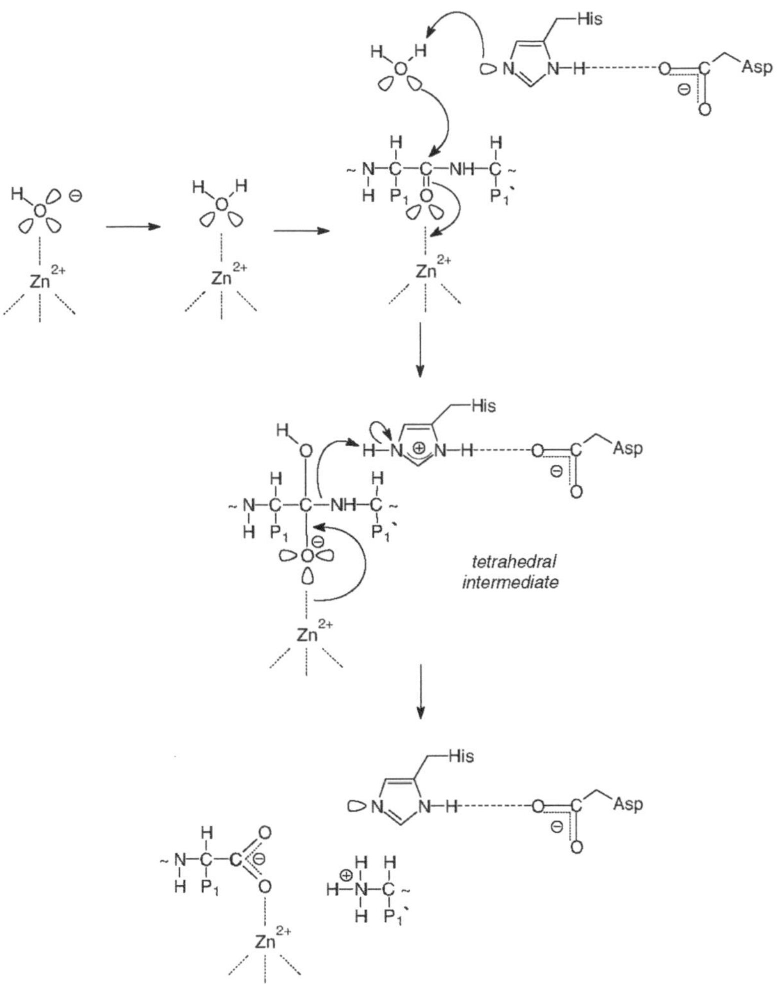

SCHEME 2.6 Catalytic mechanism for peptidases with a mononuclear metal center.

L-benzylsuccinate (2(*S*)-benzyl-3-carboxypropionic acid). The geometry around the metal ion is described as a distorted tetrahedron.

2.7.5.5.3   Water Bound by Two Metal Ions

There are four clans known to contain two metal ions in the catalytic center. The clans MF and MH contain two zinc ions, clan MG contains two cobalt or manganese ions, and clan MJ contains two nickel ions. The most striking contrast between double- and single-metal ion peptidases is the pentahedral coordination of the metal ions in double-metal ion peptidases.[100,101] The nucleophile is a water molecule bridged by the two metal ions symmetrically. The two metal ions in turn are ligated to five residues overall, which are either one or more His, Asp, Glu, His, and Lys.

At first, it seems to be impossible to give only one valid catalytic mechanism to the broad spectrum of proteases acting by metal ions to explain their activity.

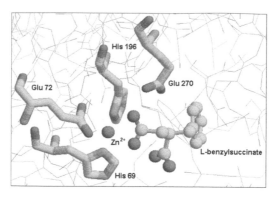

FIGURE 2.33 Representation of the active site of carboxypeptidase A in complex with the inhibitor L-benzylsuccinate. The atomic coordinates of PDB ID code 1CBX[99] were used to construct this figure. In addition to the catalytic zinc ion and the zinc ligands His 69, Glu 72, and His 196, the predicted activating residue Glu 270 is also shown.

Based on the crystallographic determination of leucine aminopeptidase, a mechanism was proposed. This reveals an important role of the catalytic metal ions for binding and activation of the substrate and also for stabilizing the transition state.[102]

The exopeptidase, leucine aminopeptidase, hydrolyzes peptides that bind adjacent to a free amino group, and it liberates amino acids from the N-terminal end of a number of proteins and polypeptides. As a consequence of the enzyme's ability to cleave leucine-containing substrates, it has been called leucine aminopeptidase. Leucine aminopeptidase has been found in plants, animals, and bacteria. In Figure 2.34, the active site of leucine aminopeptidase in complex with the inhibitor amastatin is shown. Leucine aminopeptidase is a member of the clan MF (family M17) in the MEROPS peptidase database.

2.7.5.6 Examples of Protease Inhibitors as Drugs

Since the introduction of the first affinity labels as tools to study protease mechanisms and protease active sites three decades ago, there have been numerous protease inhibitors developed as potential drugs. Among them are compounds that combat important pathophysiological states or disorders such as AIDS, high blood pressure, blood coagulation, emphysema, rheumatic arthritis, diabetes, etc. Target proteins are proteases from all the major classes: serine, cysteine, aspartate, and metalloproteases. The structure and chemical nature of the inhibitors is mainly dependent on the catalytic mechanism and the structural class the particular protease belongs to. Table 2.6 is, of course, an incomplete list of disorders, enzymes, and their inhibitors, which have been successfully introduced to the market or which are under development. Whereas first-generation compounds were initially found by producing and searching large chemical libraries for diversity, potency, and optimal pharmacokinetics, followup generation developments are mostly the result of intensive structure–activity relationship (SAR) approaches, including rational drug design based on inhibitor modeling to the active sites of target molecules.[103–105]

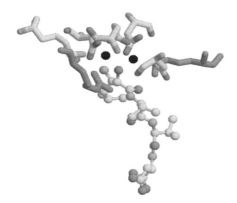

FIGURE 2.34 Representation of the active site of leucine aminopeptidase in complex with the inhibitor, amastatin. The atomic coordinates of PDB ID code 1BLL[102] were used to construct this figure. For better clarity, no numbering or labeling of the residues has been included. The catalytic zinc ions are shown as black spheres and the zinc ligands are shown in stick representation in CPK color mode (Lys 250, Asp 255, Asp 273, Asp 332, and Glu 334). In the proposed mechanism of leucine aminopeptidase, the residues Arg 336, as an additional electrophilic substrate activator and transition state stabilizer, and Lys 262, as a proton shuttle, are both shown in stick representation in gray. The inhibitor amastatin is shown in ball-and-stick representation in CPK color mode. All other residues were removed from this figure.

2.7.5.7 Interesting Websites

PDB files of all known structures of proteases are given at http://www.rcsb.org/pdb/. A database with the structural classification of proteases can be sourced at http://scop.berkeley.edu/. Another useful database that systematically classifies the known proteases into families and clans based on structural similarities is MEROPS, the peptidase database (http://merops.sanger.ac.uk/). In addition, PROCAT is a database of 3-D enzyme-active-site templates that provide facilities for the architecture of the active sites of enzymes. This database allows the user to search for 3-D enzyme-active-site template motifs within a protein structure (http://www.bio-chem.ucl.ac.uk/bsm/PROCAT/PROCAT.html). CATH is a novel hierarchical classification of protein domain structures in the Brookhaven PDB. The database is divided into four major levels: class, architecture, topology (fold family), and homologous superfamily (http://www.biochem.ucl.ac.uk/bsm/cath_new/). BRENDA, the enzyme database, has listed information on enzymes regarding their nomenclature, reaction, specificity, structure, isolation, preparation, and stability. Furthermore, cross references for the sequence and 3-D structure data banks are given (http://www.brenda.uni-koeln.de).

2.7.6 Conclusion

Proteases play an important role in triggering and regulating physiological processes as well as the catabolism of proteins. Drug development based on a knowl-

TABLE 2.6
Overview of Protease Inhibitors in the Treatment of Disorders

Disorder	Protease	Structure	Name (company)	References
AIDS	HIV-Protease (Aspartate)	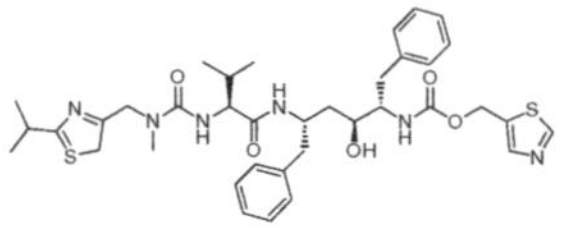	Ritonavir (Abbott Laboratories)	106
Cardiac insufficiency	ACE[a] (Metallo-)	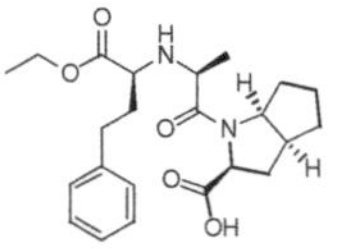	Ramipril (Aventis)	107
Cancer	MMP[b] (Metallo-)	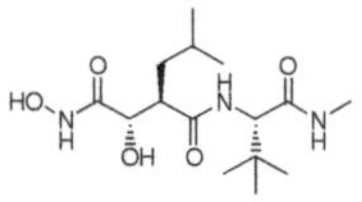	Marimastat (British Biotech)	108
Emphysema	Elastase (Serine)	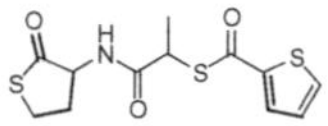	MR-889[c] (Medea Research)	109
Diabetes	DP IV (Serine)	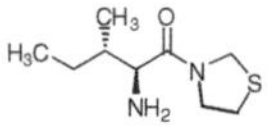	P32/98[d] (probiodrug)	110
		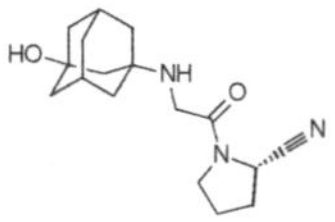	LAF 237[d] (Novartis)	111
Osteoporosis	Cathepsin K (Cysteine)	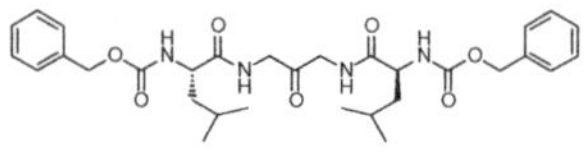	1,3-Bis(benzoyl-amino)-2-propanone (Glaxo SmithKline)	112
Hepatitis C	NS3/4A Virus Protease (Serine)	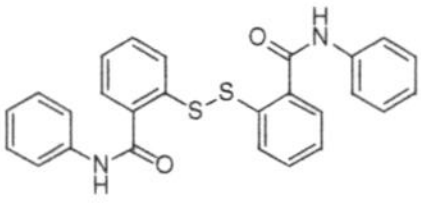	2,2′-Dithiobis[(N-phenyl)benzamide]	113

[a] Angiotensin-converting enzyme.
[b] Matrix metalloproteinase.
[c] In clinical phase III in Italy.
[d] In clinical phase II.

edge of the mechanism and the architecture of the active site has proved to be highly successful, and therefore proteases are promising targets for the future. Several mechanisms of cleaving a peptide bond have been elucidated so far. The nucleophile in serine and cysteine proteases is an integral part of the enzyme's structure, whereas metallo- and aspartic enzymes use an activated water molecule located at the active site.

ACKNOWLEDGMENT

For the creation of the Figure 2.18, Figure 2.20, and Figure 2.22 we thank W. Brandt from the IPB (Institute of Plant Biochemistry) in Halle/S.

REFERENCES

1. Ciardelli, T. and Smith, K.A. (1989). Interleukin 2: prototype for a new generation of immunoactive pharmaceuticals. *Trends in Pharmacological Science*, 10, 239–243.
2. Hopfinger, A.J. (1985). Computer-assisted drug design. *Journal of Medicinal Chemistry*, 28, 1133–1139.
3. Keseru, G.M. (2001). A virtual high throughput screen for high affinity cytochrome P450 cam substrates. Implications for in silico prediction of drug metabolism. *Journal of Computer-Aided Molecular Design*, 15, 649–657.
4. Shia, K.S., Li, W.T., Chang, C.M., Hsu, M.C., Chern, J.H., Leong, M.K., Tseng, S.N., Lee, C.C., Lee, Y.C., Chen, S.J., Peng, K.C., Tseng, H.Y., Chang, Y.L., Tai, C.L., and Shih, S.R. (2002). Design, synthesis, and structure-activity relationship of pyridyl imidazolidinones: a novel class of potent and selective human enterovirus 71 inhibitors. *Journal of Medicinal Chemistry*, 45, 1644–1655.
5. Pickett, S.D., McLay, I.M., and Clark, D.E. (2000). Enhancing the hit-to-lead properties of lead optimization libraries. *Journal of Chemical Information and Computer Science*, 40, 263–272.
6. Colmenarejo, G. (2003). *In silico* prediction of drug-binding strengths to human serum albumin. *Medicinal Research Reviews*, 23, 275–301.
7. Demuth, H.U. (1990). Recent developments in inhibiting cysteine and serine proteases. *Journal of Enzyme Inhibition*, 3, 249–278.
8. Rando, R.R. (1974). Chemistry and enzymology of k_{cat} inhibitors. *Science*, 185, 320–324.
9. Carter, P. and Wells, J.A. (1988). Dissecting the catalytic triad of a serine protease. *Nature*, 332, 564–568.
10. Carter, P. and Wells, J.A. (1987). Engineering enzyme specificity by substrate-assisted catalysis. *Science*, 237, 394–399.
11. Dunn, B.M. (2002). Structure and mechanism of the pepsin-like family of aspartic peptidases. *Chemical Reviews*, 102, 4431–4458.
12. Hartley, B.S. (1960). Proteolytic enzymes. *Annual Reviews in Biochemistry*, 29, 45–72.
13. Gerhartz, B., Niestroj, A.J., and Demuth, H.U. (2002). Enzyme classes and mechanisms. In *Proteinase and Peptidase Inhibition*, Smith, H.J. and Simons, C., Eds. Taylor & Francis, London, pp. 1–20.

14. Schechter, I. and Berger, A. (1967). On the size of the active site in proteases. I. Papain. *Biochemical and Biophysical Research Communications*, 27, 157–162.

15. Kamphuis, I.G., Kalk, K.H., Swarte, M.B., and Drenth, J. (1984). Structure of papain refined at 1.65 Å resolution. *Journal of Molecular Biology*, 179, 233–256.

16. Perona, J.J., Tsu, C.A., McGrath, M.E., Craik, C.S., and Fletterick, R.J. (1993). Relocating a negative charge in the binding pocket of trypsin. *Journal of Molecular Biology*, 230, 934–949.

17. Polgar, L. and Halasz, P. (1982). Current problems in mechanistic studies of serine and cysteine proteinases. *Biochemical Journal*, 207, 1–10.

18. Akahane, K. and Umeyama, H. (1986). Binding specificity of papain and cathepsin B. *Enzyme*, 36, 141–149.

19. Asboth, B., Majer, Z., and Polgar, L. (1988). Cysteine proteases: the S2P2 hydrogen bond is more important for catalysis than is the analogous S1P1 bond. *FEBS Letters*, 233, 339–341.

20. Liebman, J.F. and Greenberg, A. (1989). *Molecular Structure and Energetics*, Vol. 9, Mechanistic Principles of Enzyme Activity. John Wiley & Sons, London.

21. Fastrez, J. (1983). On the stability of tetrahedral intermediates within the active sites of serine and cysteine proteases. *European Journal of Biochemistry*, 135, 339–341.

22. Asboth, B., Stokum, E., Khan, I.U., and Polgar, L. (1985). Mechanism of action of cysteine proteinases: oxyanion binding site is not essential in the hydrolysis of specific substrates. *Biochemistry*, 24, 606–609.

23. Polgar, L. and Csoma, C. (1987). Dissociation of ionizing groups in the binding cleft inversely controls the endo- and exopeptidase activities of cathepsin B. *Journal of Biological Chemistry*, 262, 14448–14453.

24. Brocklehurst, K., Brocklehurst, S.M., Kowlessur, D., O'Driscoll, M., Patel, G., Salih, E., Templeton, W., Thomas, E., Topham, C.M., and Willenbrock, F. (1988). Supracrystallographic resolution of interactions contributing to enzyme catalysis by use of natural structural variants and reactivity-probe kinetics. *Biochemical Journal*, 256, 543–558.

25. Pickersgill, R.W., Goodenough, P.W., Sumner, I.G., and Collins, M.E. (1988). The electrostatic fields in the active-site clefts of actinidin and papain. *Biochemical Journal*, 254, 235–238.

26. Brocklehurst, K., O'Driscoll, M., Kowlessur, D., Phillips, I.R., Templeton, W., Thomas, E.W., Topham, C.M., and Wharton, C.W. (1989). The interplay of electrostatic and binding interactions determining active centre chemistry and catalytic activity in actinidin and papain. *Biochemical Journal*, 257, 309–310.

27. Varughese, K.I., Ahmed, F.R., Carey, P.R., Hasnain, S., Huber, C.P., and Storer, A.C. (1989). Crystal structure of a papain-E-64 complex. *Biochemistry*, 28, 1330–1332.

28. Storer, A.C., Angus, R.H., and Carey, P.R. (1988). Comparison of the kinetics of the papain-catalyzed hydrolysis of glycine- and alanine-based esters and thiono esters. *Biochemistry*, 27, 264–268.

29. Baker, B.R. (2003). *Design of Active-Site-Directed Irreversible Enzyme Inhibitors*, John Wiley & Sons, New York.

30. Harada, N. and Nakanishi, K. (1983). *Circular Dichroic Spectroscopy: Exciton Coupling in Organic Stereochemistry*, University Science Books, Mill Valley, CA.

31. Muller, D.J., Janovjak, H., Lehto, T., Kuerschner, L., and Anderson, K. (2002). Observing structure, function and assembly of single proteins by AFM. *Progress in Biophysics and Molecular Biology*, 79, 1–43.

32. Blow, D.M. and Steitz, T.A. (1970). X-ray diffraction studies of enzymes. *Annual Reviews in Biochemistry*, 39, 63–100.

33. Lipscomb, W.N. (1983). Structure and catalysis of enzymes. *Annual Reviews in Biochemistry*, 52, 17–34.
34. Johnson, L.N. and Blundell, T.L. (1976). *Protein Crystallography.* Academic Press, New York.
35. Drenth, J. (1994). *Principles of Protein X-ray Crystallography,* Springer, New York.
36. Ernst, R.R. (1975). Two-dimensional spectroscopy. *Chimica*, 29, 179–183.
37. Williamson, M.P., Havel, T.F., and Wuthrich, K. (1985). Solution conformation of proteinase inhibitor IIA from bull seminal plasma by 1H nuclear magnetic resonance and distance geometry. *Journal of Molecular Biology*, 182, 295–315.
38. Otting, G., Qian, Y.Q., Billeter, M., Muller, M., Affolter, M., Gehring, W.J., and Wuthrich, K. (1990). Protein–DNA contacts in the structure of a homeodomain–DNA complex determined by nuclear magnetic resonance spectroscopy in solution. *EMBO Journal*, 9, 3085–3092.
39. Wüthrich, K. (1986). *NMR of Proteins and Nucleic Acids,* John Wiley & Sons, New York.
40. Fischer, E. (1894). Einfluss der configuration auf die wirkung der enzyme. *Chemische Berichte*, 27, 2985–2993.
41. Koshland, D.E., Jr. (1958). Application of a theory of enzyme specificity to proteinase synthesis. *Proceedings of the National Academy of Sciences U.S.A.*, 44, 98–104.
42. Lipscomb, W.N. (1972). Structures and mechanisms of enzymes. *Proceedings of the Robert A. Welch Foundation Conferences on Chemical Research*, 15, 130–182.
43. Quiocho, F.A. and Lipscomb, W.N. (1971). Carboxypeptidase A: a protein and an enzyme. *Advances in Protein Chemistry*, 25, 1–78.
44. Christianson, D. W. and Lipscomb, W. N. (1986). X-ray crystallographic investigation of substrate binding to carboxypeptidase A at subzero temperature. *Proceedings of the National Academy of Science U.S.A.*, 83, 7568–7572.
45. Kilshtain-Vardi, A., Glick, M., Greenblatt, H.M., Goldblum, A., and Shoham, G. (2003). Refined structure of bovine carboxypeptidase A at 1.25 Å resolution. *Acta Crystallographica Section D — Biological Crystallography*, 59, 323–333.
46. Walsh, K.A. (1975). Unifying concepts among proteases. In *Proteases and Biological Control*, Reich, E., Rifkin, D.B., and Shaw, E., Eds., Cold Spring Harbor Laboratory, Cold Spring Harbor, NY, pp. 1–11.
47. Neurath, H. (1984). Evolution of proteolytic enzymes. *Science*, 224, 350–357.
48. Rose, T. and Di Cera, E. (2002). Substrate recognition drives the evolution of serine proteases. *Journal of Biological Chemistry*, 277, 19243–19246.
49. Krem, M.M., Prasad, S., and Di Cera, E. (2002). Ser(214) is crucial for substrate binding to serine proteases. *Journal of Biological Chemistry*, 277, 40260–40264.
50. Barrett, A.J. and Rawlings, N.D. (1995). Families and clans of serine peptidases. *Archives of Biochemistry and Biophysics*, 318, 247–250.
51. Barrett, A.J. and Rawlings, N.D. (2001). Evolutionary lines of cysteine peptidases. *Biological Chemistry*, 382, 727–733.
52. Webb, E.C. (1992). *Enzyme Nomenclature 1992,* Academic Press, San Diego, CA.
53. Tipton, K.F. (1994). Nomenclature Committee of the International Union of Biochemistry and Molecular Biology (NC-IUBMB). Enzyme nomenclature. Recommendations 1992. Supplement 1: corrections and additions. *European Journal of Biochemistry*, 223, 1–5.
54. Barrett, A.J. (1995). Nomenclature Committee of the International Union of Biochemistry and Molecular Biology (NC-IUBMB). Enzyme nomenclature. Recommendations 1992. Supplement 2: corrections and additions. *European Journal of Biochemistry*, 232, 1–6.

55. Barrett, A.J. (1996). Nomenclature Committee of the International Union of Biochemistry and Molecular Biology (NC-IUBMB). Enzyme nomenclature. Recommendations 1992. Supplement 3: corrections and additions. *European Journal of Biochemistry*, 237, 1–5.

56. Barrett, A.J. (1997). Nomenclature Committee of the International Union of Biochemistry and Molecular Biology (NC-IUBMB). Enzyme Nomenclature. Recommendations 1992. Supplement 4: corrections and additions. *European Journal of Biochemistry*, 250, 1–6.

57. Tipton, K. and Boyce, S. (1999). Nomenclature Committee of the International Union of Biochemistry and Molecular Biology (NC-IUBMB). Enzyme Supplement 5. *European Journal of Biochemistry*, 264, 610–650.

58. Lecaile, F., Kaleta, J., and Brömme, D. (2003). Human and parasitic papain-like cysteine proteases: their role in physiology and pathology and recent developments in inhibitor design. *Chemical Reviews*, 102, 4459–4488.

59. Yamamoto, D., Matsumoto, K., Ohishi, H., Ishida, T., Inoue, M., Kitamura, K., and Mizuno, H. (1991). Refined x-ray structure of papain E-64-c complex at 2.1-Å resolution. *Journal of Biological Chemistry*, 266, 14771–14777.

60. Joshua-Tor, L., Xu, H.E., Johnston, S.A., and Rees, D.C. (1995). Crystal structure of a conserved protease that binds DNA: the bleomycin hydrolase, Gal6. *Science*, 269, 945–950.

61. Sivaraman, J., Nagler, D.K., Zhang, R., Menard, R., and Cygler, M. (2000). Crystal structure of human procathepsin X: a cysteine protease with the proregion covalently linked to the active site cysteine. *Journal of Molecular Biology*, 295, 939–951.

62. Drenth, J., Kalk, K.H., and Swen, H.M. (1976). Binding of chloromethyl ketone substrate analogues to crystalline papain. *Biochemistry*, 15, 3731–3738.

63. Ton-That, H., Liu, G., Mazmanian, S.K., Faull, K.F., and Schneewind, O. (1999). Purification and characterization of sortase, the transpeptidase that cleaves surface proteins of Staphylococcus aureus at the LPXTG motif. *Proceedings of the National Academy of Sciences U.S.A.*, 96, 12424–12429.

64. Meldal, M., Svendsen, I., and Scharfenstein, J. (1998). Inhibition of cruzipain visualized in a fluorescence-quenched solid-phase inhibitor library assay. D-amino acid inhibitors for cruzipain, cathepsin B, and cathepsin L. *Journal of Peptide Science*, 4, 83–91.

65. Mazmanian, S.K., Liu, G., Jensen, E.R., Lenoy, E., and Schneewind, O. (2000). Staphylococcus aureus sortase mutants defective in the display of surface proteins and in the pathogenesis of animal infections. *Proceedings of the National Academy of Sciences U.S.A.*, 97, 5510–5515.

66. Mazmanian, S.K., Liu, G., Ton-That, H., and Schneewind, O. (1999). Staphylococcus aureus sortase, an enzyme that anchors surface proteins to the cell wall. *Science*, 285, 760–763.

67. Ilangovan, U., Ton-That, H., Iwahara, J., Schneewind, O., and Clubb, R. T. (2001). Structure of sortase, the transpeptidase that anchors proteins to the cell wall of Staphylococcus aureus. *Proceedings of the National Academy of Sciences U.S.A.*, 98, 6056–6061.

68. Ding, J., McGrath, W.J., Sweet, R.M., and Mangel, W.F. (1996). Crystal structure of the human adenovirus proteinase with its 11 amino acid cofactors. *EMBO Journal*, 15, 1778–1783.

69. Walker, N.P., Talanian, R.V., Brady, K.D., Dang, L.C., Bump, N.J., Ferenz, C.R., Franklin, S., Ghayur, T., Hackett, M.C., and Hammill, L.D. (1994). Crystal structure of the cysteine protease interleukin-1 beta-converting enzyme: a (p20/p10)2 homodimer. *Cell*, 78, 343–352.

70. Rotonda, J., Nicholson, D.W., Fazil, K.M., Gallant, M., Gareau, Y., Labelle, M., Peterson, E.P., Rasper, D.M., Ruel, R., Vaillancourt, J.P., Thornberry, N.A., and Becker, J.W. (1996). The three-dimensional structure of apopain/CPP32, a key mediator of apoptosis. *Nature Structural Biology*, 3, 619–625.

71. Denault, J.B. and Salvesen, G.S. (2002). Caspases: keys in the ignition of cell death. *Chemical Reviews*, 102, 4489–4500.

72. Yoneda, M., Hirofuji, T., Anan, H., Matsumoto, A., Hamachi, T., Nakayama, K., and Maeda, K. (2001). Mixed infection of Porphyromonas gingivalis and Bacteroides forsythus in a murine abscess model: involvement of gingipains in a synergistic effect. *Journal of Periodontal Research*, 36, 237–243.

73. Imamura, T., Potempa, J., and Travis, J. (2000). Comparison of pathogenic properties between two types of arginine-specific cysteine proteinases (gingipains-R) from porphyromonas gingivalis [In Process Citation]. *Microbial Pathogenesis*, 29, 155–163.

74. Manoury, B., Hewitt, E.W., Morrice, N., Dando, P.M., Barrett, A.J., and Watts, C. (1998). An asparaginyl endopeptidase processes a microbial antigen for class II MHC presentation [see comments]. *Nature*, 396, 695–699.

75. Urade, R., Oda, T., Ito, H., Moriyama, T., Utsumi, S., and Kito, M. (1997). Functions of characteristic Cys-Gly-His-Cys (CGHC) and Gln-Glu-Asp-Leu (QEDL) motifs of microsomal ER-60 protease. *Journal of Biochemistry (Tokyo)*, 122, 834–842.

76. Okudo, H., Urade, R., Moriyama, T., and Kito, M. (2000). Catalytic cysteine residues of ER-60 protease. *FEBS Letters*, 465, 145–147.

77. Hall, T.M., Porter, J.A., Young, K.E., Koonin, E.V., Beachy, P.A., and Leahy, D.J. (1997). Crystal structure of a hedgehog autoprocessing domain: homology between hedgehog and self-splicing proteins. *Cell*, 91, 85–97.

78. Perler, F.B. (1998). Protein splicing of inteins and hedgehog autoproteolysis: structure, function, and evolution. *Cell*, 92, 1–4.

79. Liebeschuetz, J.W., Jones, S.D., Morgan, P.J., Murray, C.W., Rimmer, A.D., Roscoe, J.M., Waszkowycz, B., Welsh, P.M., Wylie, W.A., Young, S.C., Martin, H., Mahler, J., Brady, L., and Wilkinson, K. (2002). PRO_SELECT: combining structure-based drug design and array-based chemistry for rapid lead discovery. 2. The development of a series of highly potent and selective factor Xa inhibitors. *Journal of Medicinal Chemistry*, 45, 1221–1232.

80. Steinmetz, A.C., Demuth, H.U., and Ringe, D. (1994). Inactivation of subtilisin Carlsberg by *N*-((tert-butoxycarbonyl)alanylprolylphenylalanyl)-*O*-benzoylhydroxylamine: formation of a covalent enzyme-inhibitor linkage in the form of a carbamate derivative. *Biochemistry*, 33, 10535–10544.

81. Toide, K., Shinoda, M., and Miyazaki, A. (1998). A novel prolyl endopeptidase inhibitor, JTP-4819 — its behavioral and neurochemical properties for the treatment of Alzheimer's disease. *Reviews in Neuroscience*, 9, 17–29.

82. Fulop, V., Bocskei, Z., and Polgar, L. (1998). Prolyl oligopeptidase: an unusual beta-propeller domain regulates proteolysis. *Cell*, 94, 161–170.

83. Tong, L., Qian, C., Massariol, M.J., Bonneau, P.R., Cordingley, M.G., and Lagace, L. (1996). A new serine-protease fold revealed by the crystal structure of human cytomegalovirus protease. *Nature*, 383, 272–275.

84. Hoog, S.S., Smith, W.W., Qiu, X., Janson, C.A., Hellmig, B., McQueney, M.S., O'Donnell, K., O'Shannessy, D., DiLella, A.G., Debouck, C., and Abdel-Meguid, S.S. (1997). Active site cavity of herpesvirus proteases revealed by the crystal structure of herpes simplex virus protease/inhibitor complex. *Biochemistry*, 36, 14023–14029.

85. Paetzel, M., Dalbey, R.E., and Strynadka, N.C. (2002). Crystal structure of a bacterial signal peptidase apoenzyme: implications for signal peptide binding and the Ser-Lys dyad mechanism. *Journal of Biological Chemistry*, 277, 9512–9519.

86. Paetzel, M., Dalbey, R.E., and Strynadka, N.C. (1998). Crystal structure of a bacterial signal peptidase in complex with a beta-lactam inhibitor. *Nature*, 396, 186–190.

87. Brannigan, J.A., Dodson, G., Duggleby, H.J., Moody, P.C., Smith, J.L., Tomchick, D.R., and Murzin, A.G. (1995). A protein catalytic framework with an N-terminal nucleophile is capable of self-activation. *Nature*, 378, 416–419.

88. Smith, J.L., Zaluzec, E.J., Wery, J.P., Niu, L., Switzer, R.L., Zalkin, H., and Satow, Y. (1994). Structure of the allosteric regulatory enzyme of purine biosynthesis. *Science*, 264, 1427–1433.

89. Duggleby, H.J., Tolley, S.P., Hill, C.P., Dodson, E.J., Dodson, G., and Moody, P.C. (1995). Penicillin acylase has a single-amino-acid catalytic centre. *Nature*, 373, 264–268.

90. Lowe, J., Stock, D., Jap, B., Zwickl, P., Baumeister, W., and Huber, R. (1995). Crystal structure of the 20S proteasome from the archaeon T. acidophilum at 3.4 Å resolution. *Science*, 268, 533–539.

91. Groll, M., Brandstetter, H., Bartunik, H., Bourenkow, G., and Huber, R. (2003). Investigations on the maturation and regulation of archaebacterial proteasomes. *Journal of Molecular Biology*, 327, 75–83.

92. Almond, J.B. and Cohen, G.M. (2002). The proteasome: a novel target for cancer chemotherapy. *Leukemia*, 16, 433–443.

93. Puehler, G., Pitzer, F., Zwickl, P., and Baumeister, W. (1994). Proteasome: multisubunit proteinases common to thermoplasma and eukaryotes. *Systematic and Applied Microbiology*, 16, 734–741.

94. Groll, M., Ditzel, L., Lowe, J., Stock, D., Bochtler, M., Bartunik, H.D., and Huber, R. (1997). Structure of 20S proteasome from yeast at 2.4 Å resolution. *Nature*, 386, 463–471.

95. Silva, A.M., Cachau, R.E., Sham, H.L., and Erickson, J.W. (1996). Inhibition and catalytic mechanism of HIV-1 aspartic protease. *Journal of Molecular Biology*, 255, 321–346.

96. Kim, E.E., Baker, C.T., Dwyer, M.D., Murcko, M.A., Rao, B.G., Tung, R.D., and Navia, M. A. (1995). Crystal structure of HIV-1 protease in complex with VX-478, a potent and orally bioavailable inhibitor of the enzyme. *Journal of the American Chemical Society*, 117, 1181–1182.

97. Christianson, D.W. and Lipscomb, W.N. (1989). Carboxypeptidase A. *Accounts of Chemical Research*, 22, 62–69.

98. Matthews, B.W. (1989). Structural basis of the action of thermolysin and related zinc peptidases. *Accounts of Chemical Research*, 21, 333–340.

99. Mangani, S., Carloni, P., and Orioli, P. (1992). Crystal structure of the complex between carboxypeptidase A and the biproduct analog inhibitor L-benzylsuccinate at 2.0 Å resolution. *Journal of Molecular Biology*, 223, 573–578.

100. Strater, N. and Lipscomb, W.N. (1995). Two-metal ion mechanism of bovine lens leucine aminopeptidase: active site solvent structure and binding mode of L-leucinal, a gem-diolate transition state analogue, by x-ray crystallography. *Biochemistry*, 34, 14792–14800.

101. Tahirov, T.H., Oki, H., Tsukihara, T., Ogasahara, K., Yutani, K., Ogata, K., Izu, Y., Tsunasawa, S., and Kato, I. (1998). Crystal structure of methionine aminopeptidase from hyperthermophile, Pyrococcus furiosus. *Journal of Molecular Biology*, 284, 101–124.

102. Kim, H. and Lipscomb, W.N. (1993). X-ray crystallographic determination of the structure of bovine lens leucine aminopeptidase complexed with amastatin: formulation of a catalytic mechanism featuring a gem-diolate transition state. *Biochemistry*, 32, 8465–8478.

103. Zollner, H. (1989). *Handbook of Enzyme Inhibitors*, VCH, Weinheim, Germany.

104. Bode, W. and Huber, R. (2000). Structural basis of the endoproteinase–protein inhibitor interaction. *Biochimica et Biophysica Acta*, 1477, 241–252.

105. Vendrell, J., Querol, E., and Aviles, F.X. (2000). Metallocarboxypeptidases and their protein inhibitors. Structure, function and biomedical properties. *Biochimica et Biophysica Acta*, 1477, 284–298.

106. Ala, P.J. and Chang, C.-H. (2002). HIV aspartate proteinase: resistance to inhibitors. In *Proteinase and Peptidase Inhibition*, Smith, H.J. and Simons, C., Eds., Taylor & Francis, London, pp. 367–382.

107. Hooper, N.M. (2002). Zinc metallopeptidases. In *Proteinase and Peptidase Inhibition*, Smith, H.J. and Simons, C., Eds., Taylor & Francis, London, pp. 352–366.

108. Renkiewicz, R., Qiu, L., Lesch, C., Sun, X., Devalaraja, R., Cody, T., Kaldjian, E., Welgus, H., and Baragi, V. (2003). Broad-spectrum matrix metalloproteinase inhibitor marimastat-induced musculoskeletal side effects in rats. *Arthritis and Rheumatism*, 48, 1742–1749.

109. Edwards, P.D. (2002). Human neutrophil elastase inhibitors. In *Proteinase and Peptidase Inhibition*, Smith, H.J. and Simons, C., Eds., Taylor & Francis, London, pp. 154–177.

110. Hoffmann, T., Glund, K., McIntosh, C.H., Pederson, R.A., Hanefeld, M., Rosenkranz, B., and Demuth, H.-U. (2001). DPPIV-inhibition as treatment of type II diabetes. In *Cell-Surface Aminopeptidases: Basic and Clinical Aspects*, Mizutani, S., Ed., Elsevier Science, Amsterdam, pp. 381–387.

111. Villhauer, E.B., Brinkman, J.A., Naderi, G.B., Burkey, B.F., Dunning, B.E., Prasad, K., Mangold, B.L., Russell, M.E., and Hughes, T.E. (2003). 1-[[(3-hydroxy-1-adamantyl)amino]acetyl]-2-cyano-(S)-pyrrolidine: a potent, selective, and orally bioavailable dipeptidyl peptidase IV inhibitor with antihyperglycemic properties. *Journal of Medicinal Chemistry*, 46, 2774–2789.

112. Rukamp, B. and Powers, J.C. (2002). Cathepsins. In *Proteinase and Peptidase Inhibition*, Smith, H.J. and Simons, C., Eds., Taylor & Francis, London, pp. 84–126.

113. Bartenschlager, R. and Koch, J.-O. (2002). The hepatitis C virus NS3 serine-type proteinase. In *Proteinase and Peptidase Inhibition*, Smith, H.J. and Simons, C., Eds., Taylor & Francis, London, 333–351.

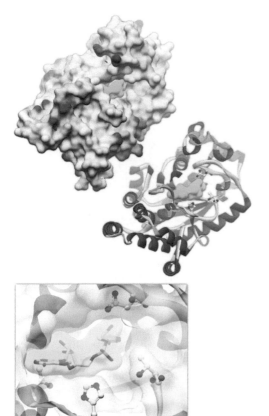

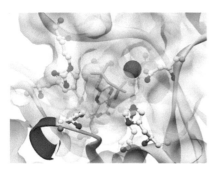

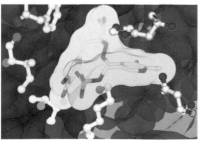

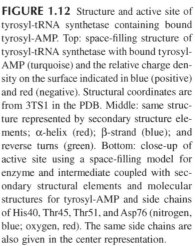

FIGURE 1.12 Structure and active site of tyrosyl-tRNA synthetase containing bound tyrosyl-AMP. Top: space-filling structure of tyrosyl-tRNA synthetase with bound tyrosyl-AMP (turquoise) and the relative charge density on the surface indicated in blue (positive) and red (negative). Structural coordinates are from 3TS1 in the PDB. Middle: same structure represented by secondary structure elements; α-helix (red); β-strand (blue); and reverse turns (green). Bottom: close-up of active site using a space-filling model for enzyme and intermediate coupled with secondary structural elements and molecular structures for tyrosyl-AMP and side chains of His40, Thr45, Thr51, and Asp76 (nitrogen, blue; oxygen, red). The same side chains are also given in the center representation.

FIGURE 1.14 Top: space-filling structure of the active site of dihydropteroate synthase bound with a close structural analog of the hydromethylpterin pyrophosphate substrate (yellow). Structural coordinates were taken from the PDB (1AD4). The molecular structure of the substrate analog is shown in the space-filling structure along with some of the amino acid side chains (Phe171, Arg52, Lys203, Arg239, His241, Asn11, Asp84, and Asp167) implicated in the active site (pyrophosphate, dark yellow; nitrogen, blue; oxygen, pink). A bound Mn^{++} is shown as a bright pink sphere. Secondary structural elements are α-helix (red); β-strand (light blue); and reverse turns (green). Bottom: space-filling model of the active site of DOPA decarboxylase with bound pyridoxal phosphate and a DOPA analog, carbidopa. Structural coordinates are from 1JS3 in the PDB. Space occupied by the bound ligands (turquoise) contains pyridoxal phosphate in a Schiff base with the hydrazine of carbidopa. Parts of two identical subunits are shown for this dimeric enzyme, one in red and the other in purple. Molecular structures of Phe101 and Ile103 on one subunit and Thr82, Asp271, and Lys303 on the other subunit are shown (nitrogen, blue; oxygen, red; phosphorus, yellow).

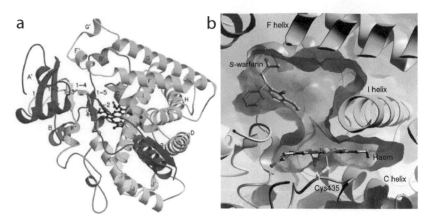

FIGURE 2.11 Crystal structure of human CYP2C9, determined recently by Williams and colleagues (2003). Panel (a) shows the structure of CYP2C9 colored from blue at the N-terminus to red at the C-terminus. The heme group is shown as a ball-and-stick model in the center of the molecule. Panel (b) depicts the active site of CYP2C9, showing a substrate, *S*-warfarin, occupying the substrate-binding site. This active site is large and could be occupied either by substrates or inhibitors larger than *S*-warfarin, or multiple compounds of comparable size to *S*-warfarin. With *S*-warfarin bound in this location, the heme group remains available to metabolize additional substrate molecules. (We gratefully acknowledge copyright permission from *Astex Technology* and *Nature Publishing Group* for these diagrams.)

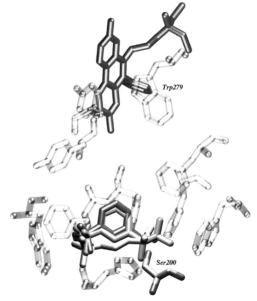

FIGURE 4.1 The active site (Ser 200) of vertebrate acetylcholinesterase is buried deep inside the enzyme molecule. White: docked substrate acetylcholine (PDB code 1ACE); red: competitive inhibitor edrophonium (PDB code 2ACK); green: transition-state analog trimethylammonio-trifluoroacetophenone (PDB code 1AMN); brown: noncompetitive inhibitor propidium (PDB code 1N5R) is bound at the entrance to the active site (Trp 279).

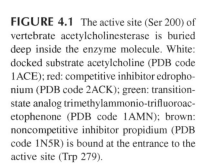

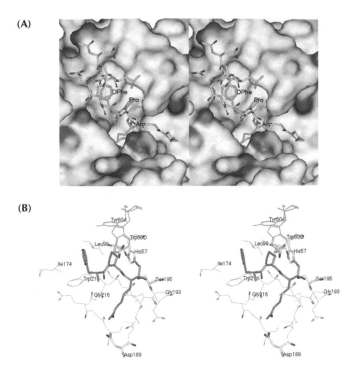

FIGURE 5.13 (A) Stereo view of the active site region of thrombin in complex with PPACK (**5.192**, green carbon atoms) superimposed with the structure of a proteolytically stable FPA mimetic Acetyl-Asp-Phe-Leu-Ala-Glu-Gly-Gly-Gly-Val-Argψ(CO–CH$_2$)Gly-Pro-OH, in which the NH of the P1'-Gly has been replaced by a CH$_2$ group (gray carbon atoms). Thrombin is represented as a solid surface with the colors showing the electrostatic surface potential (negatively and positively charged regions are shown in red and blue with different intensities, respectively). The side chains of P3-DPhe, P2-Pro, and P1-Arg in inhibitor **5.192** and P9-Phe, P2-Val, and P1-Arg in the FPA mimetic occupy similar binding sites on thrombin. The structures were generated from PDB entries 1ppb for **5.192** and 1ucy for the FPA mimetic, both obtained from the Protein Data Bank (www.rcsb.org/pdb/). (B) Stereo view of the active site region of PPACK-thrombin. The residues of the catalytic triad (Ser195, His57, Asp102) and Asp189 at the bottom of the S1-binding pocket are shown as sticks with yellow carbon atoms; all other thrombin residues are shown as lines. Tyr60A and Trp60D of the thrombin-specific insertion loop are colored in green. The irreversible inhibitor PPACK (**5.192**) forms a covalent, hemiketal-like bond to the thrombin residue Ser195 and alkylates His57. Selected thrombin residues are labeled.

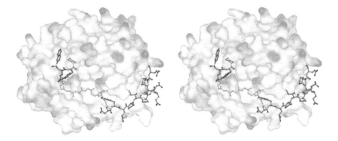

FIGURE 5.14 Stereo view of the bivalent inhibitor **5.232** in complex with thrombin generated from the PDB entry 1qur. Thrombin is represented as a transparent solid surface with the colors showing the electrostatic surface potential (for an explanation see Figure 5.13). The carbon atoms of the inhibitor's active site segment are colored in magenta, the linker is shown in green, and the hirudin-like segment is colored with gray carbons.

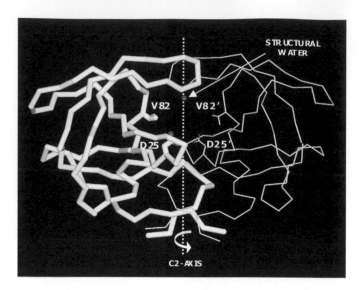

FIGURE 5.16 Structural features of HIV-1 protease that influence inhibitor design. HIV-1 protease is a C2-symmetric homodimer and its active site is located at the dimer interface. The catalytic aspartates (D25 and D25′) are located at the base of the pocket and interact with the hydroxyl group adjacent to the scissile bond. The structural water molecule (red sphere) at the top of the pocket bridges the flaps and peptidomimetic inhibitors. Val 82 and 82, which are located in the S1 and S1 subsites, are involved in key hydrophobic interactions with the P1 and P1 substituents of most inhibitors. The bound inhibitor has been omitted for clarity.

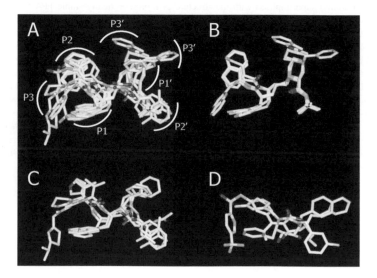

FIGURE 5.18 HIV-1 protease inhibitors interact with the same residues in the active site of the protein. A: Superposition of the bound conformations of saquinavir, nelfinavir, indinavir, L-756,423, amprenavir, ritonavir, lopinavir, CGP-73547, mozenavir, and tipranavir clearly reveals that they bind in the same subsites. B: Class IA compounds only differ at P2, P3, and P3. C: Similar binding modes are observed for classes IB and IC; D: Class II inhibitors appear to be the most diverse, except for the central core that displaces the structural water and places a hydroxyl between the catalytic aspartates.

3 Kinetics

Irene Lee and Anthony J. Berdis

CONTENTS

3.1 INTRODUCTION

Enzymes are powerful biological catalysts that are essential for the proper maintenance and propagation of any organism. These properties make them excellent candidates as therapeutic targets to combat diseases of either genetic or pathogenic origin. In this regard, one goal of molecular medicine is to develop and implement effective agents that can modulate the activity of various enzymes involved in essential biological pathways. The process of developing and characterizing these small molecules, i.e., rational drug design, often requires *a priori* knowledge of the enzyme in question. This chapter provides a rational approach to first characterizing the dynamic behavior of enzymes in the absence of potential modulators such as inhibitors or activators by studying the kinetics of enzyme-catalyzed reactions. To lay the foundation for understanding these catalysts, a brief discussion is provided regarding the principles of chemical catalysis as well as how these general principles are applied to the more complex biological catalysts. This is followed by a quantitative evaluation of single-substrate reactions, which includes sections on how to monitor enzyme activity, the various ways to analyze the generated rate data, and how to interpret the information derived from these experiments. A thorough discussion of the biologically relevant kinetic parameters V_{max}, K_m, and V_{max}/K_m is provided. Building upon the analysis for a simple single-substrate reaction, more complex scenarios involving multisubstrate reactions are discussed. Here again, the focus is on first designing the proper kinetic experiments and then interpreting the

$$S \xrightarrow{\;k_{+1}\;} P$$

$$\text{rate} = k_{+1}[S]$$

SCHEME 3.1

$$S + S \xrightarrow{\;k_{+1}\;} P \qquad \text{rate} = k_{+1}[S]^2$$

$$S + A \xrightarrow{\;k_{+1}\;} P \qquad \text{rate} = k_{+1}[S][A]$$

SCHEME 3.2

generated initial velocity data within the context of several possible kinetic mechanisms. These include distinguishing between a sequential- and a double-displacement mechanism, providing information regarding the order of substrate binding and product release, and identifying the location of the rate-limiting step along the reaction pathway. Finally, a discussion of the utility of applying pre-steady-state kinetic techniques to measure individual rate constants is provided. In general, the various kinetic sections outlined here provide the cornerstone for understanding the effects of therapeutically relevant inhibitors and inactivators of enzyme activity. Furthermore, understanding the kinetic behavior of an enzyme is essential to understanding the chemical mechanism, transition state structure, and regulation of any enzyme that again can be used for the rational design of therapeutic agents.

3.2 CHEMICAL KINETICS

The rate of a reaction is the speed at which the reacting substances are transformed to the corresponding products. For a reaction in which a single-reactant molecule S is converted to P, the rate of the reaction should be proportional to the concentration of S, as shown in Scheme 3.1. Because this reaction is dependent on the concentration of only one reactant, S, it is called a first-order reaction. The value k is defined as the first-order rate constant that relates the concentration of S to the rate of the reaction.

In an analogous manner, a reaction in which the rate depends on the concentration of two reactants is called a second-order reaction (Scheme 3.2). The rate constant k in this case is a second-order rate constant.

It is important to differentiate the rate constant from the rate of the same reaction. The rate measures the amount of substrate (S) consumption or product (P) formation per unit time, and it is the raw data obtained in a kinetic experiment. Reaction rates are dependent on the concentration of the reactants. On the other hand, rate constants are independent of substrate concentration and relate the rates of a reaction to the concentration of the reactants.

The conversion of S to P shown in Scheme 3.1 is an irreversible reaction in which only S is converted to P. In reality, however, many reactions are reversible such that P can be reconverted to S as shown in Scheme 3.3.

In the reversible reaction shown in Scheme 3.3, the rate of the reaction is a function of k_{+1} and k_{-1}. At equilibrium, the net rate of the reaction is zero because

$$S \underset{k_{-1}}{\overset{k_{+1}}{\rightleftharpoons}} P$$

SCHEME 3.3

the rate of S consumption in the forward direction equals the rate of its reformation from P in the reverse direction. The relationship between the forward and reverse reactions is shown as

$$K_{eq} = \frac{k_{+1}}{k_{-1}} = \frac{[P]}{[S]} \tag{3.1}$$

where K_{eq} is the equilibrium constant of the reaction.

3.3   ENZYME KINETICS

Enzyme kinetics is the branch of enzymology that deals with the factors affecting the rates of enzyme-catalyzed reactions. An enzyme catalyzes the rate of a reaction without changing the equilibrium concentration of the reactants and products, and it is recycled after each round of the reaction. The kinetics of an enzymatic reaction can be studied at two stages: (1) at pre-steady-state, also known as the transient state, which monitors the microscopic events along the reaction pathway during the first round of the reaction (prior to enzyme turnover); or (2) at steady state, which monitors multiple rounds or turnovers of an enzymatic reaction.

We shall begin our discussion with steady-state kinetics because this approach is usually first employed to study an enzyme. There are several reasons for this: (1) the instrumentation needed for this approach is readily accessible to most laboratories; (2) steady-state experiments require a very little amount of enzyme and are more economical in terms of experimentation; and (3) steady-state kinetics is more applicable to understanding cellular metabolism because it measures the activity of enzymes in physiologically relevant conditions.

The simplest enzymatic reaction is a single-substrate reaction adopting a minimal form as shown in Scheme 3.4.

In this example, E is the enzyme, S is the substrate (reagent), and P is the product. The first step of the reaction involves the binding of S to E to form a productive complex that leads to the formation of P in the second step. The overall reaction is irreversible because the conversion of ES to $E + P$ only proceeds in the forward direction with a rate constant of k_{+2}. The symbols k_{+1}, k_{-1}, and k_{+2} are microscopic rate constants that define forward and reverse processes along the reaction pathway. Using steady-state kinetic experiments, however, one can determine the macroscopic

$$E + S \underset{k_{-1}}{\overset{k_{+1}}{\rightleftharpoons}} ES \xrightarrow{k_{+2}} E + P$$

SCHEME 3.4

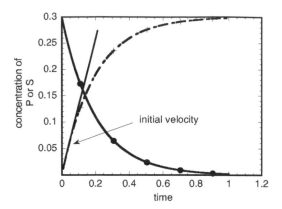

FIGURE 3.1 A time course for the disappearance of substrate (●) and the appearance of product (+) in the enzyme-catalyzed reaction shown in Scheme 3.4. The initial velocity is obtained from the linear region at the beginning of the time course.

rate constants and the kinetic parameters that summarize all the microscopic rate constants as a whole.

Figure 3.1 shows the time course of a typical enzymatic reaction exemplified in Scheme 3.4. In this type of experiment, the concentration of the substrate present in the reaction (typically in the μM to mM range) is in vast excess of the enzyme (typically in the pM to nM range) generating a pseudo–first-order condition such that the rate of the reaction, v, shows apparent dependency on [S] only. Note that the amount of substrate consumption or product formed over time is nonuniform. Closer examination of the earlier part of the time course in Figure 3.1 reveals that the amount of product formed or substrate consumed is linear. During this period, the amount of substrate consumed is less than 10% of the total substrate, $[S]_0$, present in the reaction, where substrate depletion or product inhibition is negligible. The reaction rates, defined by the slopes of the linear regions, are referred to as the initial velocities (v_0) and are used in steady-state kinetics analyses. As the reaction progresses, the velocity is affected by complicating factors such as the effect of reversible reactions, inhibition of the enzyme by the product, and substrate depletion, all of which can complicate data analysis and subsequent interpretation.

According to Scheme 3.4, both E and S participate in forming ES, and thus the rate of P formation, denoted as v, is dependent on the concentration of both E and S. In general, because the concentration of the reactant present in the reaction is constant, the initial velocity, v_0, is directly proportional to the concentration of the enzyme used. For example, the reaction rate can be doubled by increasing the enzyme concentration by twofold. However, the reverse scenario does not hold true when v_0 is measured at a fixed level of enzyme but at varying [S]. Figure 3.2 shows a graphical representation of v_0 vs. [S] for the reaction illustrated in Scheme 3.1.

At low [S], v_0 increases linearly with [S], and this region exhibits first-order kinetics in which v_0 is related to [S] by the constant V_{max}/K_m. At intermediate [S], the linear relationship between v_0 and [S] disappears until, at very high [S], v_0 attains

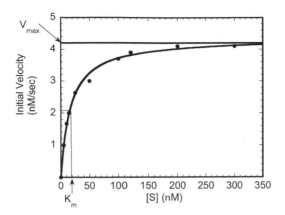

FIGURE 3.2 Plotting the initial velocities vs. [*S*] generates a hyperbolic plot. The concentration of substrate *S* is varied from 5 to 300 μ*M*. The symbol (●) represents experimental initial velocity data obtained at the specified [*S*]. According to the plot, $V_{max} = 4.2$ n*M*/sec and $K_m = 20$ μ*M*.

a limiting value (V_{max}) that exhibits apparent independency in [*S*], a condition sometimes referred to as zero-order kinetics. As illustrated in Figure 3.2, V_{max} is defined as the maximal velocity, whereas K_m is the concentration of [*S*] required to obtain $1/2\ V_{max}$. The turnover number of an enzyme, k_{cat}, is obtained by dividing V_{max} by the total amount of enzyme used in the reaction, $[E]_0$.

3.3.1 MEANING OF THE STEADY-STATE KINETIC PARAMETERS

To appreciate the meaning of the steady-state kinetic parameters introduced in Figure 3.2, we need to refer back to the reaction in Scheme 3.4. The general expression for the rate of this reaction is:

$$v_0 = \frac{d[P]}{dt} = k_{+2}[ES] \qquad (3.2)$$

and

$$\frac{d[ES]}{dt} = k_1[E][S] - k_{-1}[ES] - k_{+2}[ES] \qquad (3.3)$$

This reaction is irreversible because *P* cannot be converted back to *S* (no k_{-2} term). In analyzing the rate equation, either a rapid equilibrium or a steady-state assumption may be applied to describe the kinetics of this reaction scheme. According to the rapid-equilibrium assumption, the chemistry step k_{+2} is the slowest step along the reaction pathway. This means that the rate of *ES* formation and breakdown to *E* + *S* is much faster than the rate of *P* formation ($k_{-1} \gg k_{+2}$) such that the formation of *ES* achieves equilibrium. Thus,

$$\frac{k_{-1}}{k_{+1}} = \frac{[E]_f[S]}{[ES]} = K_s \tag{3.4}$$

where $[E]_f$ is the free enzyme concentration, and $[S] \gg [E]_f + [ES]$.

The ratio k_{-1}/k_{+1} is denoted as K_s, which is the dissociation constant of the first step of the reaction shown in Scheme 3.4, and ES is generally referred to as the Michaelis complex. Because $[E]_f = [E] - [ES]$,

$$K_s = \frac{([E]-[ES])[S]}{[ES]} \tag{3.5}$$

and, rearranging Equation 3.5 yields

$$[ES] = \frac{[E][S]}{K_s + [S]} \tag{3.6}$$

Because $v_0 = k_{+2}[ES]$, and $V_{max} = k_{+2}[E]_0$,

$$v_0 = \frac{V_{max}[S]}{K_m + [S]} \tag{3.7}$$

When $[S]$ is much larger than K_m, v_0 approximates V_{max}. Therefore, the maximum velocity, V_{max}, occurs at high substrate concentrations where the enzyme is saturated; that is, when it is entirely in the ES form. This expression, the Michaelis–Menten equation, is the basic equation of enzyme kinetics. It describes the rectangular hyperbola as plotted in Figure 3.2. Dividing V_{max} by the total enzyme concentration yields the turnover number of the enzyme, k_{cat}. The value k_{cat} defines the number of catalytic turnover events that occur per unit time, and it provides a lower limit on the first-order rate constant of the slowest or the rate-determining step of the reaction.

In the steady-state assumption (also known as the Briggs–Haldane relationship), the rate of ES formation equals its rate of conversion to $E + P$ over the time course of the reaction. Thus, the concentration of ES becomes constant, and $d[ES]/dt = 0$. Because $[E]_0 = [E] + [ES]$, using the steady-state approximation can be expressed as

$$k_{+1}([E]_0 - [ES])[S] = (k_{-1} + k_{+2})[ES],$$

$$k_{+1}[E]_0[S] = (k_{-1} + k_{+2})[ES] + k_{+1}([ES][S])$$

$$k_{+1}[E]_0[S] = [ES][(k_{-1} + k_{+2}) + k_{+1}[S]]$$

where dividing both sides by k_{+1} and solving for $[ES]$ yields

$$[ES] = \frac{[E]_0[S]}{(K_m + [S])} \tag{3.8}$$

where K_m is the Michaelis constant and is defined as

$$K_m = \frac{(k_{-1} + k_{+2})}{k_{+1}} \tag{3.9}$$

The velocity of the reaction can be related to Equation 3.8 as

$$v_0 = \frac{dP}{dt} = k_{+2}[ES] = \frac{k_2[E]_0[S]}{(K_m + [S])} \tag{3.10}$$

The Michaelis constant K_m can also be expressed as

$$K_m = \frac{k_{-1}}{k_{+1}} + \frac{k_{+2}}{k_{+1}} = K_s + \frac{k_{+2}}{k_{+1}} \tag{3.11}$$

The value K_s is the dissociation constant of the Michaelis complex and is inversely related to the enzyme's affinity for the substrate, that is, a high K_s reflects low affinity, and vice versa. The value K_m is often erroneously taken as the measure of the affinity of the enzyme for its substrate. As seen in Equation 3.11, K_m is equivalent to K_s only when $k_{+2} < k_{+1}$. If k_{+2} is significantly large, then the K_m value will be larger than K_s.

The specificity of the enzyme is defined by the ratio of the kinetic constants, k_{cat}/K_m. This value is a second-order rate constant and is generally used for comparing the efficiency of an enzyme catalyzing a reaction using a defined substrate. When $[S] \ll K_m$, very little $[ES]$ is formed. Consequently, the final enzyme concentration $[E]_f$ approximates the total enzyme concentration $[E]_0$, so that Equation 3.10 reduces to a second-order rate equation, as shown in Equation 3.12.

$$v_0 \approx \left(\frac{k_{+2}}{K_m}\right)[E]_0[S] \approx \left(\frac{k_{cat}}{K_m}\right)[E]_0[S] \tag{3.12}$$

In Equation 3.12, k_{+2} becomes k_{cat} when it is the rate-limiting step. Therefore, k_{cat}/K_m relates the reaction rate to free $[E]$ and $[S]$ such that the rate of the reaction varies directly with how often enzyme and substrate encounter each other in solution. If an enzyme can catalyze the same chemical transformation for a series of different substrates, k_{cat}/K_m can be used to assess the respective efficiency of each substrate. This is commonly referred to as the catalytic efficiency of an enzyme. In general, a higher k_{cat}/K_m value reflects a better substrate for a specific enzyme.

$$E + S \underset{k_{-1}}{\overset{k_{+1}}{\rightleftharpoons}} ES \underset{k_{-2}}{\overset{k_{+2}}{\rightleftharpoons}} E + P$$

SCHEME 3.5

3.3.2 REVERSIBLE REACTIONS

Many enzymatic reactions are reversible. Consequently, Scheme 3.4 can be expanded to include a reversible step denoted as k_2 as shown in Scheme 3.5.

The presence of k_{-2} necessitates the inclusion of the reverse step in the Michaelis–Menten equation such that

$$v = \frac{\dfrac{V_{max}^f [S]}{K_M^S} - \dfrac{V_{max}^r [P]}{K_M^P}}{1 + \dfrac{[S]}{K_M^S} + \dfrac{[P]}{K_M^P}} \tag{3.13}$$

where in the forward reaction,

$$V_{max}^f = k_s [E]_0 \tag{3.14}$$

and

$$K_M^S = \frac{k_{-1} + k_2}{k_{+1}} \tag{3.15}$$

and in the reverse reaction,

$$V_{max}^r = k_{-1} [E]_0 \tag{3.16}$$

and

$$K_M^P = \frac{k_{-1} + k_2}{k_{-2}} \tag{3.17}$$

At $[P] = 0$, $[E]_0 = [E] + [ES]$, and $v = v_0$.
At equilibrium, the observed rate v is zero (the Haldane Relationship)

$$K_{eq} = \frac{[P]}{[S]} = \frac{V_{max}^f K_M^P}{V_{max}^r K_M^S} \tag{3.18}$$

3.3.3 ENZYME ASSAYS

Because kinetic experiments measure the rates of a reaction, it is imperative that one accurately quantifies the concentration of substrate consumed, the concentration of product produced, or both during the course of the experiment. At present, two major methodologies are used to measure the rates of enzymatic reactions: a continuous assay and a discontinuous assay. In a continuous assay, the substrate and the product generate different signals such that they can be distinguished *in situ* by spectroscopic techniques. Some of the more common signals used to quantitatively distinguish substrate from product include UV-visible spectroscopy and fluorescence spectroscopy. UV-visible spectroscopy measures the change in absorbance of a molecule at a specific wavelength over time. The change is observed as a function of time defining the reaction rate. Upon determining the rates of absorbance change and the extinction coefficient of the compound of interest, one can apply the Beer–Lambert law to calculate the concentration of substrate or product at that wavelength. For example, the protease chymotrypsin hydrolyzes N-acetyl tryptophan p-nitrophenyl ester to yield N-acetyl tryptophan and p-nitrophenolate.[1] The product p-nitrophenolate absorbs strongly at 410 n*m* at values greater than pH 6. The substrate p-nitrophenyl ester, on the other hand, shows minimal absorption at 410 n*m*. The time course for p-nitrophenolate production, therefore, can be detected by monitoring the increase in absorbance at 410 n*m* as shown in Figure 3.3. Because the extinction coefficient of p-nitrophenolate is known, the initial rate v_o can be calculated from the slope in Figure 3.3.

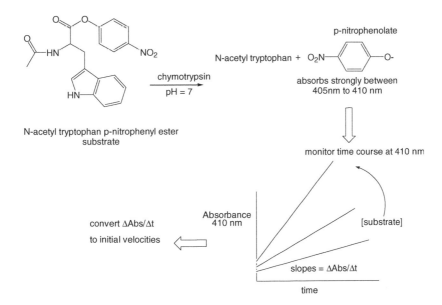

FIGURE 3.3 Chymotrypsin hydrolyzes *N*-acetyl tryptophan *p*-nitorphenyl ester to yield *p*-nitrophenolate. This reaction can be monitored continuously over time using a UV-visible spectrophotometer. The initial velocity of the reaction increases with substrate concentration present in the reaction.

Fluorescence spectroscopy utilizes a similar principle as UV-visible spectroscopy to measure reaction rates. In fluorescence spectroscopy, the compound of interest (either the substrate or the product) absorbs light at a specific wavelength and then reemits light at a longer wavelength. Therefore, it is often associated with two experimental parameters: the excitation wavelength where light is absorbed and the emission wavelength where light is emitted. Generally, the amount of light absorbed and subsequently emitted is proportional to the concentration of the compound present in the reaction. However, exceptions to this proportionality have been observed at high concentrations of the fluorescent compound due to an artifact known as the inner filter effect, and therefore care should be taken to check for linearity between the fluorescence signal and the fluorescent-compound concentration. Providing that the emission signal correlates linearly with the concentration of the target compound, one can calibrate the concentration of a compound with the intensity of light emitted, thereby converting the rates of fluorescence change to the rates of substrate consumed or product yielded.

Although the continuous-assay approach has been widely used to monitor enzyme kinetics, this method requires that the substrate and product have different spectroscopic properties. Unfortunately, not all chemical reactions yield products that are spectroscopically distinct from the substrates *in situ*. As an alternative, a discontinuous assay can be used to measure the rates of a reaction in which equal aliquots of a reaction mixture are withdrawn at different times and terminated by deactivating the enzyme. These reaction aliquots are known as the time points of a reaction, and the concentration of product formed at each time point can be derivatized to yield a chromophore or a fluorophore *in situ* that can be quantified by UV-visible spectroscopy or fluorescence spectroscopy in a manner analogous to the continuous assay described previously. For example, adenosine triphosphate (ATP) is hydrolyzed by ATPases to yield adenosine diphosphate (ADP) and inorganic phosphate (P_i). P_i can be rapidly and stoichiometrically converted into a chromophore that absorbs strongly at 660 n*m* using a Malachite Green assay. Therefore, the amount of ATP hydrolyzed at each time point can be directly quantified by determining the amount of P_i present by the Malachite Green assay, using a calibration curve that relates known [P_i] with the corresponding absorbance at 660 n*m*. Plotting the [P_i] produced vs. the time of quenched reaction provides a time course for the reaction, from which the initial rates, v_0, are determined from the slopes within the linear regions of the plot (Figure 3.4).

If the substrate and the product cannot be distinguished by chemical derivatization *in situ*, they must be physically separated from each other. Chromatography techniques are routinely used to separate the substrate from the product. A common way to detect and quantify the concentrations of substrate and product, or either, in a reaction time course is to use a radiolabeled substrate. For example, radiolabeled adenosine triphosphate ($\alpha^{32}P$-ATP) is hydrolyzed by an ATPase to yield inorganic phosphate and radiolabeled adenosine diphosphate ($\alpha^{32}P$-ADP). The radiolabeled $\alpha^{32}P$-ADP and $\alpha^{32}P$-ATP can be separated by thin-layer chromatography such that the relative amount of product generated at each time point can be quantified by the amount of radioactivity present at each spot. Plotting the concentration of $\alpha^{32}P$-ADP produced over time allows the determination of v_0. In addition to thin-layer chro-

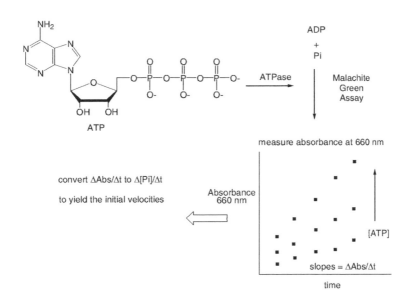

FIGURE 3.4 The inorganic phosphate produced from ATPase-catalyzed hydrolysis of adenosine triphosphate (ATP) is quantified colorimetrically using the Malachite Green assay. Plotting the absorbance at 660 n*m* at different reaction time points generates a linear plot that allows the determination of the initial velocities at increasing [ATP].

matography, separation techniques such as high-performance liquid chromatography, electrophoresis, and gel filtration have been used to separate substrate from product. Once separated, the product can be quantified by chemical derivation in conjunction with detection by an optical signal or radioactivity counting.

3.3.4 EXPERIMENTAL MEASUREMENT OF V_{MAX}, K_{CAT}, AND K_M

The kinetic constants V_{max} and K_m are determined graphically with initial velocity (v_0) measurements obtained at varying substrate concentrations. The most straightforward way of presenting the data is to directly plot the reaction rates as a function of [*S*]. Providing that the Michaelis–Menten kinetics is obeyed, one should be able to fit the Michaelis–Menten equation (Equation 3.7) to the direct plot to yield the V_{max} and K_m values. The data-fitting process can be accomplished by using a software program that provides nonlinear regression-fitting capability. Programs such as Kaleidagraph (Synergy, Inc.) allow the user to input a defined fitting function for routine data-fitting purposes. Alternatively, programs such as Enzfitter (Biosoft, Inc.) that contain a predefined library of equations pertinent to enzyme kinetics and protein–ligand binding studies can also be used. An example of fitting a direct velocity vs. [*S*] plot using nonlinear regression is shown in Figure 3.5.

Besides fitting the data by nonlinear regression, one can also transform the rate data to generate linear plots to determine the kinetic constants. The most commonly used linear plot is the Lineweaver–Burk plot, also known as a double-reciprocal plot. Figure 3.6 shows an example of a Lineweaver–Burk plot in which reciprocal

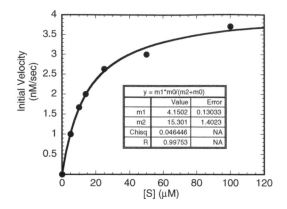

FIGURE 3.5 The initial velocity saturation curve is fitted with Equation 3.7 to determine the V_{max} and K_m values. The concentration of substrate S is varied from 5 to 100 μM. The computer program Kaleidagraph (Synergy, Inc.) was used to fit Equation 3.7 in the form of y = m1*m0/m2+m0, where m0 is [S], to the rate data represented by (●). The V_{max} determined from the fit is 4.15 ± 0.13 nM/sec and the K_m is 15.3 ± 1.4 μM.

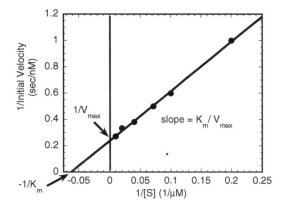

FIGURE 3.6 The reciprocals of initial velocities are plotted against the reciprocals of [S] to yield a Lineweaver–Burk plot. The concentration of substrate S is varied from 5 to 100 μM. The V_{max} determined from the y-intercept is 4.17 nM/sec. The K_m value determined from the slope (K_m/V_{max}) or the x-intercept is 17 μM.

reaction rates ($1/v_0$) are plotted against reciprocal substrate concentrations ($1/[S]$) to yield a straight line that is fitted with the linear function y = mx + b, where y is $1/v_0$; x is $1/[S]$; b is the y-intercept, $1/V_{max}$; and m is the slope, K_m/V_{max}. The mathematical basis for the Lineweaver–Burk plot is derived from algebraic rearrangement of the Michaelis–Menten equation (Equation 3.7) to yield Equation 3.19.

$$v = \frac{V_{max}[S]}{K_m + [S]} = \frac{V_{max}}{1 + \dfrac{K_m}{[S]}} \tag{3.19}$$

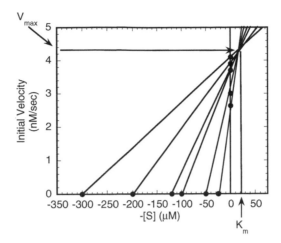

FIGURE 3.7 In the Eisenthal–Cornish–Bowden plot, the initial velocity data are plotted on the y-axis, whereas the substrate concentrations [S] are plotted on the negative x-axis. The substrate concentration S is varied from 25 to 300 μM. The convergence of the lines reveals graphically that $V_{max} = 4.3$ nM/sec and the $K_m = 13$ μM.

Taking the reciprocal of Equation 3.19 yields

$$\frac{1}{v} = \left(\frac{K_m}{V_{max}}\right)\frac{1}{[S]} + \frac{1}{V_{max}} \tag{3.20}$$

Therefore, plotting $1/v_0$ as a function of $1/[S]$ allows the evaluation of K_m/V_{max} and $1/V_{max}$ by measuring the slope and the y-intercept of the plot, respectively.

Although the Lineweaver–Burk plot has been widely used as a linear transformation method to determine steady-state kinetic parameters from experimental data, it has one major limitation. Because the line of the plot is primarily defined by the data points at low [S], which are more error prone, the kinetic parameters determined from linear regression may not accurately reflect the true values. Consequently, other linear graphical methods have been developed to alleviate such errors. For example, the Eisenthal–Cornish–Bowden plot entails plotting v_0 values along the y-axis and the corresponding [S] values along the negative x-axis to yield a series of lines that intersect at $[S] = K_m$ and $v_0 = V_{max}$ as shown in Figure 3.7.

The V_{max} and the K_m values are deduced from the convergence of the plot. Some transformed plots are also derived from the rearrangement of the Michaelis–Menten equation or the Lineweaver–Burk equation to generate linear plots that can be fitted with a defined linear function. For example, the Eadie–Hofstee plot is derived from the Michaelis–Menten equation such that $v_0 = V_{max} - K_m (v_0/[S])$. By plotting v_0 as a function of $v_0/[S]$, a straight line is obtained with its slope defining K_m and its y-intercept defining V_{max}. Multiplying the Lineweaver–Burk equation by [S] yields a function that relates $[S]/v_0$ linearly with [S] to yield

$$\frac{[S]}{v} = \left(\frac{1}{V_{\max}}\right)[S] + \frac{K_m}{V_{\max}} \tag{3.21}$$

In this case, plotting $[S]/v_0$ vs. $[S]$ yields the Hanes–Wolff plot in which the slope defines $1/V_{\max}$, and the y-intercept defines $K_m/V_{\max}$.

3.3.5 Multisubstrate Reaction Mechanisms

The preceding section provided a fundamental understanding toward quantifying enzymatic activity using Michaelis–Menten kinetics. Although this technique has been widely applied to study various enzyme systems, its utility is somewhat limited because it is most appropriately applied to enzymes that utilize only a single substrate. These types of enzymes are best represented by isomerases such as epimerase and mutases such as phosphoglyceromutase that convert a single substrate into a single product. However, Michaelis–Menten kinetics can be applied to certain multisubstrate enzymes. For example, hydrolytic reactions that are catalyzed by proteases have two substrates — a protein or peptide that undergoes hydrolysis, and water. For these enzymes, it is rather easy to vary the concentration of the protein or peptide substrate to generate $V_{\max}$, K_m, and $V_{\max}/K_m$ values. However, it is rather difficult to vary the concentration of water because it remains fairly constant at 55 M. Thus, Michaelis–Menten kinetics can be applied to these types of enzymes because the concentration of the second substrate remains fixed and saturated.

The vast majority of enzymes found in nature utilize two or more distinct substrates. When discussing these higher-order enzymatic reactions, it is important to first understand the nomenclature originally developed by W.W. Cleland that has been utilized by kineticists to comprehend the theory and experimental approaches behind steady-state kinetics. Substrates are designated A, B, C, and D in the order in which they bind to the enzyme, whereas products are designated P, Q, R, and S in the order in which they dissociate from the enzyme after catalysis. Enzyme forms are denoted as either E, the free enzyme, or F, which is another stable enzyme form typically associated with double-displacement reactions (see the following text). As the number of substrates utilized has increased, the number of enzyme–substrate, enzyme–product, and enzyme–substrate–product complexes has also expanded. For example, substrate A can bind to the free enzyme E to form the binary EA complex. This transitory complex can then bind to substrate B to form the transitory, ternary EAB complex. These complexes are defined as being transitory because their "lifetime" is relatively short. That is, they can either proceed toward catalysis to yield a product or decompose back to free enzyme. Despite the increase in the number of substrates, the terminology and mechanistic deductions for kinetic parameters such as $V_{\max}$, K_m, and $V_{\max}/K_m$ discussed for unisubstrate reactions are still valid and applicable. In nearly all cases, the Michaelis constant for substrate A is denoted as K_a, whereas that for substrate B is denoted as K_b.

In the case of bisubstrate enzyme reactions, simple Michaelis–Menten kinetics have been applied by maintaining the concentration of substrate B fixed while varying the concentration of substrate A. Although this approach can provide values for $V_{\max}$,

K_m, and V_{max}/K_m for substrate A, it should not be employed because it runs the risk of obtaining erroneous kinetic parameters for substrate A as $[B]$ has been arbitrarily set. This will become clearly evident in Section 3.3.6, demonstrating that the kinetic parameters for one substrate are most often sensitive to the concentration of the other substrate. In these instances, artificially low V_{max} and V_{max}/K_m values for substrate A are obtained because $[B]$ was arbitrarily fixed at a subsaturating concentration.

A more standard approach to studying the kinetic behavior of multisubstrate enzymes is to measure the rate of product formation while varying the concentration of both substrates. Before describing the details of experimental design and interpretation, it is important to emphasize the advantages associated with this type of analysis. Although varying the concentration of both substrates adds complexity to data acquisition and analysis, this technique can be easily applied to accurately determine the kinetic parameters V_{max}, K_m, and V_{max}/K_m for both substrates. Furthermore, the equilibrium dissociation constants, K_{ia} and K_{ib}, for the substrates can be measured although the relevance of these terms is dependent upon the kinetic mechanism. This is unique because this information cannot be obtained using the modified Michaelis–Menten approach discussed above. Perhaps more important, however, is how the implementation of this technique can be used to distinguish between a sequential reaction mechanism and a double-displacement or "ping-pong" kinetic mechanism. In some instances, one can obtain preliminary evidence to distinguish among the various sequential kinetic mechanisms, i.e., rapid-equilibrium ordered and rapid-equilibrium random kinetic mechanisms, which aids in assigning the potential rate-limiting step of the enzyme reaction as well as the order of substrate binding and product release.

For convenience, Section 3.3.6 will be limited to the analysis of bisubstrate enzymatic reactions. The same approach can also be applied to higher-order reaction mechanisms such as ter- and quad-reactant mechanisms, although the complexity of the experiments and data analyses increases significantly. For those interested, an excellent overview of this technique is provided in a paper written by Viola and Cleland. The overall approach for studying multisubstrate enzymes is similar to that previously described for analyzing unisubstrate reactions. The key exception, of course, is that $[A]$ is now varied at several different fixed $[B]$. Figure 3.8 provides representative data plotting the initial rate in product formation vs. $[A]$ at the differing $[B]$. At each fixed concentration of substrate B, saturation kinetics is observed for the variable substrate A as expected. It is important to note that the K_m and V_{max} values for substrate A change as the concentration of substrate B increases. This reemphasizes that the kinetic parameters for substrate A are dependent upon the concentration of substrate B, and vice versa.

At this point, one is interested in obtaining the kinetic parameters for the enzymatic reaction. The data can be directly fitted to the rate equation best defining the initial velocity data. However, the rate equation is dependent upon the kinetic mechanism of the enzyme in question. This leads to the apparent paradox of steady-state kinetics — how can the proper rate equation be used to define the mechanism if the kinetic mechanism is not known? Before proceeding with the assignment of rate equations, it is most instructive to first evaluate initial velocity data in order to distinguish between the two general classes of enzyme mechanisms: sequential and

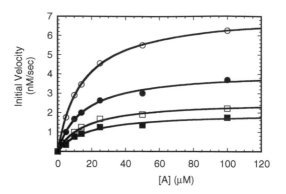

FIGURE 3.8 Initial velocity patterns measuring rate as a function of substrate A at several different concentrations of substrate B. The concentration of substrate A is varied from 5 to 100 μM. The concentrations of substrate B are 2 μM (■), 2.9 μM (□), 5 μM (●), and 20 μM (○).

double-displacement. A sequential mechanism is one in which both substrates must be bound to the enzyme in order for catalysis to occur. This can occur in either an ordered or random fashion. As the name implies, an ordered mechanism depicted in Scheme 3.6a involves the obligatory binding of substrate A prior to the binding of substrate B. In a random mechanism (Scheme 3.6b), substrates can bind in any order to generate the productive ternary EAB complex.

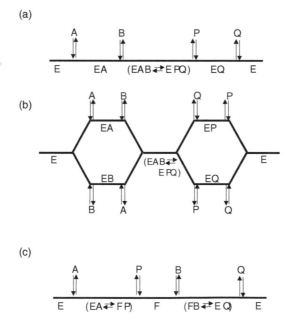

SCHEME 3.6 Cleland representations for (a) an ordered reaction mechanism (b) a random reaction mechanism, and (c) a double-displacement reaction mechanism.

Each type of sequential mechanism can be further subdivided into different classes that are differentiated by the location of the rate-limiting step. A rapid-equilibrium mechanism, for example, indicates that chemistry is the rate-limiting step along the reaction pathway, whereas in a steady-state ordered mechanism, the rate-limiting step can be any step other than chemistry. One specialized mechanism is the Theorell–Chance mechanism in which release of the second product, Q, is the slowest step along the reaction pathway. Another specialized sequential mechanism that will not be discussed extensively here is the steady-state random mechanism. Briefly, a steady-state random mechanism allows for the binding of substrate in any order. However, because the rate-limiting step can be any step other than chemistry, the overall rate equation becomes much more complex as it possesses many squared terms in both the numerator and denominator to account for the increased number of potential enzyme forms that can exist along the reaction pathway.

In a double-displacement mechanism, a catalytic event can occur when only one of the two substrates is bound to the enzyme. As depicted in Scheme 3.6c, a catalytic event occurs after the binding of substrate A but prior to the binding of substrate B. After the first catalytic event, a portion of substrate A remains in the active site of the enzyme, whereas the other portion denoted as product P dissociates. The first catalytic event produces a different stable enzyme form denoted as F to which substrate B can bind and proceed with catalysis to generate product Q and regenerate free enzyme E.

Figure 3.9A and Figure 3.9B show the initial velocity data for an enzyme possessing either a sequential or a ping-pong mechanism, respectively. Visual inspection of these plots does not reveal a significant difference between the two; this makes it nearly impossible to unambiguously assign the proper rate equation. However, simply replotting the data in the form of a double-reciprocal plot, i.e., $1/v_0$ vs. $1/[A]$ at the differing $[B]$, allows easy distinction between a sequential or ping-pong mechanism. Specifically, the double-reciprocal plot in Figure 3.10A shows a series of lines intersecting to the left of the y-axis, which is indicative of a sequential mechanism. By contrast, the double-reciprocal plot in Figure 3.10B reveals a series of parallel lines that are indicative of a ping-pong mechanism. Although exceptions to these generalized rules do exist, it is clear that plotting the data in double-reciprocal form has a distinct advantage in providing a reasonable distinction between kinetic mechanisms. It is now appropriate to fit the data to the rate equation to define the kinetic parameters.

We shall first focus on the random sequential reaction mechanism displayed in Scheme 3.6b. The overall rate equation for this mechanism is given by Equation 3.22.

$$v_0 = \frac{V_{max}[A][B]}{K_{ia}K_b + K_a[B] + K_b[A] + [A][B]} \tag{3.22}$$

Although this is similar in form and composition to the Michaelis–Menten equation, there are some obvious exceptions worthy of discussion. In Equation 3.22, K_{ia} is the intrinsic dissociation constant for substrate A, whereas K_a and K_b are the Michaelis constants for substrates A and B, respectively. Furthermore, the numerator,

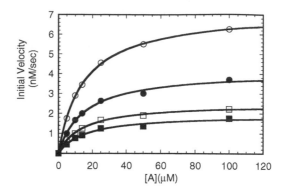

FIGURE 3.9A Initial velocity pattern for a sequential reaction mechanism in which the rate of product formation is plotted as a function of substrate A at several different concentrations of substrate B. The concentration of substrate A is varied from 5 to 100 μM. The concentrations of substrate B are 2 μM (■), 2.9 μM (□), 5 μM (●), and 20 μM (○).

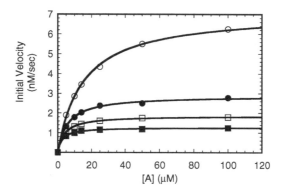

FIGURE 3.9B Initial velocity pattern for a double-displacement reaction mechanism in which the rate of product formation is plotted as a function of substrate A at several different concentrations of substrate B. The concentration of substrate A is varied from 5 to 100 μM. The concentrations of substrate B are 2 μM (■), 2.9 μM (□), 5 μM (○), and 20 μM (●).

representing the thermodynamic driving force for the reaction, must contain terms corresponding to V_{max} and both substrates. The denominator corrects for the slowing down of the reaction due to the distribution of substrates among the various enzyme forms such as E, EA, EB, and EAB that can exist along the reaction pathway. As such, the denominator is also expanded to account for the terms corresponding to the concentrations of both substrates, the Michaelis constants for both substrates, and the equilibrium-dissociation constant for substrate A. In double-reciprocal form, the rate equation becomes Equation 3.23.

$$\frac{1}{v_0} = \frac{K_{ia}K_b}{V_{max}[A][B]} + \frac{K_b}{V_{max}[A]} + \frac{K_a}{V_{max}[B]} + \frac{1}{V_{max}} \tag{3.23}$$

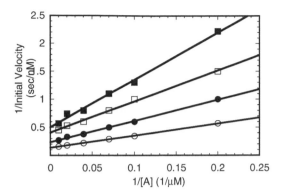

FIGURE 3.10A Double-reciprocal plot for a sequential reaction mechanism. The plot of 1/initial velocity vs. 1/[A] at several different concentrations of substrate B provides a series of lines that intersect to the left of the y-axis. The concentration of substrate A is varied from 5 to 100 μM. The concentrations of substrate B are 2 μM (■), 2.9 μM (□), 5 μM (○), and 20 μM (●).

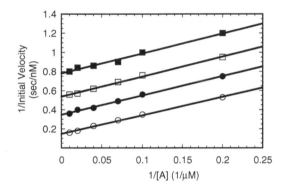

FIGURE 3.10B Double-reciprocal plot for a double-displacement reaction mechanism. The plot of 1/initial velocity vs. 1/[A] at several different concentrations of substrate B provides a series of parallel lines. The concentration of substrate A is varied from 5 to 100 μM. The concentrations of substrate B are 2 μM (■), 2.9 μM (□), 5 μM (○), and 20 μM (●).

Although this appears to make the rate equation more complex, it actually simplifies it by converting it into a linear form that can be directly analyzed through various plotting regressions. For example, when A is varied at different levels of B, the equation becomes

$$\frac{1}{v_0} = \frac{1}{[A]}\left(\frac{K_{ia}K_b}{V_{max}[B]} + \frac{K_b}{V_{max}}\right) + \frac{K_a}{V_{max}[B]} + \frac{1}{V_{max}} \tag{3.24}$$

and gives rise to the plot previously shown in Figure 3.10A. Visual inspection of this plot reveals that both the slope and intercept decrease as the concentration of B increases and gives rise to what is referred to as the "slope" and "intercept" effect, respectively. This observation is apparent because the double-reciprocal equation now has the form of $y = mx + b$ in which the slope term $[(K_{ia}K_b/V_{max}[B] + K_b/V_{max})]$ and intercept term $[(K_d/V_{max}[B] + 1/V_{max})]$ show a dependency on the concentration of substrate B. Note that both the slope and intercept values are in the form of linear equations and can thus be plotted to give values corresponding to V_{max}, Michaelis constants, and equilibrium-dissociation constants. Working through this algebraic problem, we find that the intercept value from the primary plot is defined as Equation 3.25.

$$y = \frac{1}{[B]}\left(\frac{K_a}{V_{max}}\right) + \frac{1}{V_{max}} \tag{3.25}$$

When plotted as intercept vs. $1/[B]$, one obtains the plot provided in Figure 3.11A. This is referred to as a secondary plot in which the intercept of this plot (the "intercept of the intercept") is $1/V_{max}$, whereas the slope of this plot (the "slope of the intercept") is equivalent to K_d/V_{max}. Thus, two kinetic parameters of interest, V_{max} and K_a, can be directly measured by implementing this approach. It is also important to emphasize how this replot reflects the different enzyme forms that exist as a function of substrate concentration. The intercept values from the primary double-reciprocal plot are obtained under conditions in which $[A]$ is high but $[B]$ is varied. Thus, the intercept of the intercept is defined optimally when the concentration of both substrate A and substrate B are maintained high. Under these conditions, the enzyme form that must predominate is the EAB form and is represented by the $1/V_{max}$ term. By contrast, the slope of the intercept is obtained when $[A]$ is high but $[B]$ is low. Thus, the law of mass action predicts that the enzyme form that should predominate is the EA form that is represented by K_d/V_{max}.

Similar analyses can be performed to evaluate the replots for the slope effect. The slope value from the primary plot is defined as

$$y = \frac{1}{[B]}\left(\frac{K_{ia}K_b}{V_{max}}\right) + \frac{K_b}{V_{max}} \tag{3.26}$$

and when plotted as the slope vs. $1/[B]$, yields the plot provided in Figure 3.11B. The intercept (intercept of the slope) is K_b/V_{max}, whereas the slope of this plot (slope of the slope) is equivalent to $K_{ia}K_b/V_{max}$. This allows the other two kinetic parameters of interest, K_b and K_{ia}, to be graphically and empirically measured. As before, this replot reflects the different enzyme forms that exist as a function of substrate concentration. The slope values from the primary double-reciprocal plot are obtained under conditions in which $[A]$ is low, and $[B]$ is varied. Thus, the intercept of the slope replot is obtained under conditions in which $[A]$ is maintained low and that for substrate B is maintained high. Under these conditions, EB is the enzyme form that predominates and is represented by the K_b/V_{max} term. By contrast, the slope of the slope is obtained when the concentrations of both substrates are low. Under these

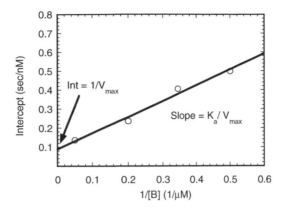

FIGURE 3.11A Secondary replot of intercept vs. $1/[B]$. The intercept of this plot provides a V_{max} value of 11.1 nM/sec and the slope of this plot provides a K_a value of 9.4 μM.

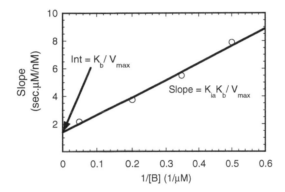

FIGURE 3.11B Secondary replot of slope vs. $1/[B]$. The intercept of this plot provides a K_b value of 16.7 μM and the slope of this plot provides a K_{ia} value of 8.3 μM. Note that the nearly identical values for K_{ia} and K_a (determined from Figure 3.11A) are consistent with a rapid-equilibrium random mechanism.

conditions, the predominant enzyme form is the free enzyme, E, and is represented by the term $K_{ia}K_b/V_{max}$.

It should be noted that the same rate equation discussed can be applied to the rapid-equilibrium random, steady-state ordered, and Theorell–Chance mechanisms. This means, of course, that the initial velocity patterns and double-reciprocal plots for these three mechanisms are indistinguishable. However, these mechanisms do differ in the location of the rate-limiting step along the reaction pathway, and thus other kinetic techniques such as product and dead-end inhibition studies, isotope effects, and pre-steady-state kinetics studies can be used to differentiate among them. Despite the identity in rate equations for these unique mechanisms, it should be noted that the enzyme forms are represented differently, as summarized in Table 3.1. Again, these differences do not affect the initial velocity patterns or the graphical analyses used to measure the kinetic parameters. Rather, they indicate which enzyme

TABLE 3.1
Summary of Common Bisubstrate Kinetic Mechanisms

Mechanism	Rate Equation[a]	Double-Reciprocal Plots	Enzyme Forms
Rapid-equilibrium random	$v = VAB/(K_{ia}K_b + K_aB + K_bA + AB)$	Both intersect to the left of the y-axis	$K_{ia}K_b = E$ $K_bA = EA$ $K_aB = EB$ $AB = EAB$
Steady-state ordered	$v = VAB/(K_{ia}K_b + K_aB + K_bA + AB)$	Both intersect to the left of the y-axis	$K_{ia}K_b, K_aB = E$ $K_bA = EA$ $AB = EAB$
Theorell–Chance	$v = VAB/(K_{ia}K_b + K_aB + K_bA + AB)$	Both intersect to the left of the y-axis	$K_{ia}K_b + K_aB = E$ $K_bA = EA$ $AB = EAB$
Rapid-equilibrium ordered	$v = VAB/(K_{ia}K_b + K_bA + AB)$	Vary A: intersects to the left of the y-axis; Vary B: intersects on the y-axis	$K_{ia}K_b = E$ $K_bA = EA$ $AB = EAB$
Steady-state random	Complex	Intersection to the left of the y-axis, curvature near y-axis possible	Complex
Double displacement	$v = VAB/(K_aB + K_bA + AB)$	Both provide parallel lines	$K_aB = E$ $K_bA = F$

[a] AB = All possible binary and stable enzyme forms.

forms can exist and which enzyme forms do exist under any set of reaction conditions. This information has practical applications for understanding the dynamic behavior of dead-end inhibitors that may be of therapeutic importance.

A unique sequential mechanism is the rapid-equilibrium ordered mechanism. Although similar in form to the steady-state ordered mechanism depicted in Scheme 3.6a, this mechanism differs from the former because the chemistry step is the slowest step along the reaction pathway. This distinction has ramifications with regards to the various kinetic terms because the $K_a[B]$ form does not exist in this mechanism. Thus, the overall rate equation for an equilibrium-ordered mechanism is Equation 3.27.

$$v_0 = \frac{V_{max}[A][B]}{K_{ia}K_b + K_b[A] + [A][B]} \tag{3.27}$$

This asymmetric equation does not give the identical double-reciprocal patterns observed for the other sequential mechanisms, as previously described. For example, in double-reciprocal form, the rate equation becomes

$$\frac{1}{v_0} = \frac{K_{ia}K_b}{V_{max}[A][B]} + \frac{K_b}{V_{max}[A]} + \frac{1}{V_{max}} \tag{3.28}$$

that again has a linear form that can be directly analyzed through various plotting regressions. When A is varied at different levels of B, the equation becomes

$$\frac{1}{v_0} = \frac{1}{A}\left(\frac{K_{ia}K_b}{V_{max}[B]}\right) + \frac{K_b}{V_{max}[B]} + \frac{1}{V_{max}} \tag{3.29}$$

and gives rise to the plot provided in Figure 3.12A, which looks identical to the double-reciprocal plots discussed above. As before, both the slope ($K_{ia}K_b/V_{max}[B]$) and intercept ($K_b/V_{max}[B] + 1/V_{max}$) terms reveal a dependency on the concentration of substrate B. In this instance, the intercept value from the primary plot is defined as Equation 3.30.

$$y = \frac{1}{[B]}\left(\frac{K_b}{V_{max}}\right) + \frac{1}{V_{max}} \tag{3.30}$$

When plotted as the slope vs. $1/[B]$, a linear plot is obtained from which the slope of the slope is equal to K_b/V_{max} and the intercept of the slope is equal to $1/V_{max}$. By contrast, the slope value from the primary plot is defined as $y = 1/[B](K_{ia}K_b/V_{max})$ and contains only one term. Thus, when plotted as the slope vs. $1/[B]$, a line that passes directly through the origin is obtained. This indicates that a term corresponding to K_a/V_{max} does not exist because the intercept of the slope is zero. However, the slope-of-the-slope value reflects free enzyme and is equivalent to $K_{ia}K_b/V_{max}$.

As with all bisubstrate reactions, it is useful to plot the data in both double-reciprocal forms, i.e., $1/v_0$ vs. $1/[A]$ at various levels of B and vice versa. In the case of the rapid-equilibrium ordered mechanism, the plot of $1/v_0$ vs. $1/[B]$ at several fixed $[A]$ displays a series of lines that converge on the y-axis (Figure 3.12B). This pattern is distinctive from all the other double-reciprocal plots and is diagnostic for a rapid-equilibrium ordered mechanism. In general, it is wise for the experimenter to plot the data both ways because this not only provides a way to distinguish various kinetic mechanisms but can also provide evidence for the order in which substrates bind to the enzyme, at least in the case of the rapid-equilibrium mechanism.

Having considered various sequential mechanisms, let us revisit the double-displacement or ping-pong mechanism that represents the other type of bireactant mechanism (Scheme 3.6). The rate equation for this class of mechanism is Equation 3.31.

$$v_0 = \frac{V_{max}[A][B]}{K_a[B] + K_b[A] + [A][B]} \tag{3.31}$$

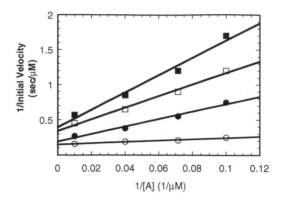

FIGURE 3.12A Double-reciprocal plot for a rapid-equilibrium ordered reaction mechanism. The plot of 1/initial velocity vs. 1/[A] at several different concentrations of substrate B provides a series of lines that intersect to the left of the y-axis. The concentration of substrate A is varied from 5 to 100 μM. The concentrations of substrate B are 2 μM (■), 2.9 μM (□), 5 μM (○), and 20 μM (●).

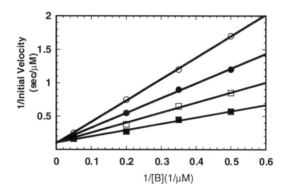

FIGURE 3.12B Double-reciprocal plot for a rapid-equilibrium ordered reaction mechanism. The plot of 1/initial velocity vs. 1/[B] at several different concentrations of substrate A provides a series of lines that intersect on the y-axis. The concentration of substrate A is varied from 2 to 20 μM. The concentrations of substrate A are 10 μM (■), 14 μM (□), 25 μM (○), and 100 μM (●).

This rate equation is symmetrical as it can be broken into two separate equations that reflect each half reaction of the complete reaction. For example, at saturating concentrations of substrate B, the equation can be simplified to

$$v_0 = \frac{V_{max}[A]}{K_a + [A]} \tag{3.32}$$

whereas at saturating levels of substrate A, the rate equation simplifies to

$$v_0 = \frac{V_{max}[B]}{K_b + [B]} \tag{3.33}$$

As discussed earlier, the diagnostic double-reciprocal pattern for a double-displacement mechanism is a series of parallel lines (see Figure 3.10B). Again, secondary replots can be used to empirically measure the kinetic parameters of interest. When varying the concentration of A, the double-reciprocal equation becomes

$$\frac{1}{v_0} = \frac{K_a}{V_{max}}\frac{1}{[A]} + \frac{1}{V_{max}}\left(1 + \frac{K_b}{[B]}\right) \tag{3.34}$$

and again provides both a slope and an intercept term. When plotted as the intercept vs. 1/[B], the linear transformation provides the slope of the intercept equal to K_b/V_{max} and the intercept of the intercept equal to $1/V_{max}$. By contrast, the slope value from the primary plot is defined as $y = (K_a/V_{max})$ and contains a single term that is independent of [B]. When plotted as the slope vs. 1/[B], a line with only an intercept value equal to K_a/V_{max} is obtained because a value corresponding to $K_{ia}K_b/V_{max}$ does not exist.

The reader should be cautious in unambiguously defining a reaction mechanism based solely on initial velocity data because exceptions to these generalized rules do exist. One such example is the degeneration of a rapid-equilibrium random mechanism into a rapid-equilibrium ordered mechanism. This can occur under conditions in which the value of K_{ia} is much greater than that of K_a. At the molecular level, this indicates that substrate A binds to the EB complex with higher affinity as compared to its affinity to free enzyme, E. This is commonly referred to as synergism in binding. The algebraic formalism again reveals that this is indeed the case. If the value of K_{ia} is much greater than that of K_a, the rate equation for the rapid-equilibrium random mechanism shown in Equation 3.22 reduces to Equation 3.35 because the $K_a[B]$ term becomes insignificant.

$$v_0 = \frac{V_{max}[A][B]}{K_{ia}K_b + K_a[B] + K_b[A] + [A][B]} \tag{3.22}$$

$$v_0 = \frac{V_{max}[A][B]}{K_{ia}K_b + K_b[A] + [A][B]} \tag{3.35}$$

The initial velocity patterns form a series of lines that converge on the y-axis when the data are plotted as $1/v_0$ vs. 1/[B] and lead the experimenter to believe that the mechanism was rapid-equilibrium ordered rather than rapid-equilibrium random. Another example is the degeneration of a sequential mechanism in the appearance of a ping-pong mechanism. This can occur if the value of K_{ia} is substantially less than that of K_a such that the $K_{ia}K_b$ term becomes negligible, and the rate equation degenerates into

$$v_0 = \frac{V_{max}[A][B]}{K_a[B] + K_b[A] + [A][B]} \tag{3.36}$$

This section was provided to point out the major pitfall of steady-state kinetics: Initial velocity is deductive in nature such that data should only be used to eliminate certain kinetic mechanisms and never be used as sole justification in assigning a mechanism. As will be discussed in Chapter 4, other techniques such as product and dead-end inhibition studies as well as heavy atom isotope effect studies should be used to complement initial velocity data when attempting to assign a kinetic mechanism.

3.3.6 PRE-STEADY-STATE OR TRANSIENT KINETICS

The reaction shown in Scheme 3.4 involves the formation of a single binary complex, [ES], and it is an oversimplification of a real enzymatic reaction. In fact, more complex reaction mechanisms can exist that contain two or more binary complexes along the reaction pathway. For example, both reaction Scheme 3.7 and Scheme 3.8 show an enzymatic reaction pathway containing more than one distinct binary complex, i.e., ES and EP, and a transiently formed intermediate EX in the latter.

The steady-state kinetic analyses described above cannot distinguish a reaction undergoing Scheme 3.7 or Scheme 3.8, as both schemes can also be described by Equation 3.7 and thus give rise to the same Michaelis–Menten plot shown in Figure 3.2. This is because the steady-state kinetic constants k_{cat} and K_m are complex functions that combine rate constants from multiple steps in the overall enzymatic reaction. However, these reaction pathways can be distinguished by employing pre-steady-state kinetic analyses to identify the number of reaction intermediates that exist along the reaction pathway. As the name implies, pre-steady-state kinetics can measure physical and chemical events at an enzyme's active site that occur prior to attainment of the steady state, i.e., the first turnover of the enzyme. As such, this approach can monitor the rate of formation of the transient reaction intermediates. In the case of Scheme 3.7, the formation of EP depends on only one step — the formation of ES. On the contrary, the formation of EP in Scheme 3.8 depends on two steps — the formation of ES followed by the formation of EX. By determining the kinetics of EP formation, one can distinguish between a one-step and a two-step mechanism. Because a reaction intermediate is only transiently formed during each cycle of the reaction, one can only monitor its formation during the first round of

$$E + S \underset{k_{-1}}{\overset{k_{+1}}{\rightleftharpoons}} ES \underset{k_{-2}}{\overset{k_{+2}}{\rightleftharpoons}} EP \xrightarrow{k_{+3}} E + P$$

SCHEME 3.7

$$E + S \underset{k_{-1}}{\overset{k_{+1}}{\rightleftharpoons}} ES \underset{k_{-2}}{\overset{k_{+2}}{\rightleftharpoons}} EX \xrightarrow{k_{+3}} EP \xrightarrow{k_{+4}} E + P$$

SCHEME 3.8

the reaction, that is, before an enzyme undergoes steady-state turnover. In general, reaction intermediates are formed on a millisecond time scale, and thus special instrumentation such as stopped-flow apparatus or chemical quench-flow instruments are needed to accurately monitor the pre-steady-state kinetic reactions.

To illustrate the pre-steady-state approach, we will use the reaction shown in Scheme 3.7 and Scheme 3.8 as an example. We tentatively assume that the reaction intermediates, ES or EX, are detectable by the change in tryptophan fluorescence intensity inherent to the enzyme due to conformational changes associated with enzyme intermediate formation. Therefore, the rate constant associated with the formation of the reaction intermediate (k_{obs}) can be determined from the time course of tryptophan fluorescence changes upon mixing E with S. In Scheme 3.7, the formation of ES is dependent on $[E]$ and $[S]$ as well as the forward and reverse rate constants k_{+1} and k_{-1}, respectively. If $[S]$ is maintained in sufficient excess over $[E]$ to ensure pseudo-first-order conditions, then the time course in ES formation will approximate a single exponential process, which can be fitted with

$$[ES]_t = [E]_0(1\ \exp^{-kt}) \tag{3.37}$$

where $[ES]_t$ is the concentration of the enzyme–substrate complex at time t, and apparent k is the pseudo-first-order rate constant of the reaction. The rate constants for ES formation can then be measured by varying $[S]$ and maintaining $[E]$ constant. For a one-step reaction mechanism, the plot of k_{obs} vs. $[S]$ will be linear and equal to $k_{+1}[S] + k_{-1}$. Linear regression analysis can again provide values corresponding to k_{+1} (the slope) and k_{-1} (the y-intercept), as illustrated in Figure 3.13A.

If the reaction proceeds via Scheme 3.8 and involves the formation of EX prior to chemical transformation of S to yield EP, one would detect a nonlinear functional relationship between k_{obs} for the formation of EX and the concentration of $[S]$. In this case, the concentration dependence of the rate of EX formation follows the hyperbolic function

$$k_{obs} = \frac{K_1 k_{+2}[S]}{K_1[S]+1} + k_{-2} = \frac{k_{+2}[S]}{[S]+\dfrac{1}{K_1}} \tag{3.38}$$

where K_1 equals k_{+1}/k_{-1}. Therefore, the reaction Scheme 3.7 and Scheme 3.8 can be distinguished by the dependency of k_{obs} towards $[S]$. A linear plot of k_{obs} vs. $[S]$ (Figure 3.13A) is indicative of a one intermediate mechanism (Scheme 3.7), whereas a hyperbolic plot (Figure 3.13B) is indicative of a two intermediate mechanism (Scheme 3.8).

3.4 SUMMARY

This chapter has defined much of the basic theory and practical applications of chemical kinetics and its application to more complex biological catalysts. The reader should now possess a clear understanding of the kinetic parameters V_{max}, K_m, and

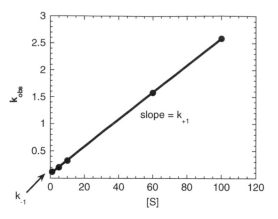

FIGURE 3.13A A linear plot relating the observed rates (k_{obs}) with increasing [S] at constant [E] is consistent with Scheme 3.4.

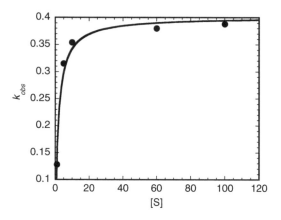

FIGURE 3.13B A hyperbolic plot relating the observed rate (k_{obs}) with increasing [S] at constant [E] is consistent with Scheme 3.7.

V_{max}/K_m both in terms of how these values are obtained and how they are interpreted with respect to enzymatic behavior. Likewise, more complex reactions involving multisubstrate reactions were examined to provide evidence for discriminating among various kinetic mechanisms. It should be emphasized, however, that steady-state studies represent a deductive form of kinetic analysis in which the experimenter is advised to use the generated data to eliminate possibilities rather than come to specific (and possibly erroneous) conclusions. Other chapters evaluating product and dead-end inhibition as well as the use of heavy atom isotope effects should be used to augment initial velocity data to rationally deduce the kinetic mechanism of an enzyme. Collectively, these chapters should provide the researcher with the necessary tools to evaluate the effects of therapeutic agents on biological targets as well as aid in their design and implementation.

FURTHER READING

Cleland, W.W. (1967). *Advances in Enzymology and Related Areas of Molecular Biology,* 29, pp. 1–32.

Cleland, W.W. (1979). *Methods in Enzymology,* 63, pp. 103–138.

Copeland, R.A. (1996). *Enzymes — A Practical Introduction to Structure, Mechanism, and Data Analysis,* VCH, New York.

Fersht, A. (1985). *Enzyme Structure and Mechanism,* W.H. Freeman, New York.

Gilbert, S.P. and Mackey, A.T. (2000). *Methods,* 22, pp. 337–354.

Johnson, K.A. (1992). *The Enzymes,* 20, pp. 1–61.

Lanzetta, P.A., Alvarez, L.J., Reinach, P.S., and Andia, O.A. (1979). *Anal. Biochem.,* 269, 16493–16501.

Segel, I.H. (1993). *Enzyme Kinetics,* John Wiley & Sons, New York.

Viola, R.E. and Cleland, W.W. (1982). *Methods in Enzymology,* 87, pp. 353–366.

Zerner, B., Bond, R.P., and Bender, M.L. (1964). *J. Am. Chem. Soc.,* 86, 3674–3677.

4 Enzyme Inhibitors

Jure Stojan

CONTENTS

4.1 INTRODUCTION

Compounds that influence the rates of enzyme-catalyzed reactions are called modulators, moderators, or modifiers. Usually, the effect is to reduce the rate, and this is called *inhibition*. Sometimes the enzyme reaction is increased, and this is called *activation*. Accordingly, the compounds are termed *inhibitors* or *activators*.

When studying these phenomena, one has to understand the molecular events leading to the experimentally observed effects. A number of techniques are available for reaching a basic explanation for the so-called mechanism or mode of action of a substance on the enzymic reaction under investigation. Although the methods for solving the 3-D structure of enzyme-inhibitor complexes result in clear steric presentations, and quantum mechanical calculations provide data consistent with basic physicochemical laws, the great practical value of classical kinetic information cannot and must not be neglected. Additionally, it should be stressed that not only the mode of action of an inhibitor is important but its effects on the substrate turnover, in many cases, provide information that cannot be obtained from studies of pure enzyme-substrate systems. On the basis of such kinetic observations, inhibitors are usually divided into two groups. The first consists of reversible inhibitors that form nonco-valent interactions with various parts of the enzyme surface, which can be easily reversed by dilution or dialysis. The second group comprises irreversible inhibitors that interact with different functional groups on the enzyme surface by forming strong covalent bonds that often persist even during complete protein breakdown.

4.2 TYPES OF INHIBITORS

4.2.1 REVERSIBLE INHIBITORS

Traditionally, three types of reversible inhibition are distinguished by the relation between the velocities of the inhibited and uninhibited reactions. The degree of inhibition is unaffected by change in the substrate concentration in the pure noncompetitive type, reduced by increasing substrate concentration if the inhibitor competes for the active site with the substrate (for competitive inhibitor), and increased at higher substrate concentrations in the so-called coupling, anticompetitive, or uncompetitive behavior.

Although the different types of inhibition imply ideas about the mechanism of inhibitor action, they cannot necessarily be interpreted in terms of the molecular events described by the particular type. Nevertheless, it is expected that structural similarities between the substrate and the inhibitor exclude simultaneous binding, leading to a reduction of the degree of inhibition with increasing substrate concentration and, thus, to competitive inhibition. On the other hand, if the two can become attached to the enzyme simultaneously, their structures are unlikely to be similar so that they do not compete, and the inhibition is termed noncompetitive (Figure 4.1). Unfortunately, such a pictorial representation is directly correlated only with a classical, single intermediate, Michaelis–Menten reaction mechanism, but unexpected types of inhibition can be observed in multisubstrate or multiintermediate reactions.

Considering only the framework of an enzymic reaction exposed to the action of a reversible inhibitor, the degree of inhibition is defined as the reduction of the rate divided by the rate of uninhibited reaction:

$$i = \frac{\upsilon_0 - \upsilon_i}{\upsilon_0} \tag{4.1}$$

Indeed, the question is, what influences the particular rates in Equation 4.1? Although the relationships that describe velocities (both in the absence and presence of the inhibitor) are often rather complex, there is a simple equation that applies to many systems. It is operative even if the mechanism is not just a single intermediate Michaelis–Menten type. If S stands for the substrate and I for the inhibitor, its general form clearly reveals that the basic types of inhibition are only determined by the relative magnitudes of the two inhibitor dissociation constants (K_{ia} and K_{ib}):

$$\upsilon = \frac{V[S]}{Ks\left(1 + \dfrac{[I]}{K_{ia}}\right) + [S]\left(1 + \dfrac{[I]}{K_{ib}}\right)} \tag{4.2}$$

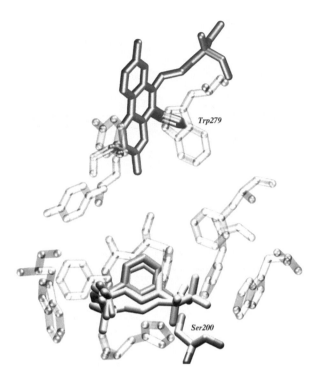

FIGURE 4.1 (See color insert following page 176.) The active site (Ser 200) of vertebrate acetylcholinesterase is buried deep inside the enzyme molecule. White: docked substrate acetylcholine (PDB code 1ACE); red: competitive inhibitor edrophonium (PDB code 2ACK); green: transition-state analog trimethylammonio-trifluoroacetophenone (PDB code 1AMN); brown: noncompetitive inhibitor propidium (PDB code 1N5R) is bound at the entrance to the active site (Trp 279).

Competitive, reversible inhibition is seen when K_{ib} is much larger than K_{ia}, so that the term $\dfrac{[I]}{K_{ib}}$ can be neglected. If the opposite is found, we get an uncompetitive inhibition, and a noncompetitive one when the dissociation constants are equal (Figure 4.2 and Figure 4.3). In practice, however, the two dissociation constants usually differ, but to such an extent that neither of the terms in the denominator, $\dfrac{[I]}{K_{ia}}$ or $\dfrac{[I]}{K_{ib}}$, can be neglected. We call the inhibition mixed, and the interpretation in terms of reaction mechanism becomes more complicated.

General mathematical representation with regard to the observations of reversible inhibition can be further clarified by considering the two dissociation constants such that both denominator terms $\dfrac{[I]}{K_{ia}}$ and $\dfrac{[I]}{K_{ib}}$ can be neglected. Indeed, the substance is now a very poor inhibitor, and Equation 4.2 reduces to the original Michae-

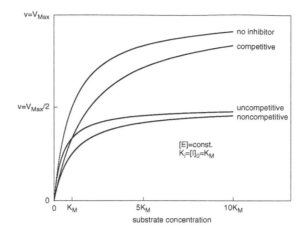

FIGURE 4.2 Theoretical direct plots (v vs. [S]) for a Michaelian enzyme in the absence and in the presence of three classical types of inhibitors.

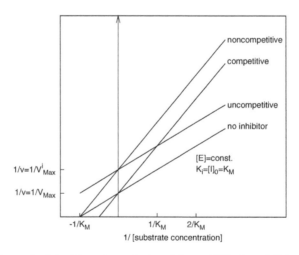

FIGURE 4.3 Theoretical double-reciprocal plots ($1/v$ vs. $1/[S]$) for a Michaelian enzyme in the absence and in the presence of three classical types of inhibitors.

lis–Menten form. Of course, an equivalent reduction occurs if the inhibitor is not present at all.

The above considerations are only valid if the enzyme–substrate–inhibitor system is in the steady state. As already discussed in Chapter 3, this means that the enzyme is present at such a low relative concentration that the depletion of other reactants throughout the entire experiment can be neglected. This, again, holds only when initial reaction rates are taken into account and when all initial complexes are formed virtually instantaneously. Such a behavior is observed with inhibitors showing the affinities described by dissociation constants down to 100 nM, but those with still higher affinities (lower value of dissociation constant) are named *tight binders*.

4.2.2 Transition-State Analogs

In 1898, Emil Fisher proposed the "key and lock" theory of specificity in enzymes. According to J.B.S. Haldane[1] and L. Pauling,[2] optimal complementarity occurs between the enzyme active site and the substrate in the transition state. Because the maximal number of possible weak interactions is required for the substrate to reach transition-state destabilization, the idea came about to synthesize a molecule complementary to the active site surface without it needing to be deformed (Figure 4.1). Such transition-state substrate analogs proved to be extremely effective competitive inhibitors, exerting dissociation constants as low as femtomolar. Usually, the reason for such tight binding is the very slow dissociation rate constant (k_{off}) rather than the high rate of association (k_{on}). The latter is limited by diffusion, and in the case of small organic inhibitors, seldom exceeds values above $10^9 M^{-1} sec^{-1}$. The kinetic properties of tight binders, however, prevent classical steady-state studies of initial rates. In reality, the pre-steady-state is prolonged when the enzyme is in stoichiometric amounts in proportion to the inhibitor. The enzyme becomes instantaneously and completely inhibited if the inhibitor is in great excess. Indeed, pre-steady-state data are much more informative, but the conditions under which they have to be gathered require special analytical treatment. However, tight-binding transition-state analogs are very convenient titrating agents, especially when their target enzymes are not pure.

4.2.3 Irreversible Inhibitors

Compounds that interact with the enzyme in such a way as to cause permanent loss of activity are irreversible inhibitors. They make stable covalent bonds, mainly with the enzyme active-site residues, and so they are also termed *catalytic poisons*. If such binding occurs, the inhibitors are needed only in trace amounts because each inhibitor molecule eventually finds its own enzyme molecule and abolishes the latter's activity. In practice, the formation of a stable covalent bond is so slow that the rate of inactivation can usually be determined by following the time course of residual enzyme activity, after various preincubation times with the inhibitor, even under conditions when the inhibitor is in great excess. Consequently, the initial activity would be unaffected by the presence of such an inhibitor if the reaction is started by the addition of the enzyme. Sometimes, however, every encounter of the inhibitor with the enzyme does not lead to successful bonding. In such cases, irreversible inactivation is preceded by a reversible step, similar to the classical reversible inhibition.

4.3 KINETICS OF INHIBITED ENZYME REACTIONS

Analysis of the effects of enzyme inhibitors by kinetic means is the oldest and the most thoroughly elaborated functional evaluation. Testing the action of an inhibitor on a target enzyme is not only the most fundamental functional check but also a probe for resolving and influencing the catalytic mechanism. The aim is to figure out an appropriate mathematical formulation, i.e., an equation that, within the bounds of experimental error, reproduces all the available experimental data. If the equation is consistent with the basic events that are expected to occur during the catalysis, it becomes a hypothesis. When, however, a new piece of evidence arises, and the equation is no longer satisfied, it should be modified, usually enlarged. So, in practice, the kinetic analysis of inhibited enzyme reactions is a multistep procedure towards clarifying our understanding of the particular catalytic process.

The first step after establishing that the substance influences the enzymic reaction is to test whether it acts instantaneously or slowly, by following the activity after various preincubation times. Subsequently, the (ir)reversibility must be established by one of the available techniques: dilution, dialysis, chromatography, or electrophoresis. On the basis of these pilot experiments, a decision can be made on further procedures.

4.3.1 CLASSICAL INSTANTANEOUS INHIBITION MECHANISMS

4.3.1.1 Types of Mechanisms

If an inhibitor acts reversibly and instantaneously, it is convenient to perform an array of initial rate measurements by changing the concentration of both the substrate and the inhibitor. The simplest and the easiest characterization is done by plotting the data using Lineweaver–Burk diagrams of $1/v$ against $1/[S]$ at each $[I]$. If the plots are linear, the lines can cross on either of the axes in the second quadrant, or they can be parallel (Figure 4.3). The interpretation of each pattern is done in terms of reaction schemes and by the corresponding equations derived from them.

The classical reaction scheme (Scheme 4.1) representing competitive inhibition is as follows:

$$EI$$
$$k_{+1} \updownarrow k_{-i}$$
$$S+E \underset{k_{-1}}{\overset{k_1}{\rightleftharpoons}} ES \xrightarrow{k_2} E+P$$
$$+$$
$$I$$

SCHEME 4.1

In the steady state, the rates of formation and disappearance of the complexes are identical:

$$k_1[S][E] = (k_{-1}+k_2)[ES] \tag{4.3}$$

$$k_i[I][E] = k_{-i}[EI] \tag{4.4}$$

The expressions for the total enzyme concentration and the reaction rate are:

$$[E]_0 = [E]+[ES]+[EI] \tag{4.5}$$

$$v = k_2[ES] \tag{4.6}$$

Insertion of the expressions for E and EI, in terms of ES, into the expression for the total enzyme concentration gives

$$[E]_0 = \left(\frac{k_{-1} + k_2}{k_1[S]} + 1 + \frac{k_{-1} + k_2 k_i[I]}{k_1[S]k_i} \right)[ES] \tag{4.7}$$

and the derived steady-state rate equation is

$$v = \frac{k_2[E]_0[S]}{[S] + \dfrac{k_{-1} + k_2}{k_1}\left(1 + \dfrac{k_i[I]}{k_{-i}} \right)} \tag{4.8}$$

Because $\dfrac{k_{-1} + k_2}{k_1}$ is the Michaelis constant for the reaction in the absence of

an inhibitor, and $\dfrac{k_{-i}}{k_i}$ is the true equilibrium constant for the dissociation of the

complex EI into E and I, the rate equation may be written in the form

$$\upsilon = \frac{V_{max}[S]}{[S] + K_M\left(1 + \dfrac{[I]}{K_i} \right)} \tag{4.9}$$

which in reciprocal form becomes

$$\frac{1}{v} = \frac{1}{[S]}\frac{K_M}{V_{max}}\left(1 + \frac{[I]}{K_i} \right) + \frac{1}{V_{max}} \tag{4.10}$$

It is evident from this equation that only the slope of Lineweaver–Burk lines is influenced by the changing $[I]$ and, consequently, the lines cross on the y-axis (Figure 4.3). In other words, the crossing of the lines on the y-axis in the double reciprocal diagram is diagnostic for pure competitive inhibition.

A similar derivation, starting from the reaction scheme (Scheme 4.2) for the uncompetitive type of inhibition, leads to the rate equation

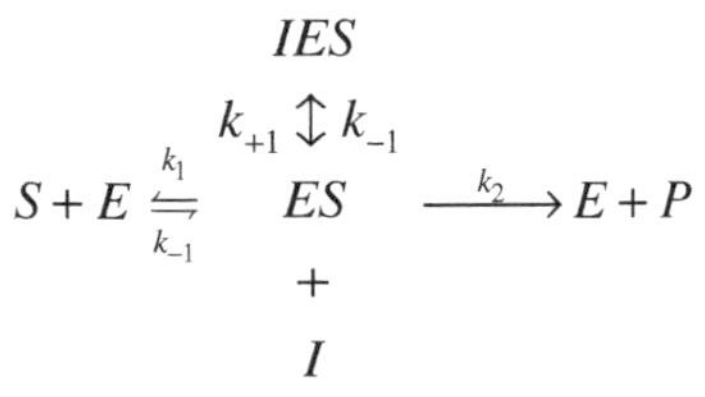

$$IES$$
$$k_{+1} \updownarrow k_{-1}$$
$$S + E \underset{k_{-1}}{\overset{k_1}{\rightleftharpoons}} ES \xrightarrow{k_2} E + P$$
$$+$$
$$I$$

SCHEME 4.2

$$\upsilon = \frac{V_{max}[S]}{[S]\left(1 + \dfrac{[I]}{K_i}\right) + K_M} \tag{4.11}$$

and, from its reciprocal form

$$\frac{1}{\upsilon} = \frac{1}{[S]} \frac{K_M}{V_{max}} + \frac{1}{V_{max}}\left(1 + \frac{[I]}{K_i}\right) \tag{4.12}$$

it is clear that only the intercepts of Lineweaver–Burk lines are influenced by the changing inhibitor concentration (Figure 4.3). Again, parallel lines in a double reciprocal diagram are diagnostic for pure uncompetitive inhibition.

The third and final pure classical inhibition pattern is the noncompetitive type, represented by the following reaction scheme (Scheme 4.3).

$$
\begin{array}{ccc}
IE & & IES \\
k_{+1} \updownarrow k_{-1} & & k_{+1} \updownarrow k_{-1} \\
S + E & \underset{k_{-1}}{\overset{k_1}{\rightleftharpoons}} & ES \xrightarrow{k_2} E + P \\
+ & & + \\
I & & I
\end{array}
$$

SCHEME 4.3

The rate equation and its reciprocal form are:

$$\upsilon = \frac{V_{max}[S]}{([S] + K_M)\left(1 + \dfrac{[I]}{K_i}\right)} \tag{4.13}$$

$$\frac{1}{\upsilon} = \frac{1}{[S]} \frac{K_M}{V_{max}}\left(1 + \frac{[I]}{K_i}\right) + \frac{1}{V_{max}}\left(1 + \frac{[I]}{K_i}\right) \tag{4.14}$$

In this pure type of noncompetitive inhibition, the equilibrium constants for the dissociation of the complexes, i.e., IE and IES, are considered to be equal, and so, changing the inhibitor concentration affects both the slope and the intercept of Lineweaver–Burk lines. This means that lines in the double reciprocal plot crossing at the x-axis (Figure 4.4) are diagnostic for pure noncompetitive inhibition.

4.3.1.2 Determination of Inhibition Constants

After qualitative characterization of the mechanism of action of an inhibitor, it is necessary to evaluate its relative potency. One, historically very popular, approach

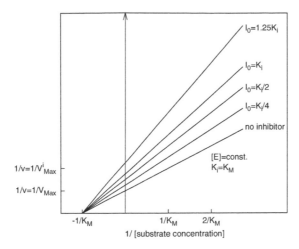

FIGURE 4.4 Theoretical double-reciprocal plots ($1/v$ vs. $1/[S]$) for a Michaelian enzyme in the absence and in the presence of various concentrations of a classical noncompetitive inhibitor.

was to determine the concentration of the inhibitor that produces 50% inhibition. The so-called pI_{50} can be obtained by taking the negative logarithm of this concentration, just as the concentration of protons is used for the calculation of pH. Just a few measurements, with different concentrations of the inhibitor, should be taken to enable extrapolation to the required concentration. However, the concentration of the inhibitor needed to achieve such a degree of inhibition may depend significantly on the amount of substrate used in the measurement. It should be recalled from previous considerations that only the action of a noncompetitive reversible inhibitor is independent of substrate concentration, yielding the concentration at pI_{50} numerically identical to the value of the dissociation constant K_i. Unfortunately, a pure noncompetitive mechanism is a rather rare situation. Therefore, it is much more reliable to evaluate the corresponding inhibition constant to describe the relative effectiveness of a drug.

The easiest way to estimate the inhibition constants in classical mechanisms is to analyze the initial rates at various substrate concentrations in the absence and in the presence of one inhibitor concentration (Figure 4.2 and Figure 4.3). From each double reciprocal plot (Figure 4.3), the values for $V_{\max}$ and K_M are determined graphically or by linear regression. Subsequently, the calculation is performed by using the appropriate relation between K_M in the presence of the inhibitor and K_i from Equation 4.9, Equation 4.11, and Equation 4.13. In the competitive mechanism, where the slope of the double-reciprocal plot changes, the Michaelis constant in the presence of the inhibitor becomes K_M^i and

$$K_i = \frac{K_M}{K_M^i - K_M}[I] \qquad (4.15)$$

In an uncompetitive mechanism, the intercept changes so that the Michaelis constant as well as the catalytic constant ($V_{max}/[E]_0$) are inhibitor dependent and

$$K_i = \frac{K_M}{K_M - K_M^i}[I] \tag{4.16}$$

or

$$K_i = \frac{V_{max}^i}{V_{max} - V_{max}^i}[I] \tag{4.17}$$

whereas in the noncompetitive mechanism, both the slope and intercept change. Therefore, only the catalytic constant ($V_{max}/[E]_0$) depends on the presence of the inhibitor

$$K_i = \frac{V_{max}^i}{V_{max} - V_{max}^i}[I] \tag{4.18}$$

Another conventional, but more reliable, way for determining the type of inhibition and the corresponding equilibrium constant is to plot intercepts $\left(\dfrac{1}{V_{max}}\right)$ and slopes $\left(\dfrac{K_M}{V_{max}}\right)$ of double-reciprocal graphs in the absence and in the presence of several inhibitor concentrations as a function of the inhibitor concentration (Figure 4.5 and Figure 4.6).

The so-called secondary plots are linear for classical inhibition mechanisms, and extrapolation of the line to the intercept with the x-axis yields the value of the

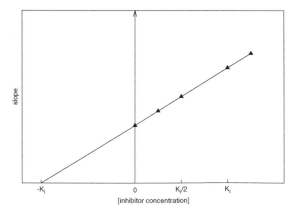

FIGURE 4.5 Theoretical secondary plot of the slopes of double-reciprocal plots from Figure 4.3 at various inhibitor concentrations (*slope* vs. [*I*]) for a Michaelian enzyme.

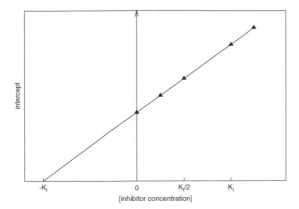

FIGURE 4.6 Theoretical secondary plot of the intercepts of double-reciprocal plots from Figure 4.3 at various inhibitor concentrations (*intercept* vs. [*I*]) for a Michaelian enzyme.

TABLE 4.1
Summary of the Equations for Intercepts and Slopes of Double-Reciprocal Plots for the Three Classical Inhibition Mechanisms

	Competitive	Uncompetitive	Noncompetitive
Intercept	Independent	$\dfrac{1}{V_{max}}\left(1+\dfrac{[I]}{K_i}\right)$	$\dfrac{1}{V_{max}}\left(1+\dfrac{[I]}{K_i}\right)$
Slope	$\dfrac{K_M}{V_{max}}\left(1+\dfrac{[I]}{K_i}\right)$	Independent	$\dfrac{K_M}{V_{max}}\left(1+\dfrac{[I]}{K_i}\right)$

inhibition constant. Again, the determination can be graphical or by linear regression. The equations for each type of inhibition can be obtained from Equation 4.10, Equation 4.12, and Equation 4.14, and are summarized in Table 4.1.

Several other methods for estimating the inhibition constant have been commonly used for many years. One of the most frequently applied was introduced by Dixon,[3] where reciprocals of initial rates at fixed substrate concentrations are plotted against inhibitor concentrations (Figure 4.7). For a classical inhibition mechanism, such a diagram is linear and yields reliable estimates, but it does not discriminate between the various inhibition types.

A different approach to analyzing kinetic results utilizes a nonlinear computer program to estimate the parameters of the rate equation by fitting the derived equation directly to experimental data. A simple program that can handle simultaneously multiple sets of data, involving two independent variables ([*S*] and [*I*]), was developed by R.G. Duggleby,[4] exactly for these purposes. The program also lists the complete statistical analysis, which proves very convenient when searching for the best kinetic model for a complex reaction.

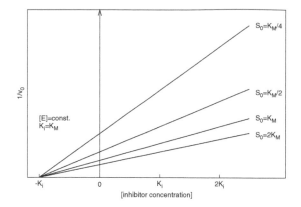

FIGURE 4.7 Theoretical Dixon plots ($1/v$ vs. $[I]$) for a Michaelian enzyme in the presence of a classical noncompetitive inhibitor at various substrate concentrations.

4.3.1.3 Nonproductive Binding and Substrate Inhibition

Two special cases of classical inhibition mechanisms arise where in fact no inhibitor is present. The first is nonproductive binding of a substrate to a nonspecific enzyme. If an enzyme can catalyze the cleavage of different substrates, as for instance digestive proteases, which bind polymers of various length quite specifically, it might interact with small artificial substrates in many ways. Usually, the detection of enzyme activity as well as the interpretation of the results using such sterically stable molecules is much easier and straightforward, and so they are used as probes for revealing reaction mechanisms. However, if a significant amount of substrate forms a stable complex with the enzyme that cannot undergo further conversion into products, the reaction rate is diminished but the kinetics show no deviations from the Michaelian type. The reaction scheme (Scheme 4.4) for nonproductive binding is the classical pure competitive reaction scheme, where I is substituted by another S:

$$ES$$
$$k_{+s} \updownarrow k_{-s}$$
$$S + E \quad \underset{k_{-1}}{\overset{k_1}{\rightleftharpoons}} \quad ES \xrightarrow{k_2} E + P$$
$$+$$
$$S$$

SCHEME 4.4

Unfortunately, the derivation leads to an equation that is indistinguishable from the equation for the normal situation. Therefore, special attention should be paid when studying rather unspecific enzymes.

The second special situation is the so-called substrate inhibition arising from the binding of an additional substrate molecule to the ES complex. The reaction scheme (Scheme 4.5) is analogous to the scheme for uncompetitive inhibition and

$$SES$$

$$k_{+s} \updownarrow k_{-s}$$

$$S + E \underset{k_{-1}}{\overset{k_1}{\rightleftharpoons}} \quad ES \quad \xrightarrow{k_2} E + P$$

$$+$$

$$S$$

SCHEME 4.5

the kinetics clearly show deviation from the Michaelian hyperbola: At high substrate concentrations, the activity decreases and might, in some cases, even approach zero. Indeed, the Lineweaver–Burk diagram is nonlinear and, unless more information is available, the determination of the inhibition constant is quite uncertain. Inhibition by substrate at relatively high concentrations is expected with enzymes where the active site is buried deep inside the core of the molecule and might be interpreted as a sort of "traffic jam."

4.3.2 MORE GENERAL REACTION MECHANISMS

In the preceding paragraphs, an attempt was made to clarify some elementary connections between molecular events taking place during the inhibited enzymic reaction, and the mathematical formulation and evaluation. In practice, however, the systems are much more complex, although they can often be conducted under conditions that permit simplifications to the level of classical mechanisms. Sometimes, a small correction in the starting assumptions can dramatically improve the accord between theoretical equation and experimental data. For instance, by allowing the two inhibition constants in the pure noncompetitive inhibition to be different (i.e., different inhibitor affinities for free enzyme and for the enzyme–substrate complex), the noncompetitive inhibition can be turned into a mixed one. This seems to be much more natural. On the other hand, if the species *EI* and *ESI* can undergo further reaction, the derivation of rate equation and analysis become more complicated, double-reciprocal plots will be nonlinear, and the type of inhibition will not be obvious. Moreover, there is evidence for some systems in which a number of intermediates are involved in the reaction pathway.[5] This can only be established by using a particular type of inhibitor in a precisely conducted experiment.

4.3.3 SLOW INHIBITION

Theoretically, an equilibrium in the reaction between an enzyme and an inhibitor is reached at the end of a preequilibration phase. The length of this phase varies considerably from reaction to reaction. It can, in fact, last from a few fractions of a nanosecond to hours or even days. However, the establishment of an equilibrium is said to be instantaneous if it is completed before the actual measurement can be started. Of course, by using more powerful rapid-reaction methods, the very early stages of enzymic reactions with ligands can be monitored. So, it depends very much on the laboratory equipment employed when distinguishing between rapid and slow reactions.[6]

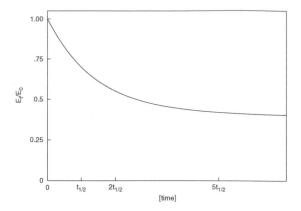

FIGURE 4.8 Theoretical time course of residual enzyme activity after various preincubation times with a slow binding inhibitor ($[E]_t/[E]0$ vs. [*time*]).

In practice,[7] if an inhibitor abolishes the enzyme activity completely when applied in a concentration range of the Michaelis–Menten constant for its specific substrate, its affinity is likely to be very high. Thus, it is worth testing its action at lower concentrations, after various preincubation times before adding the substrate. In the case where the remaining activity is time dependent (Figure 4.8) and the concentration used is at least 10 times higher than the concentration of the enzyme, the substance is a slow inhibitor. In particular, slow inhibition is a consequence of slow establishment of the equilibrium between the enzyme and the ligand. It is, therefore, possible from preequilibrium data, for such systems, to determine individual rate constants for the association of the inhibitor with the enzyme, and for the dissociation of the enzyme–inhibitor complex. The ratio of the two constants obtained gives the equilibrium constant, analogous to that for instantaneous inhibitors.

After measuring the time course of activity remaining at various inhibitor concentrations, the easiest analysis is to plot the data on a semilogarithmic diagram (Figure 4.9). If the plot is linear, the equilibration process proceeds as a pseudo-first-order reaction. The kinetic scheme (Scheme 4.6) for such a reaction is very simple:

$$E + I \overset{k_{+1}}{\underset{k_{-1}}{\rightleftharpoons}} EI$$

SCHEME 4.6

The time course of residual free enzyme concentration is

$$[E]_t = [E]_{eq} + ([E]_0 - [E]_{eq})e^{-(kt)} \qquad (4.19)$$

where E_t is the concentration of free enzyme at any time, E_0 and E_{eq} are the initial and equilibrium enzyme concentrations, and the overall pseudo-first-order rate constant

$$k = k_{+1}[I] + k_{-1} \tag{4.20}$$

The remaining activity is proportional to the amount of free enzyme (E_t), and a classical procedure for evaluating the individual association and dissociation constants is performed as follows: the slope of initial portions of semilogarithmic diagrams, representing k, plotted against the corresponding inhibitor concentrations is linear (see Equation 4.20). The value of the intercept on the y-axis equals k_{-1} and the slope equals the second-order rate constant k_{+1}. So, for the classical determination, two types of linear graphs must be constructed. First, in a set of graphs at various inhibitor concentrations, the logarithm of the remaining activity is plotted against the preincubation time to yield slopes (Figure 4.9), which are subsequently plotted, in a second diagram, against the corresponding inhibitor concentrations. Indeed, the analogy with the diagnostics of classical inhibition patterns, using Lineweaver–Burk graphs and secondary plots, is obvious.

It is, however, much more convenient and reliable to estimate the rate and equilibrium constants by fitting Equation 4.19 directly to the remaining activity data at various inhibitor concentrations, using a nonlinear regression computer program.

Sometimes the reaction between the enzyme and inhibitor is completed just before the first remaining activity measurement is technically possible. It is said that the reaction occurs during the dead time of a particular data-acquisition technique. After adding the substrate, the system is practically in steady state and the inhibitor acts virtually as a classical one. To access preequilibrium information in such cases, the reaction can be started by the addition of enzyme and measured on an apparatus equipped with a rapid-reaction accessory. Several techniques exist for measuring early reaction stages.[6] They differ not only in the principle of how the reaction system is perturbed but also in the dead time. For instance, remaining-activity experiments using a stopwatch to time removal of samples from a reaction vessel cannot be performed with a dead time shorter that 5 sec, whereas stopped-flow measurements can start as early as 0.7 msec after mixing the components. It

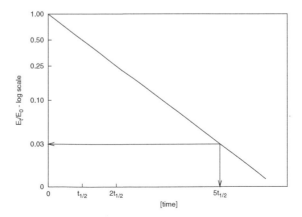

FIGURE 4.9 Theoretical time course of residual enzyme activity after various times of preincubation with a slow binding inhibitor {log($[E]_t/[E]_0$) vs. [$time$]}.

is clear, therefore, that an instantaneous reaction might appear slow if a faster technique was available.

Unfortunately, experiments that are started by the addition of enzyme do not include the substrate just as a detecting aid. It might act as a reactant whose concentration changes during the measurement and, still more important, it may influence sterically or allosterically the reaction between the enzyme and inhibitor. Data collected in such measurements are named *progress curves*.[8] Virtually, they may be very similar to classical initial-rate progress curves, but in reality, they are always curved. The curvature is not just a consequence of substrate depletion but rather, of a slow enzyme–inhibitor interaction. Although many mechanisms can account for various types of nonlinear progress curves, two of them have proved to be very common, and are illustrated by Scheme 4.7 and Scheme 4.8. The first one represents the slow binding of a ligand to the enzyme:

Mechanism A

Mechanism A

$$E \quad \overset{k_{1}*S}{\underset{k_{-1}}{\rightleftharpoons}} ES \xrightarrow{k_2} products$$

$$k_{+3} * I \updownarrow k_{-3}$$

$$EI *$$

SCHEME 4.7

and the second involves the rapid formation of an initial addition complex that subsequently undergoes a slow isomerization reaction:

Mechanism B

Mechanism B

$$E \quad \overset{k_{1}*S}{\underset{k_{-1}}{\rightleftharpoons}} ES \xrightarrow{k_2} products$$

$$k_{+3} * I \updownarrow k_{-3}$$

$$EI \quad \overset{k_{+4}}{\underset{k_{-4}}{\rightleftharpoons}} EI *$$

SCHEME 4.8

Both mechanisms, as well as many others,[9] are described by the same general integrated equation of the following form:

$$[P]_t = v_s t + (v_0 - v_s)(1 - e^{-(kt)}) / k \qquad (4.21)$$

where v_0, v_s, and k represent, respectively, the initial velocity, the final steady-state velocity, and the apparent overall first-order rate constant for establishing equilibrium between the free enzyme and inhibitor, and $EI*$ complex. All three parameters can easily be recognized from the diagrams in Figure 4.10 and Figure 4.11: The tangents to the curves through the origin reflect either initial rates, v_0, in the absence of the

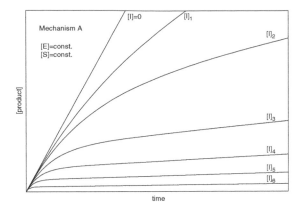

FIGURE 4.10 Theoretical progress curves for the inhibition of a Michaelian enzyme by a slow-binding inhibitor acting according to Mechanism A. The line represents the reaction in the absence of the inhibitor, and the curves the reaction in the presence of increasing inhibitor concentrations. The substrate concentration is always set to the same value. Note the identical initial parts of the curves and sloped asymptotes of the curves at infinite time.

inhibitor in Mechanism A, or instantaneous inhibition stage in Mechanism B. The asymptotes at infinite time, v_s, reflect the equilibrium in the presence of the inhibitor, whereas the curvature reflects the transition stage characterized by the overall rate constant, k. Even without going into the details of kinetic theory, the general validity of this equation is obvious. In the case when v_s equals v_0, the inhibitor is ineffective, according to Mechanism A, or classical, as predicted by Mechanism B.

The analysis of progress curves describing slow inhibition is performed by consecutive steps. To begin with, careful inspection of the available sets of progress

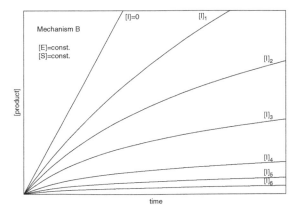

FIGURE 4.11 Theoretical progress curves for the inhibition of a Michaelian enzyme by a slow-binding inhibitor acting according to Mechanism B. The line represents the reaction in the absence of the inhibitor, and the curves the reaction in the presence of increasing inhibitor concentrations. The substrate concentration is always set to the same value. Note the changing initial parts of the curves and sloped asymptotes of the curves at infinite time.

curves helps to discriminate between Mechanism A and Mechanism B, i.e., if the initial velocity does or does not vary as a function of inhibitor concentration. Next, for each progress curve obtained at varying concentrations of substrate and inhibitor, the parameters v_0, v_s, and k from Equation 4.21 should be determined by nonlinear regression. Initial and steady-state velocities can now be analyzed using classical diagnostic graphs such as double-reciprocal and secondary plots whereas the dependence of k is checked at varying inhibitor and substrate concentrations. At this stage, the information gathered usually enables the construction of one or a few most probable reaction mechanisms for which the significance of the parameters v_0, v_s, and k has to be derived. For illustration, such a significance is given for Mechanism A:

$$v_0 = \frac{V_M[S]}{K_m + [S]} \tag{4.22}$$

$$k = k_{-3} + \frac{k_3 K_m [I]}{K_m + [S]} \tag{4.23}$$

$$v_s = v_0 \frac{k_{-3}}{k} \tag{4.24}$$

$$v_{0\ preincubation} = v_0 \frac{k_{-3}}{k_{+3}[I] + k_{-3}} \tag{4.25}$$

Finally, the adequacy of the chosen mechanism should be checked by comparing theoretically calculated progress curves, using Equation 4.21 (for Mechanism A, v_0, v_s, and k should be substituted with Equation 4.22 to Equation 4.24 or Equation 4.25), with experimental data at all substrate and inhibitor concentrations.

4.3.4 SLOW TIGHT-BINDING INHIBITION

Very high affinity of an inhibitor for an enzyme may require experiments to be conducted at such low inhibitor concentrations that they become comparable to that of the enzyme. Under such conditions, called tight-binding conditions, it is necessary in the analysis to allow for the depletion of inhibitor concentration during the course of the reaction. Usually, tight-binding conditions permit values for the uni- and bimolecular rate constants for the dissociation and formation of enzyme–inhibitor complex to be obtained by remaining-activity experiments. The analysis may, therefore, be performed according to Scheme 4.6. However, due to progressively decreasing inhibitor concentration, the semilogarithmic plot of remaining activity vs. preincubation time is nonlinear, thus suggesting a distinction from simple slow inhibition. If such a behavior is found, then Equation 4.19 is not appropriate. An analytical solution for Scheme 4.6 under tight-binding conditions is possible, but the equation is very complex.[8] Sometimes, when the dissociation rate is extremely slow, the reaction can be treated as essentially irreversible (Scheme 4.9) and a bimolecular rate constant can be estimated by analyzing the remaining activities, according to the equation:

$$E + I \xrightarrow{k} EI$$

SCHEME 4.9

$$[EI]_t = \frac{[E]_0[I]_0(ek([E]_0 - [I]_0)t - 1)}{[E]_0 ek([E]_0 - [I]_0)t - [I]_0} \qquad (4.26)$$

Of course, such a simplification is only justified if the dissociation rate, i.e., the reactivation after the removal of the remaining free inhibitor, is practically immeasurable.

Finally, it should be mentioned that when a slow conformational change is preceded by rapid complex formation (as in Mechanism B), the instantaneous part may only be detected at significantly higher inhibitor concentrations, using rapid reaction techniques. The progress curves obtained will be similar in shape to those in Figure 4.11, and the tight binding condition will most probably not apply. However, the detection might be difficult because complete loss of activity will occur very rapidly.

4.3.5 IRREVERSIBLE INHIBITION

Compared to transition-state analogs, irreversible inhibitors often form complexes with their target enzymes much more slowly even if they are in great excess. Therefore, the equation that describes the reaction (Scheme 4.9) can be obtained under simplified conditions. Instead of the general solution, given by Equation 4.26, the derivation under the assumption $[E] \ll [I]$ yields the equation for the classical first-order decay

$$E_t = E_0 e^{-k[I]t} \qquad (4.27)$$

and the slope of the semilogarithmic plot, log E_t/E_0 vs. time, represents the product between the rate constant and the corresponding concentration of the inhibitor, k_{app} = $k[I]$.

If, however, a reversible complex is initially formed rapidly prior to covalent bonding of a catalytic poison with the active site residue, as in the following reaction scheme (Scheme 4.10), the derived residual activity equation is of the same form as Equation 4.27:

$$E + I \overset{K_i}{\rightleftharpoons} EI \xrightarrow{k_{+2}} EI*$$

SCHEME 4.10

$$E_t = E_0 e^{-\frac{k_{+2}[I]}{[I]+K_i}t} \tag{4.28}$$

Because the dependence of the overall second-order inactivation rate constant k_{app} on the concentration of the inhibitor is now hyperbolic, rather than linear as in Equation 4.27, this is a diagnostic plot for discriminating between reaction mechanisms in Scheme 4.9 and Scheme 4.10. The kinetic constants K_i and k_{+2} are evaluated essentially the same way as for the determination of K_M and k_{cat}, according to Lineweaver–Burk:

$$k_{app} = \frac{k_{+2}[I]}{[I]+K_i} \tag{4.29}$$

$$\frac{1}{k_{app}} = \frac{K_i}{k_{+2}} \cdot \frac{1}{[I]} + \frac{1}{k_{+2}} \tag{4.30}$$

Of course, the kinetic constants can always be estimated by direct nonlinear regression fitting of Equation 4.27 or Equation 4.28 to a set of residual activity data.

Yet another type of irreversible enzyme inhibition arises with compounds that bear a high degree of structural similarity to the substrate or product but are chemically unreactive in the absence of the target enzyme. When bound to the active site, however, they undergo conversion through a normal catalytic mechanism either to a product that dissociates from the active site or to a very reactive intermediate that subsequently forms a stable covalent bond with the enzyme. Consequently, this type of inhibitor is termed a *mechanism-based inactivator* or, occasionally, a *suicide substrate*. The reaction with the target enzymes of such inhibitors can be formulated by the following scheme (Scheme 4.11):[10]

$$E + I \overset{K_i}{\rightleftharpoons} EI \xrightarrow{k_{+2}} EI' \xrightarrow{k_{+3}} E + I'$$
$$\downarrow k_{+4}$$
$$E - I^*$$

SCHEME 4.11

Because turnover and inactivation can occur in different proportions, their ratio is called the partition ratio $\left(\dfrac{k_{+3}}{k_{+4}}\right)$. An effective inactivator must, in its activated form, react more rapidly with the enzyme rather than it dissociates (i.e., $k_{+4} \gg k_{+3}$). Under such conditions, the estimation of an overall inactivation rate constant k_{inact}, consisting of k_{+2} and k_{+4} as a complex relation, can be done by the analysis of residual activity measurements according to Equation 4.28 to Equation 4.30.

4.3.6 INHIBITED TWO-SUBSTRATE REACTIONS

Careful kinetic evaluation of enzyme reactions involving two substrates can reveal several different reaction mechanisms.[11] The use of inhibitors to discriminate reliably between various rival mechanisms for a particular reaction may sometimes become unavoidable.[12] This is especially true in cases where several mechanisms lead to identical rate laws in the absence of inhibitors. In principle, the analytical procedure is analogous to that of inhibited single-substrate reactions. The array of experiments, however, is larger and the derivation of appropriate equations is usually more complex. After performing measurements in the absence of an inhibitor, they must be repeated with varying concentrations of each ligand at constant concentration of all the others. When many sets of such data are collected, they should be represented graphically using all the conventional approaches as described previously. Diagrammatic representation compels close inspection, which often, per se, reduces a number of candidate mechanisms to a reasonable number. Subsequently, equations must be obtained for each of them and tested statistically, beginning with each individual set of data and finishing with the simultaneous fit of an appropriate equation to all available sets.

REFERENCES

1. Haldane, J.B.S. (1930). *Enzymes.* Longmans, Green, p. 182. Cambridge, MA: M.I.T. Press (1965).
2. Pauling, L. (1946). *Chemical Engineering News,* 24, 1375; (1948) *American Scientific,* 36, 51.
3. Dixon, M. (1953). *Biochemical Journal,* 55, 170.
4. Duggleby, R.G. (1984). Regression analysis of nonlinear Arrhenius plots: an empirical model and a computer program. *Computers in Biology and Medicine,* 14, 447–455.
5. Cha, S. (1968). A simple method for derivation of rate equation for enzyme-catalyzed reactions under the rapid equilibrium assumption or combined assumptions of equilibrium and steady state. *Journal of Biological Chemistry,* 243, 820–825.
6. Gibbson, Q. (1983). Rapid reaction methods in biochemistry. In *Modern Physical Methods in Biochemistry, Part B.* Eds. A. Neuberger and L.L.M. Van Deenen, pp. 65–84. Amsterdam: Elsevier.
7. Golinik, M. and Stojan, J. (2002). Multi-step analysis as a tool for kinetic parameter estimation and mechanism discrimination in the reaction between tight-binding fasciculin 2 and electric eel acetylcholinesterase. *Biochimica et Biophysica Acta,* 1597, 164–172.
8. Morrison, J.F. and Walsh, C.T. (1988). The behavior and significance of slow-binding enzyme inhibitors. *Advances in Enzymology,* 61, 201–301.
9. Stojan, J. (1998). Equations for progress curves of some kinetic models of enzyme-single substrate-single slow binding modifier system. *Journal of Enzyme Inhibition,* 13, 161–176.
10. Silverman, R.B. (2002). *The Organic Chemistry of Enzyme-Catalyzed Reactions.* San Diego, CA: Academic.
11. Laidler, K. and Bunting, J. (1973). *Chemical Kinetics of Enzyme Action.* Oxford, U.K.: Calderon Press.
12. Cleland, W.W. (1977). Determining the chemical mechanisms of enzyme-catalyzed reactions by kinetic studies. *Advances in Enzymology,* 45, 273–387.

5 Development of Enzyme Inhibitors as Drugs

H. John Smith and Claire Simons

CONTENTS

0-415-33402-0/05/$0.00+$1.50
© 2005 by CRC Press LLC

5.1 INTRODUCTION

The majority of drugs used clinically exert their action in one of two ways: (1) by interfering with a component (agonist) in the body, preventing interaction with its site of action (receptor), i.e., receptor antagonist, or (2) by interfering with an enzyme normally essential for the well-being of the body or involved in bacterial or parasitic or fungal growth causing disease and infectious states, where the removal of its activity by treatment is necessary, i.e., enzyme inhibitors. In recent years, the proportion of current drugs described as enzyme inhibitors has increased, and this chapter gives an account of the steps taken for designing and developing such inhibitors — from identification of the target enzyme to be blocked in a particular disease or infection to the marketplace.

As has been described in previous chapters, enzymes catalyze the reactions of their substrates by initial formation of a complex (ES) between the enzyme (E) and the substrate (S) at the active site of the enzyme. This complex then breaks down, either directly or through intermediary stages, to give the product (P) of the reaction with regeneration of the enzyme (Equation 5.1 and Equation 5.2):

$$E + S \rightleftharpoons \underset{\substack{\text{enzyme-substrate} \\ \text{complex}}}{ES} \xrightarrow{\ k_{cat}\ } E + \text{products} \tag{5.1}$$

$$E + S \rightleftharpoons ES \xrightarrow{\ k_2\ } \underset{\text{intermediate}}{E'} + P_1 \xrightarrow{\ k_3\ } E + P_2 \tag{5.2}$$

where k_{cat} is the overall rate constant for decomposition of ES into products; k_2 and k_3 are the respective rate constants for formation and breakdown of the intermediate E' [i.e., $k_{cat} = k_2 k_3/(k_2 + k_3)$].

Chemical agents known as inhibitors modify the ability of an enzyme to catalyze the reaction of its substrates, a term that is usually restricted to chemical agents, other modifiers of enzyme activity such as pH, ultraviolet light, high salt concentrations, organic solvents, and heat being known as denaturizing agents.

5.1.1 BASIC CONCEPTS

The body contains several thousand different enzymes, each catalyzing a reaction of a single substrate or group of substrates. An array of enzymes is involved in a metabolic pathway each catalyzing a specific step in the pathway up to final metabolite production (Equation 5.3). These actions are integrated and controlled in various ways to produce a coherent pattern governed by the requirements of the cell. Alternatively, the enzyme may not be part of a pathway and operates in a single-step reaction (AB).

$$A \xrightarrow{\ E_1\ } B \xrightarrow{\ E_2\ } C \xrightarrow{\ E_3\ } \ldots \xrightarrow{\ E_n\ } \text{metabolite} \tag{5.3}$$

The use of enzyme inhibitors as drugs is based on the rationale that inhibition of a suitably selected target enzyme leads first to an accumulation of the substrates and, second, to a corresponding decrease in concentration of the metabolites; one of these features leads to a useful clinical response.

5.1.1.1 Substrate (Agonist) Accumulation or Preservation

Where the substrate gives a required response (i.e., agonist), inhibition of its metabolizing enzyme leads to accumulation of the intact substrate and accentuation of that response. Several examples follow:

Accumulation of the neurotransmitter acetylcholine (**5.1**) by inhibition of the metabolizing enzyme acetylcholinesterase using neostigmine (**5.2**) is used for the treatment of myasthenia gravis and glaucoma (Equation 5.4).

$$CH_3CO_2CH_2CH_2\overset{+}{N}(CH_3)_3 \xrightarrow{\text{acetylcholinesterase}} CH_3CO_2H +$$

$$\text{(5.1)} \qquad\qquad HOCH_2CH_2\overset{+}{N}(CH_3)_3$$

$$(5.4)$$

Anticholinesterases, e.g., donepezil (**5.3**), rivastigmine (**5.4**), and galantamine (**5.5**), capable of penetrating the blood–brain barrier and thereby exerting an effect on the central nervous system, are used in the treatment of Alzheimer's disease for increasing cognitive functions.

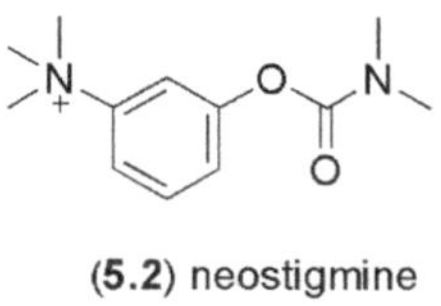

(**5.2**) neostigmine

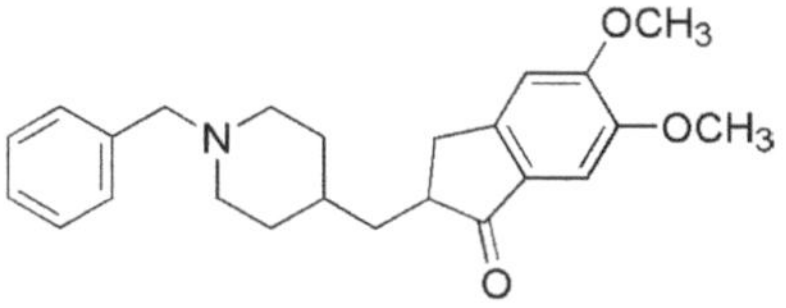

(**5.3**) donepezil

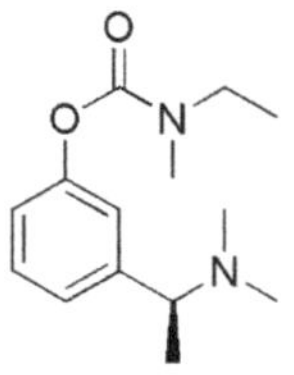

(**5.4**) rivastigmine

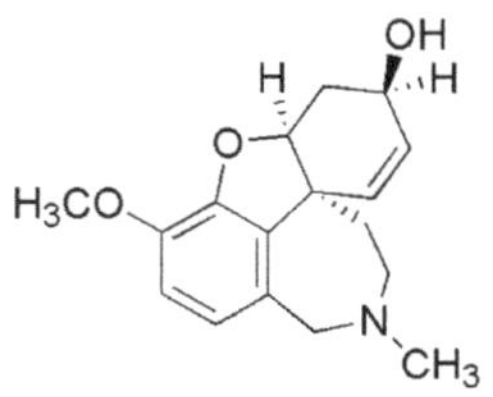

(**5.5**) galantamine

Inhibitors have been used (see Equation 5.5) as codrugs to protect an administered drug with the required action from the effects of a metabolizing enzyme. Inhibition of the metabolizing target enzyme permits higher plasma levels of the

administered drug to persist, thus prolonging its biological half-life and either pre-
serving its effect or resulting in less frequent administration.

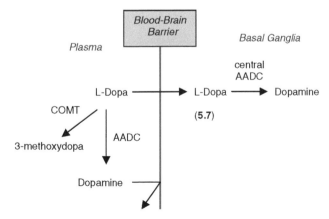

(5.6) clavulanic acid

Clavulanic acid (**5.6**), an inhibitor of certain β-lactamase enzymes produced by
bacteria for protection purposes, when administered in conjunction with a β-lacta-
mase-sensitive penicillin, preserves the antibacterial action of the penicillin towards
these bacteria.

$$\text{Drug} \xrightarrow{\text{metabolizing enzyme}} \text{Inert product(s)}$$
$$\text{or} \qquad\qquad \uparrow \qquad\qquad\qquad\qquad (5.5)$$
$$\text{agonist} \quad \text{Codrug (inhibitor)}$$

Parkinson's disease is due to degeneration in the basal ganglia, which leads to
reduction in dopamine levels that control muscle tension. Effective treatment for
considerable periods involves administration of the drug L-dopa (**5.7**), which is
decarboxylated after passage into the brain by a central acting amino acid decarboxylase (AADC).

Because L-dopa is readily metabolized by peripheral AADCs (see Figure 5.1),
it is administered with a peripheral AADC inhibitor, i.e., benzserazide (**5.8**) and
carbidopa (**5.9**) (which cannot penetrate the brain), to decrease this metabolism and
reduce the necessary administered dose.

FIGURE 5.1 Peripheral and central metabolism of L-Dopa (**5.7**).

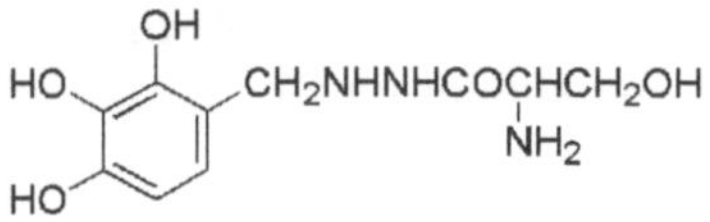

(5.8) benzserazide

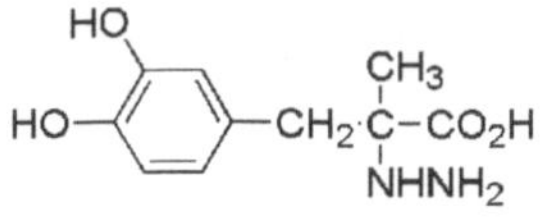

(5.9) carbidopa

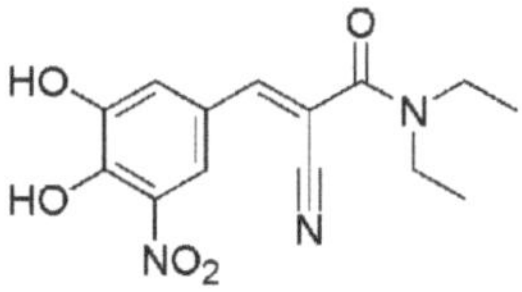

(5.10) entacapone

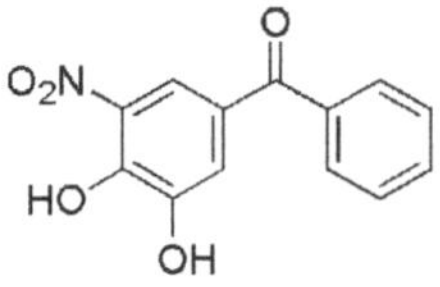

(5.11) tolcapone

A further adjuvant to the above combinations is a catechol-*O*-methyltransferase (COMT) inhibitor. COMT peripherally converts L-dopa to 3-methoxydopa with loss of potency. Entacapone (**5.10**) (COMTESS) is the inhibitor currently available for this purpose; tolcapone (**5.11**) (Tasma), previously used, led in a few instances to fatal hepatic effects and has been discontinued in the U.K.

5.1.1.2 Decrease in Metabolite Production

When the metabolite has an action judged to be clinically undesirable or too pronounced, inhibition of a relevant enzyme reduces its concentration with a decreased (desired) response.

Allopurinol is an inhibitor of xanthine oxidase and is used for the treatment of gout. Inhibition of the enzyme reduces the formation of uric acid from the purines xanthine and hypoxanthine, from the external precursors; otherwise, the uric acid deposits and produces irritation in the joints (Equation 5.6).

$$\text{Xanthine} \xrightarrow[\uparrow]{\substack{\text{xanthine} \\ \text{oxidase}}} \text{uric acid}$$

Allopurinol

(5.6)

In the above example, an enzyme acting in isolation was targeted, but additional strategies may be used with enzyme inhibitors to produce an overall satisfactory clinical response.

(1) Where the target enzyme is part of a biosynthetic pathway consisting of a sequence of enzymes with their specific substrates and coenzymes (Equation 5.7), inhibition of a carefully selected target enzyme in the pathway (see Section 5.2.1) would lead to prevention of overall production of a metabolite that either clinically gives an unrequired response or is essential to bacterial or cancerous growth.

$$A \xrightarrow{E_1} B \xrightarrow{E_2} C \xrightarrow{E_3} D \ldots \xrightarrow{E_n} E \text{ (metabolite)} \qquad (5.7)$$

$$\uparrow$$

inhibitor

(2) Sequential chemotherapy involves the use of two inhibitors simultaneously on a metabolic chain (Equation 5.8) with the aim of achieving a greater therapeutic effect than by application of either alone.

$$A \xrightarrow{E_1} B \xrightarrow{E_2} C \xrightarrow{E_3} D \ldots \xrightarrow{E_n} E \text{ (metabolite)} \qquad (5.8)$$

$$\uparrow \qquad\qquad\qquad\qquad \uparrow$$

inhibitor 1   inhibitor 2

This situation arises when dosage with a single inhibitor is limited by host toxicity or resistant bacterial strains have emerged. The best-known combination is the antibacterial mixture cotrimoxazole, consisting of trimethoprim (**5.12**) (dihydrofolate reductase [DHFR] inhibitor) and the sulfonamide sulfamethoxazole (**5.13**) (dihydropteroate synthetase inhibitor), although the usefulness of the latter in the combination has been queried.

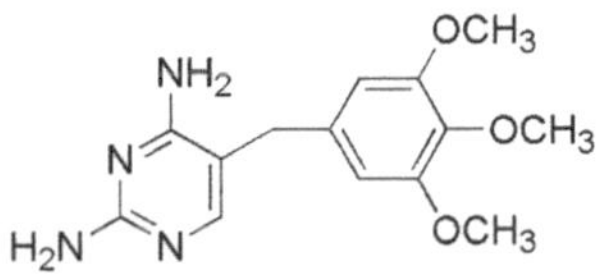

(**5.12**) trimethoprim 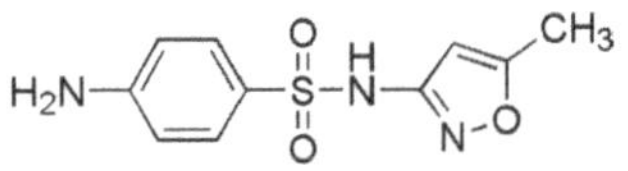   (**5.13**) sulfamethoxazole

(3) A rare example of metabolic pathway inhibition is shown in Equation 5.9 in which inhibition of an enzyme occasionally leads to formation of a "dead-end" complex between the enzyme, coenzyme, and inhibitor, rather than straightforward interaction between the inhibitor and the enzyme. 5-Fluorouracil (**5.14**) inhibits thymidylate synthetase to form a dead-end complex with the enzyme and coenzyme, tetrahydrofolate, thus preventing bacterial growth (Equation 5.9).

$$A \xrightarrow{E_1} B \xrightarrow{E_2} C \xrightarrow{E_3} D \xrightarrow{E_4} E \text{ (metabolite)} \qquad (5.9)$$

Cofactor Z E_2Z' Inhibitor
+ Inhibitor   (Dead-end complex)

Topoisomerases I and II are nuclear enzymes that catalyze the concerted breaking and rejoining of DNA strands to produce the necessary topological and conformational changes in DNA critical for many cellular processes such as replication, recombination, and transcription. The antitumor drugs doxorubicin (**5.15**) and amsacrine (**5.16**) exert their action by binding to the enzyme-(broken)DNA complex in a nonproductive ternary dead-end complex.

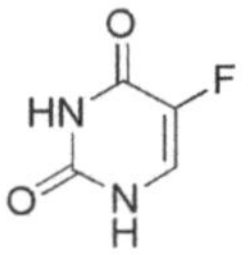

(**5.14**) 5-fluorouracil

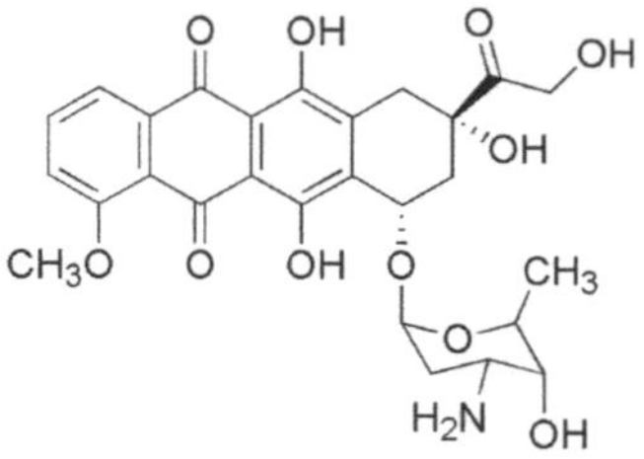

(**5.15**) doxorubicin

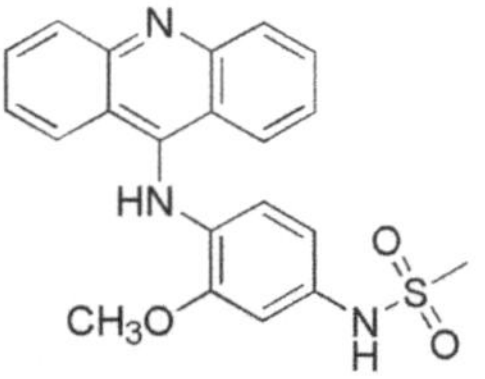

(**5.16**) amsacrine

5.2 RATIONAL SELECTION OF SUITABLE TARGET ENZYME AND INHIBITOR

5.2.1 TARGET ENZYME

Selection of a suitable target enzyme for a particular disease or infection may be aided by (1) fortuitous discovery of the side effects noted for an existing drug being used for another purpose where its main target enzyme is known, (2) drugs introduced into therapy after detection of biological activity in screening experiments in the anticancer and antibacterial setting where the target enzyme was subsequently searched for and found, (3) examination of the biochemical pathways involved either in the normal physiological functioning of the cellular processes that may have been affected in the disease or growth requirements of the bacterial or parasitic infections and requirements for viral multiplication and spread.

Drugs in current use for one therapeutic purpose have occasionally exhibited side effects indicative of potential usefulness for another, subsequent work establishing that the newly discovered drug effect is due to inhibition of a particular enzyme. Although the drug may possess minimal therapeutic usefulness in its newly found role, it does constitute an important "lead" compound for the development of analogs with improved clinical characteristics.

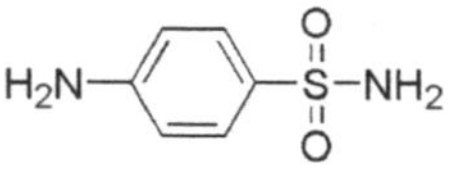

(**5.17**) sulfanilamide 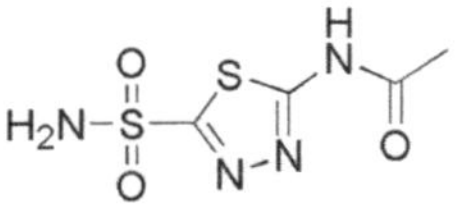   (**5.18**) acetazolamide

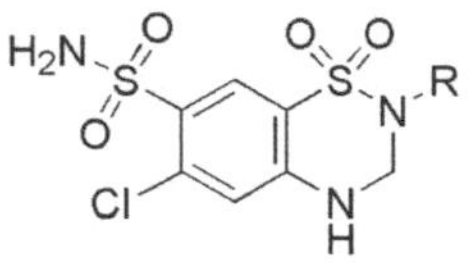

(5.19) R = H; chlorothiazide
(5.20) R = CH₃; methylchlorothiazide

The use of sulfanilamide (**5.17**) as an antibacterial drug was associated with acidosis in the body due to its inhibition of renal carbonic anhydrase (CA). This observation led to the development of the currently used potent inhibitor acetazolamide (**5.18**) as an antiglaucoma agent and subsequently the important chlorothiazide group of diuretics [e.g., chlorothiazide (**5.19**) and methylchlorothiazide (**5.20**)] although these have a different mode of action. Further developments with carbonic anhydrase have shown the presence of 14 isoenzyme forms of CA and that CA IX in particular aids hypoxia (oxygen deficiency) and thus growth in solid cancerous tumors by creating an acidic environment; specific inhibitors of CA would add to the anticancer armory.

The anticonvulsant aminoglutethimide (**5.112**) was withdrawn from the market due to inhibition of steroidogenesis (steroid hormone synthesis) and an insufficiency of 11β-hydroxy steroids. Aminoglutethimide, in conjunction with supplementary hydrocortisone, is now in clinical use for the treatment of estrogen-dependent breast cancer in postmenopausal women due to its ability to inhibit aromatase, the terminal enzyme in the pathway, which is responsible for the production of estrogens from androstenedione. Other much more potent aromatase inhibitors free of depressive side effects have subsequently been developed (see Section 5.6 examples).

Iproniazid (**5.21**), initially used as a drug in the treatment of tuberculosis, was observed to be a central nervous system stimulant due to a mild inhibitory effect on MAO. This observation, with eventual identification of the enzyme target, led to the discovery of more potent inhibitors of MAO, such as phenelzine (**5.22**), tranylcypromine (**5.23**), selegiline ((-)-deprenyl) (**5.24**), and chlorgyline (**5.25**).

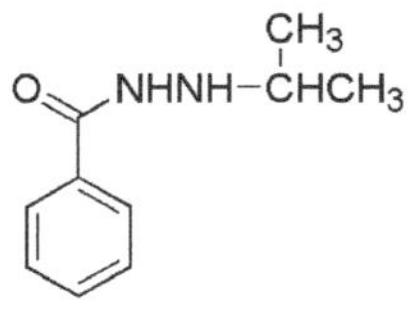

(5.21) iproniazid

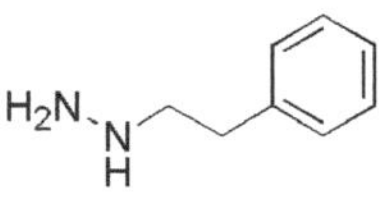

(5.22) phenelzine

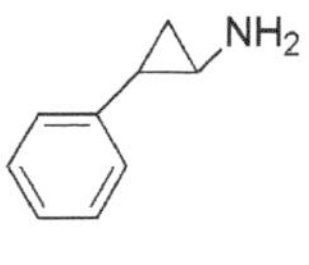

(5.23) tranylcypromine

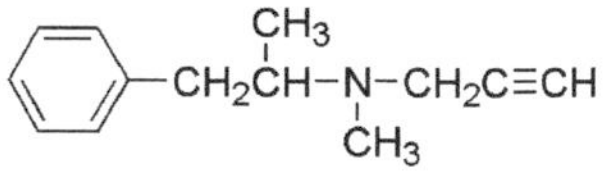

(5.24) (-)-deprenyl

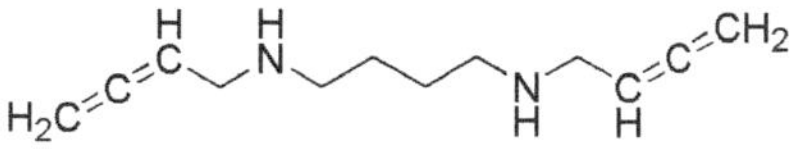

(5.25) chlorgyline

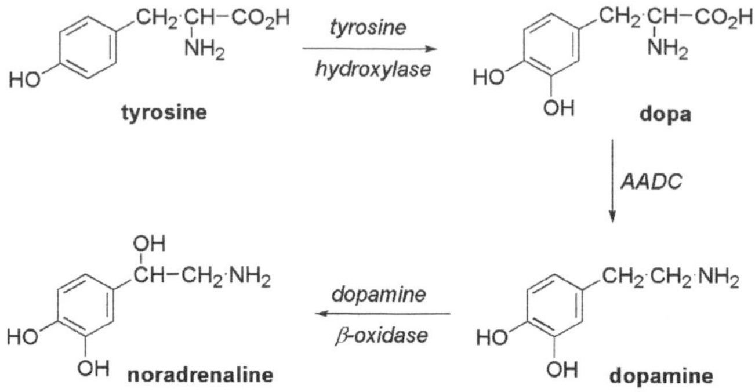

FIGURE 5.2 Conversion of dopa to dopamine by the action of AADC.

Many drugs introduced into therapy following detection of biological activity by cell culture or microbiological screening experiments have subsequently been shown to exert their action by inhibiting a specific enzyme in the tumor cell culture or parasite. This knowledge has helped in the development of clinically more useful drugs by limiting screening tests to involve only the isolated pure or partially purified target enzyme concerned and thus introducing a more rapid screening protocol.

A priori examination of the biochemical or physiological processes responsible for a disease or condition in which these are known or can be guessed at, may point to a suitable target enzyme in its biochemical environment, the inhibition of which would rationally be expected to lead to alleviation or removal of the disease or condition.

Inhibitors of the noradrenaline biosynthetic pathway were intended to decrease production of noradrenaline at the nerve–capillary junction in hypertensive patients, with an associated reduction in blood pressure. The selected target enzyme, aromatic amino acid decarboxylase (AADC), catalyzes the conversion of dopa to dopamine in the second step of the biosynthesis of noradrenaline from tyrosine (Figure 5.2). Many reversible inhibitors, although active *in vitro* against this enzyme, fail to lower noradrenaline production *in vivo*; however, in an isolated scenario, they may slow down decarboxylation of dopa in peripheral tissues. Irreversible inhibitors of AADC successfully lower noradrenaline levels (see Section 5.2.2.2). A possible explanation for the inability of the AADC inhibitors to produce a satisfactory response in a metabolic chain is as follows.

In a metabolic chain of reactions with closely packed enzymes in a steady state (see Equation 5.10) in which the initial substrate (A) does not undergo a change in concentration as a consequence of changes effected elsewhere in the chain, any type of reversible inhibitor that inhibits the first step of the chain effectively blocks that sequence of reactions.

$$A \xrightarrow[v_1]{E_1} B \xrightarrow[v_2]{E_2} C \xrightarrow[v_3]{E_3} D \xrightarrow[v_4]{E_4} \text{metabolite} \qquad (5.10)$$

Inhibitors acting at later points in the chain of closely bound enzymes may not block the metabolic pathway. If the reaction B → C (Equation 5.10) is considered, competitive inhibition of E_2 initially decreases the rate of formation of C, but eventually the original velocity (v_2) of the step is attained as the concentration of B rises due to the difference between its rates of formation and consumption. However, these changes relating to an increase in concentration of B may have secondary effects on the chain due to product inhibition (B on E_1) or product reversal (A ← B); either of these effects can reduce v_1, thus leading to a slowing down of the overall pathway, i.e., here, inhibition of the second enzyme has a successful outcome. There is a general misconception that the overall rate in a linear chain can be depressed only by inhibiting the rate-limiting reaction, i.e., the one with the lowest velocity at saturation with its substrate. Because individual enzymes cannot be saturated with their substrates, the overall rate is determined largely by the concentration of the initial substrate, so that the first enzyme will often be rate limiting, irrespective of its potential rate due to a low concentration of its substrate.

A knowledge of the structure, life cycle, and replication of the human immunodeficiency virus (HIV) has led to the development of inhibitors of the virally encoded protease essential for maturation of the virus and hence production and spread. Aspects of this work are discussed in Section 5.9.

5.2.2 Types of Inhibitor for Selected Target Enzyme

As described in detail in Chapter 4, enzyme-inhibiting processes may be divided into two main classes, reversible and irreversible, depending upon the manner in which the inhibitor (or inhibitor residue) is attached to the enzyme.

5.2.2.1 Reversible Inhibitors

Reversible inhibitors may be competitive, noncompetitive, or uncompetitive, depending upon their point of entry into the enzyme–substrate reaction scheme. In either case, the inhibitor is bound to the enzyme through a suitable combination of forces; van der Waal's, electrostatic, hydrogen bonding, and hydrophobic (see Chapter 2). The extent of the binding is determined by the equilibrium constant K_I for breakdown of the EI or EIS complex for classical inhibitors. However, on rare occasions a covalent bond may be formed with an active site residue, as in the case of a hemiacetal or hemiketal bond with the catalytic serine in serine proteases with a polypeptide aldehyde or ketone-based inhibitor, but the EI complex readily dissociates back into free enzyme and inhibitor as the free inhibitor concentration falls due to dilution, excretion, metabolism, etc.

Competitive inhibitors, as their name suggests, compete with the substrate for the active site of the enzyme, and by forming an inactive enzyme–inhibitor complex, decrease the rate of catalysis by the enzyme of the substrate (Equation 5.11):

$$E + S \xrightarrow{\ K_S\ } ES \xrightarrow{\ k_2\ } E + P$$
$$I \updownarrow K_I$$
$$EI \quad \text{inactive enzyme-} \tag{5.11}$$
$$\text{inhibitor complex}$$

The Michaelis–Menten equation for the rate (v) of an enzyme-catalyzed reaction in the presence of an inhibitor is given by

$$\upsilon = \frac{V_{max}}{1 + \dfrac{K_m}{[S]}\left(1 + \dfrac{[I]}{K_I}\right)} \tag{5.12}$$

where it is seen that, in the presence of the inhibitor, the extent to which the reaction is slowed is dependent on the inhibitor concentration $[I]$ and the dissociation constant, K_i, for the EI complex. A small value for K_i ($\approx 10^6$ to 10^8 M) indicates strong binding of the inhibitor to the enzyme. With this type of inhibitor, the inhibition may be overcome, for a fixed inhibitor concentration, by increasing the substrate concentration as seen in Equation 5.12. With competitive inhibition, only substrate binding, i.e., K_m, is affected because the inhibitor competes with the substrate for the same binding site, i.e., $K_{m'} = K_m (1 + [I]/K_I)$. Determination of K_I has been previously described in Chapter 4 and is a parameter for comparing the potency of inhibitors because it is independent of substrate and inhibitor concentration.

Noncompetitive inhibitors combine with the enzyme–substrate complex and prevent the breakdown of the complex to products (Equation 5.13).

$$
\begin{array}{ccc}
E + S & \rightleftharpoons & ES \quad \longrightarrow E + P \\[2pt]
{}_{+I}\uparrow\downarrow \,{}^{-I}_{K_I} & {}_{+I}^{K_I}\downarrow\uparrow\,{}_{-I} & \\[2pt]
EI + S & \rightleftharpoons & EIS
\end{array}
\tag{5.13}
$$

These inhibitors do not compete with the substrate for the active site; they only change the V_{max} parameter for the reaction. The binding strength of the inhibitor to either E or ES is identical so that there is a single value for K_I. The kinetics for this type of inhibitor are given by

$$\upsilon = \frac{\dfrac{V_{max}}{\left(1 + \dfrac{[I]}{K_I}\right)}\cdot[S]}{[S] + K_m} = \frac{\dfrac{V_{max}}{(1 + [I]/K_I)}}{(1 + K_m/[S])} \tag{5.14}$$

The extent of the inhibition by a fixed concentration of inhibitor is not reversed by increasing the substrate concentration (in contrast to competitive inhibition) because substrate and inhibitor bind at different sites.

Uncompetitive inhibition is the third type of reversible inhibitor and is rare in single-substrate catalysis.

$$
\begin{array}{ccc}
E + S \rightleftharpoons & ES & \longrightarrow E + P \\[2pt]
& {}_{+I}^{K_I}\downarrow\uparrow\,{}_{-I} & \\[2pt]
& EIS &
\end{array}
\tag{5.15}
$$

This type of inhibitor binds only to the enzyme–substrate complex; perhaps substrate binding produces a conformation change in the enzyme, which reveals an inhibitor-binding site. The modified Michaelis–Menten equation is shown in Equation 5.16 where it could be seen that both K_m and V_{max} are modified.

$$\upsilon = \frac{\dfrac{V_{max}}{\left(1+\dfrac{[I]}{K_I}\right)}}{1+\dfrac{K_m}{[S]\left(1+\dfrac{[I]}{K_I}\right)}} = \frac{V'_{max}}{1+\dfrac{K'_{max}}{[S]}} \tag{5.16}$$

Noncompetitive and uncompetitive inhibitors are uncommon and have only recently made their appearance in drug discovery studies as a result of random screening of chemical libraries by the pharmaceutical industry for inhibitors of a new drug target. These types of inhibitors are impossible to design (without a lead) because the characteristics of their binding sites in the ES complex are not complementary to the structure of the known substrate on which the great majority of competitive inhibitors can be readily modeled.

As previously discussed in Chapter 4, occasionally the Lineweaver–Burk plot, used to determine the inhibitor type, shows a pattern that can lie between either (1) competitive and noncompetitive inhibition, or (2) noncompetitive and uncompetitive inhibition. This form of inhibition is termed mixed inhibition and arises because the inhibitor binds to both E and the ES complex but with different binding constants (K_i and K_I).

There are two special types of competitive inhibitors that bind very strongly to the target enzyme; the transition-state analog and tight-binding inhibitor.

A *transition-state analog* is a *stable compound* that resembles in structure the substrate portion of the enzymic transition state for chemical change; it differs in this respect from the transition-state structure formed after reaction between, for example, a serine moiety at the active site of a serine protease and a peptidyl ketone inhibitor, i.e., the oxyanion-containing tetrahedral intermediate (see Chapter 3).

An organic reaction between two types of molecules is considered to proceed through a high-energy-activated complex known as the transition state, which is formed by the collision of molecules with greater kinetic energy than the majority present in the reaction. The energy required for the formation of the transition state is the activation energy for the reaction and is the barrier to the reaction occurring spontaneously. The transition state for the reaction between hydroxyl ion and methyl iodide involves both the commencement of formation of a C–OH bond and the breaking of the C–I bond; it may break down to give either the components from which it was formed or the products of the reaction. Enzymes catalyze organic reactions by lowering the activation energy for the reaction, and one view is that they accomplish this by straining or distorting the bound substrate towards the transition state (see Chapter 2). Equation 5.17 shows a single-substrate enzymatic reaction and the corresponding nonenzymatic reaction in which $ES^{\neq}$ and $S^{\neq'}$ represent the transition states for the enzymatic and nonenzymatic reaction, respectively, and $K_N^{\neq}$ and $K_E^{\neq}$ are equilibrium constants, respectively, for their formation. K_s is the association constant for formation of ES from E and S, and K_T is the association constant for the hypothetical reaction involving the binding of $S^{\neq}$ to E. Analysis of the relationships between these equilibrium constants shows that $K_T K_N^{\neq} = K_s K_E^{\neq}$. Because the equilibrium constant for a reaction is equal to the rate constant mutiplied by h/kT, where h is Planck's constant and k is the Boltzmann's constant, $K_T = K_s(k_E/k_N)$, where k_E and k_N are the first-order rate constants for breakdown of the ES complex and the nonenzymatic reaction, respectively. Because the ratio k_E/k_N is usually of the order 10^{10} or greater, it follows that $K_T \gg K_s$. This means that the transition state $S^{\neq}$ is considered to bind to the enzyme at least 10^{10} times more tightly than the substrate.

$$
\begin{array}{ccccc}
& K_N^{\neq} & & & \\
E + S & \rightleftharpoons & E' + S^{\neq'} & \longrightarrow & E + P \\
K_S \downarrow\uparrow & & \downarrow\uparrow K_T & & \downarrow\uparrow \\
ES & \rightleftharpoons & ES^{\neq} & \longrightarrow & EP \\
& K_N^{\neq} & & &
\end{array}
\tag{5.17}
$$

A transition-state analog is a stable compound that structurally resembles the substrate portion of the unstable transition state of an enzymatic reaction. Because the bond-breaking and bond-making mechanism of the enzyme-catalyzed and nonenzymatic reactions are similar, the analog resembles $S^{\neq}$ and has an enormous affinity for the enzyme, and binds more tightly than the substrate. It would not be possible to design a stable compound that mimics the transition state closely because the transition state itself is unstable by possessing partially broken and partially made covalent bonds. Even crude transition-state analogs of substrate reactions would be expected to be sufficiently tightly bound to the enzyme to be excellent reversible inhibitors, an expectation that has been borne out.

As will be seen in Section 5.4.2.1.3, the design of a transition-state analog for a specific enzyme requires knowledge of the mechanism of the enzymatic reaction. Fortunately, the main structural features of the transition states for the majority of enzymatic reactions are either known or can be predicted with some confidence.

Tight-binding inhibitors bind tightly to the enzyme either noncovalently or covalently and are released very slowly from the enzyme because of the tight interaction. The slow binding is a time-dependent process and is believed to be due either to an enforced conformational change in the enzyme structure or reversible covalent bond formation or, more probably, simply the very low inhibitor concentration used during measurement to allow observation of a residual activity. The drugs coformycin, methotrexate, and allopurinol belong to this class and are useful drugs. Tight binding, in which the dissociation from the complex takes days, is not distinguishable in effect from covalent bonding, and this type of inhibitor may be classed as an irreversible inhibitor.

5.2.2.1.1 Parameters for Determining Relative Inhibitory Potency

In the initial screening of inhibitors, it is convenient to compare potencies within tested compounds as percentage inhibition. However, the relationship between the

TABLE 5.1
Relationship between Percentage Competitive Reversible Inhibition of an Enzyme and Relative Inhibitor Concentration

Percentage Inhibition	Relative Inhibitor Concentrations
10	0.1
50	1[a]
67	3
76	5
90	10
99.01	100
99.90	1000
99.99	10000

[a] IC_{50} value

percentage inhibition of an enzyme and inhibitor concentration ([S] = constant) is not linear, and the relative concentration to inhibit enzyme activity by 50, 90, and 99% under standard conditions increases logarithmically, i.e., 10^0, 10^1, and 10^2, respectively (see Table 5.1). In screening tests, although the potencies of different inhibitors may appear similar within the range 90 to 95% inhibition, their potencies may be very different when IC_{50} values from further experiments are compared.

Expression of potency as an IC_{50} value (concentration of inhibitor required to inhibit enzyme activity by 50%) is a convenient measure of potency within a laboratory, but this value should be used with care when comparing interlaboratory results for competitive inhibition because it is dependent on the concentration of substrate used (Equation 5.18), which may vary between laboratories.

$$IC_{50} = K_I\left(1 + \frac{S}{K_m}\right) \qquad (5.18)$$

For noncompetitive and uncompetitive inhibitors, $IC_{50} = K_I$ and is independent of substrate concentration. K_I is independent of substrate and inhibitor concentrations for all classes of reversible inhibitors.

5.2.2.2 Irreversible Inhibitors

Compounds producing irreversible enzyme inhibition fall into two groups: active site–directed (affinity labeling) inhibitors and mechanism-based inactivators (k_{cat} inhibitors, suicide substrates).

5.2.2.2.1 Active Site-Directed Irreversible Inhibitors

These resemble the substrate sufficiently to form a reversible enzyme–inhibitor complex, analogous to the enzyme–substrate complex, within which reaction occurs between functional groups of the inhibitor (e.g., $-COCH_2Cl$, $-COCHN_2$, $-OCONHR$, $-SO_2F$) and enzyme ($-SH$, $=N-$, $-NHR$, $-OH$). A stable covalent bond is formed with irreversible inhibition of the enzyme. Active site-directed irreversible inhibitors are designed to exhibit specificity towards their target enzymes because they are structurally modeled on the specific substrate of the enzyme concerned. These inhibitors are termed affinity-labeling agents when used to probe the nature of the functional groups present in the active site.

Irreversible inhibitors progressively reduce enzyme activity with time, and the reaction follows pseudo-first-order kinetics as described in Chapter 4. The biochemical environment of the enzyme (see Section 5.2.1) is unimportant so that any step in a biosynthetic pathway may be inhibited with decrease in overall metabolite production. However, because these compounds belong to a group that mainly consists of alkylating and acylating agents, they have not been developed as drugs as they would be expected to react with a range of tissue constituents containing amino or thiol groups besides the target enzyme, with potentially serious side effects. However, they have been used successfully as affinity labeling agents.

5.2.2.2.2 Mechanism-Based Enzyme Inactivators

Many irreversible inhibitors of certain enzymes have previously been recognized, among which the range of electrophilic centers normally associated with active site-directed irreversible inhibitors, e.g., $-COCH_2Cl$, $-COCHN_2$, $-OCONHR$, $-SO_2F$, were absent, and therefore the means by which they inhibited the enzyme was unclear. The action of these inhibitors has, in more recent years, become understandable because they have been categorized as mechanism-based enzyme inactivators. Mechanism-based enzyme inactivators bind to the enzyme through the K_s parameter to form a complex and are modified by the enzyme in such a way as to generate a reactive group that irreversibly inhibits the enzyme by forming a covalent bond with a functional group present at the active site. Occasionally, catalysis leads not to a reactive species but an enzyme–intermediate complex that is partitioned away from the catalytic pathway to a more stable complex by bond rearrangement (e.g., β-lactamase inhibitors).

These inhibitors are substrates of the enzyme, as suggested by their alternative name k_{cat} inhibitors, where, as explained earlier, k_{cat} is the overall rate constant for the decomposition of the enzyme–substrate complex in an enzyme-catalyzed reaction. Mechanism-based inactivators do not generate a reactive electrophilic center until acted upon by the target enzyme. Reaction may then occur with a nucleophile on the enzyme surface, or alternatively the species may be released and either react with an external nucleophile or decompose (Equation 5.19)

$$E + I \underset{k_{-1}}{\overset{k_{+1}}{\rightleftharpoons}} EI \xrightarrow{\ k_{+2}\ } EI^* \xrightarrow{\ k_{+3}\ } E - I$$
$$\downarrow k_{+4}$$
$$E + P$$

$$(5.19)$$

The biochemical environment of the target enzyme is unimportant (see Section 5.2.1). For example, in the noradrenaline biosynthetic pathway, α-monofluoromethyldopa (**5.26**), a mechanism-based inactivator of AADC, produces a metabolite that irreversibly inhibits and decreases the level of the enzyme by >99%. This leads to a near-complete depletion of catecholamine levels in brain, heart, and kidney, despite the occurrence of the enzyme in the second step of the noradrenaline biosynthetic pathway, as discussed earlier.

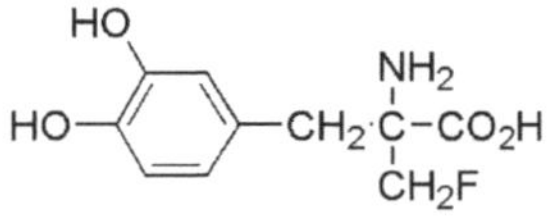

(**5.26**) α-monofluoromethyldopa

Mechanism-based inactivators do not possess a biologically reactive functional group until after they have been modified by the target enzyme and, consequently, would be expected to demonstrate high specificity of action and low incidence of adverse reactions. It is these features that have encouraged their active application in inhibitor design studies.

5.2.2.2.3 Parameters for Determining Relative Potency of Irreversible Inhibitors

Previously, for reversible inhibitors, the potency of an inhibitor was shown to be reflected in the K_I value, which is characteristic of the inhibitor and independent of inhibitor concentration. Similarly, the potency of an irreversible inhibitor is given by the binding and kinetic rate constants, both of which are independent of inhibitor concentration (Equation 5.20). This allows a precise comparison of the relative potency of inhibitors, which is necessary in the design and development of more effective inhibitors of an enzyme.

$$E + I \overset{K_I}{\leftrightharpoons} (E)(I) \xrightarrow{k_{+2}} E - I \tag{5.20}$$
$$\text{complex} \qquad \text{inhibited} \qquad \text{enzyme}$$

K_I and k_{+2} can be determined by the method previously described in Chapter 4, and inhibitor potency can be related to the ratio k_{+2}/K_I for comparative purposes. The potency increases as binding increases (K_I decreases) and the rate constant for the reaction increases (k_{+2} increases). In a similar manner, the binding and reaction rate constant for a mechanism-based enzyme inactivator can be determined, but another aspect, the partition ratio, enters into the efficiency of these agents.

The ratio of the rate constants, i.e., k_{+4}/k_{+3}, gives the partition ratio (r) for the reaction, and when this approaches zero, the mechanism-based inactivation will proceed with a single turnover of the inhibitor, where the noncovalent enzyme–inhibitor complex (EI) is transformed into an activated species (EI*) that then irreversibly inhibits the enzyme (Equation 5.21).

$$E + I \underset{K_I}{\leftrightharpoons} EI \xrightarrow{k_{+2}} EI* \xrightarrow{k_{+3}} E - I \tag{5.21}$$

For mechanism-based inactivators, the turnover rate of the enzyme is important because of enzyme resynthesis, and this rate may be 10^3 to 10^5 times slower than for natural substrates. The partition ratio of the reaction should ideally be close to zero, in which case every turnover results in inhibition; the reactive electrophilic species, by not being free to react with other molecules in the biological media, has a high degree of specificity for its target enzyme and exhibits low toxicity.

5.3   SELECTIVITY AND TOXICITY

Inhibitors used in therapy must show a high degree of selectivity towards the target enzyme. The term selectivity is preferred to specificity because the latter is unobtainable within a group of closely related enzymes.

Inhibition of closely related enzymes with different biological roles (e.g., trypsin-like enzymes such as thrombin, plasmin, and kallikrein), or reaction with constituents essential for normal functioning of the body (e.g., DNA glutathione, liver P450-metabolizing enzymes) could lead to serious side effects. An inhibitor being developed for potential clinical use is put through a spectrum of *in vitro* tests against other realistic potential enzyme targets to ascertain that it is suitably selective towards the intended target. An inhibitor with high potency, e.g., $IC_{50} = 5$ n*m*, would be screened at 1 μ*M* against other targets, and a small percentage inhibition would rate as a demonstration of acceptable selectivity. The aromatase inhibitor fadrozole (**5.27**) at higher doses than likely to be achieved clinically showed inhibition of the 18-hydroxyase in the steroidogenesis pathway, which could affect aldosterone production in the clinical setting. With the further developed compound letrozole (**5.28**) the observed selectivity between the two enzymes noted with fadrozole (10-fold) was widened by at least an order (100-fold).

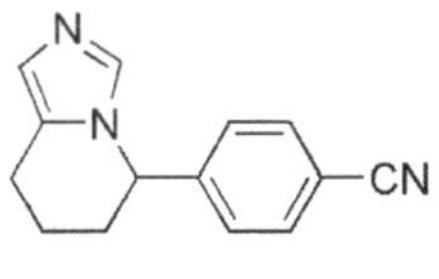

(**5.27**) fadrozole 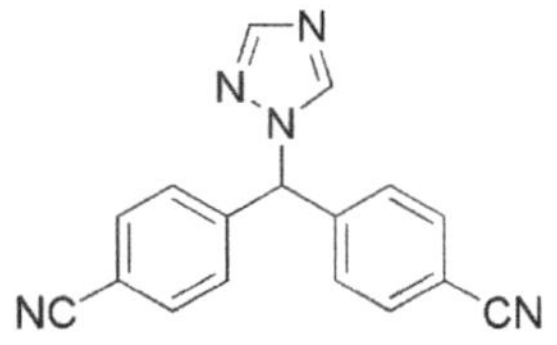   (**5.28**) letrozole

Where the target enzyme is common to the host's normal cells as well as to cancerous or parasitic cells, chemotherapy can be successful when host and parasitic cells contain different isoenzymes, e.g., dihydrofolate reductase (DHFR), with those of the parasite being more susceptible to carefully designed inhibitors.

On very rare occasions, the target enzyme may be absent from the host cell. Sulfonamides [e.g., sulfamethoxazole (**5.13**)] are toxic to bacterial cells by inhibiting dihydropteroate synthetase, an enzyme on the biosynthetic pathway to folic acid essential for bacterial growth. The host cell is unaffected because it utilizes preformed folic acid from the diet, which the susceptible bacteria is unable to do.

Another example relates to the carbonic anhydrase (CA) isoforms CA IX and CA XII that predominate in cancer cells (but are absent in normal cells) and are concerned in maintaining an acidic environment leading to hypoxia essential for solid tumor growth. Inhibitors of CA IX and XII, as antitumor agents, would need to be very selective because up to 12 other CA isoforms are known in man and are also involved in the interconversion between CO_2 and HCO_3^-, critical for many physiological processes (especially, CA I, CA II, and CA IV).

Normal and cancerous cells contain the same form of the target enzyme, DHFR, but the faster rate of growth of tumor cells makes them more susceptible to the effects of an inhibitor. Although side effects occur, these are acceptable due to the life-threatening nature of the disease.

5.4 RATIONAL APPROACH TO THE DESIGN OF ENZYME INHIBITORS

Once the target enzyme has been identified, then usually a lead inhibitor has previously been reported, or can be predicted from studies with related enzymes, or has appeared, in more recent years, from the rapid screening methods now available of industrial chemical collections (libraries). The design process is then initially concerned with optimizing the potency and selectivity of action of the inhibitor to the target enzyme using *in vitro* biochemical tests; nowadays, this has the advantage that the pure enzyme from recombinant DNA technology may be available for such studies. Candidate drugs are then examined by *in vivo* animal studies for oral absorption, stability to the body's metabolizing enzymes, and toxic side effects. Because many candidates may fall at this stage, the whole design cycle recommences because further design is then necessary to maintain desirable features and eliminate undesirable features from the *in vivo* profile. Because an *in vivo* profile in animal studies is not directly translatable to the human situation, studies with human volunteers are also required before a drug enters clinical trials.

5.4.1 LEAD INHIBITOR DISCOVERY

A lead inhibitor is usually a compound of low potency and selectivity whose structure can be used as a scaffold for structural modification to other compounds whose potency and selectivity are enhanced, together with many other desirable features, thereby leading to a compound that may eventually reach the marketplace. Discovery of a lead inhibitor for a new target enzyme may arise in conjunction with the discovery of the target enzyme (see Section 5.2.1) through (1) discovery of side effects noted for an existing drug (used for another purpose with a different target) where the side effect is the clinical effect required and expected to be shown on inhibition of the new proposed target enzyme, (2) compounds showing low potency in pharmacological, antibacterial, antiparasitic, or antiviral effects from screening experiments in which subsequent examination has revealed that it acts on the proposed target enzyme.

5.4.1.1 Modification of the Lead

The design of a novel inhibitor of a new target enzyme takes into account a combination of several different design approaches based on (1) modification of the structural scaffold of a lead inhibitor, if this is available, (2) a knowledge of the substrate and the mechanism of the catalytic reaction and perhaps a lead inhibitor (which may be from a closely related enzyme with a known structure), and (3) use of computer-assisted molecular graphics (molecular modeling).

Once a lead inhibitor has been identified, from whatever combination of the design aspects described above, a process of optimization of its potency and selectivity using *in vitro* tests is undertaken. This process involves chemical manipulation of the lead, and although many general structural approaches are available, an appropriate strategy will suggest itself for a particular lead. The principal concepts followed are generalized as follows: (1) Replace existing groups with groups comparable in size (volume) and charge, an approach well known in medicinal chemistry for receptor agonists as well as enzyme inhibitors as "isosteric replacement" (see Section 5.4.1.1.1). (2) Because the binding of peptide substrates to enzymes occurs through a hydrogen-bonding network between hydrogen bond acceptors ($R_2C=O$, $-O-$, R_3N:) and donors (=NH, bound H_2O, $-OH$) along the two surfaces (as for α–helix and β–pleated sheet structures in proteins), as well as the interaction of oppositely charged groups (COO^-, R_3NH^+), increasing the potential hydrogen-bonding groups in the inhibitor could increase potency if they are correctly positioned. (3) Many enzymes have a hydrophobic (water-repelling) cavity containing alkyl or aromatic or heterocyclic residues as part of the active site for binding of peptide amino acid residues. Introduction of similar residues in the inhibitor if correctly positioned will improve binding.

However, concepts (2) and (3) will affect the hydrophilic–hydrophobic balance in the drug, necessary for the penetration of membranes in its transport to the site where its target is situated (see Section 5.5.1). The size of the hydrophobic cavity may limit the size of any hydrophobic groups introduced. (4) There is restriction in the lead compound of extended alkyl and carbon chains because these are not only susceptible to metabolic hydroxylation by liver-metabolizing P450 enzymes (see Section 5.5.2) but can also be freely rotated and therefore may not be correctly positioned for binding in the required position of the enzyme; the energy gained on binding will be depleted by the energy required to reposition the chain, i.e., lower binding energy. Ring formation between flexible chains restricts flexibility and is a technique frequently used. (5) Remove, where possible, stereochemical aspects from the inhibitor, i.e., asymmetric carbon centers leading to (*R*)- and (*S*)-isomers and cis or trans (*Z/E*) isomerization around a C=C double bond. Such stereochemistry complicates drug development because activity usually resides with one isomer (see Section 5.5.4).

Potent, selective drug candidates from *in vitro* studies may fail in an *in vivo* animal model and require further structural manipulation to improve their *in vivo* profile when the whole of the synthetic and *in vitro* testing cycle recommences. Even then, for human testing there is usually a backup compound available to replace the main candidate drug, should it fail.

TABLE 5.2
Isosterically Related Atoms and Groups

Electronic Configurations	2(4)	2(5)	2(6)	2(7)
	=C=	–N=	–O–	–F
	–CH=	–NH–	–OH	
			–CH$_2$–	–NH$_2$
				–CH$_3$

5.4.1.1.1 Isosterism

A receptor agonist or antagonist, or enzyme inhibitor, has a skeleton complementary to the respective target protein site so as to present binding groups in the correct orientation to complementary sites on the protein for specific hydrogen bond, ionic, dipolar, and nonspecific hydrophobic interactions. Isosteric replacement of groups or atoms in the skeleton was an attempt in the very early days of drug design to maintain the structurally specific requirements of a drug before much was known concerning drug–target-protein interactions. The concept is that isosteric modification is the replacement of an atom, or group of atoms, in a molecule by another group with similar electronic and steric configurations. Isosterism was an attempt to apply to molecules or molecular fragments the premise that similarities to properties of elements within vertical groups of the Periodic Table were due to identical valence electronic configurations. Thus, two molecular fragments containing an identical numbered arrangement of electrons should have similar properties, and were termed isosteres. Further, the early recognition that benzene and thiophene were alike in many of their physical properties led to the term *ring equivalent* to describe the interchanging of –CH=CH– and –S–, which distinguishes their structures. Table 5.2 shows isosterically related atoms or groups (where the presence of hydrogen atoms is ignored).

The outer electrons (numbers in parentheses) are calculated as follows: For –N=, the lone pair is unshared, and the other 6 electrons of the three bonds are shared giving a total of (2 + 6/2) = 5 outer electrons.

The isosterism concept has been applied with great success in the past and is invariably referred to in lead manipulations in the current development of drugs, but needs to be applied with caution. Preservation of the template for correct orientation of specific atom or group bonding (i.e., the pharmacophore) by isosteric substitution should not be confused with (1) substitution of nonspecific functions that are hydrophobic and bind to suitably adjacent hydrophobic amino acid residue side chains on the target protein but outside the pharmacophore, thus increasing binding and potency, or (2) hydrophobic or hydrophilic functions introduced to adjust the balance of such functions in the molecule to improve membrane penetration or introduce metabolic stability. The isosteric concept does not distinguish between the structurally specific requirements of a drug, and its nonspecific requirements outside the pharmacophore area for improving inherent drug potency and its ability to reach its target protein.

Furthermore, the concept of isosteric replacement does not take into account ionization, and the $-NH_2$ $>N-$ function replacements mentioned here are weak bases of the phenylamine type where protonation does not occur at physiological pH.

5.4.1.1.1.1   Replacement of Univalent Atoms or Groups

The analogous groups considered here $-F$ (or $-Cl$), $-OH$, $-NH_2$, and $-CH_3$ are seen in Table 5.2, and such replacements have been successfully used in the development of hypoglycemic agents — development of inhibitory activity to dihydrofolate reductase when the $-OH$ group of folic acid (**5.29**) was replaced by $-NH_2$ [aminopterin (**5.30**)], 6 $-OH$ of hypoxanthine (**5.31**), and guanine (**5.32**) was replaced by $-SH$ to give the anticancer drugs 6-mercaptopurine (**5.33**) and 6-thioguanine (**5.34**), and barbiturates to short-acting thiobarbiturates ($-N=C-OH$ $\rightarrow -N=C-SH$).

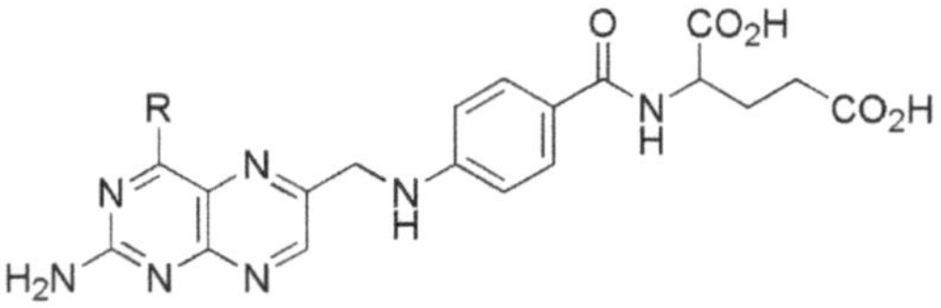

(**5.29**) R = OH; folic acid
(**5.30**) R = NH₂; aminopterin

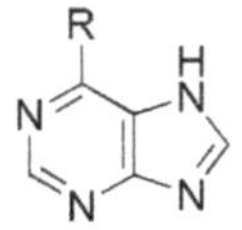

(**5.31**) R = OH; hypoxanthine   (**5.32**) R = OH; guanine
(**5.33**) R = SH; 6-mercaptopurine   (**5.34**) R = SH; 6-thioguanine

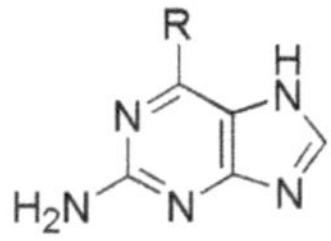

5.4.1.1.1.2   Replacement of Divalent Atoms or Groups

In esters, the rotation of C–O–C bonds is restricted by resonance (**5.35**), and aliphatic esters exist, predominantly, in the cis configuration (**5.35**) rather than the trans (**5.36**).

(**5.35**) 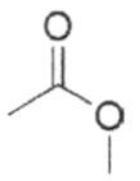   (**5.36**)

Studies on amides have also revealed similar planar structures and a predominant configuration (**5.37**) analogous to the cis ester. Conversion of the ester function in local anesthetics to amide prolongs their action in the body by preventing esterase actions; here polarity is maintained.

(5.37)

Interchange of –O–, –S–, –NH–, and –CH$_2$– have been used in the development of several groups of drugs, e.g., antihistamines, anticholinergic spasmolytics, antidepressant drugs, and antiulcer drugs.

5.4.1.1.1.3   Interchange of Trivalent Atoms and Groups
The substitution of –CH= by –N= in aromatic rings has been one of the most successful applications of isosterism. One of the most potent antihistamines, mepyramine (**5.38**), has evolved from replacement of a phenyl group in antergan (**5.39**) by pyridyl. The π-electron deficiency of the pyridine nucleus enables the nitrogen electron pair to hydrogen-bond with a water molecule, causing an increase in hydrophilicity that is significant in determining the high level of biological activity. This is a nonspecific replacement.

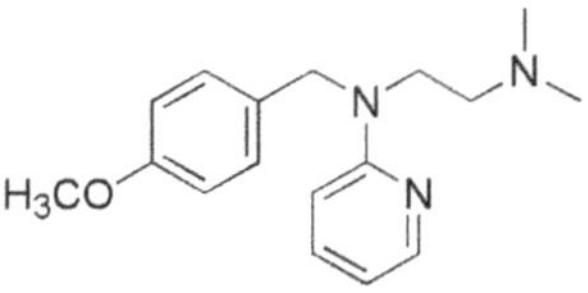

(5.38) mepyramine 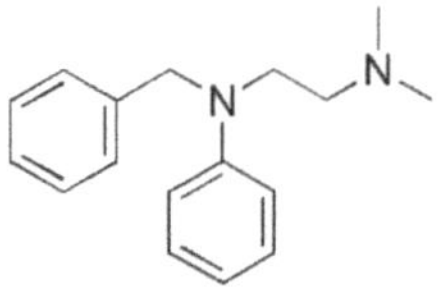   **(5.39)** antergan

The substitution of a benzene ring by pyridine has also resulted in improved activity in the tricyclic, antihistaminic, and neuroleptic (major tranquilizing) drugs with the introduction of isothipendyl (**5.40**) cf. promethazine (**5.41**) and prothipendyl (**5.42**) cf. promazine (**5.43**).

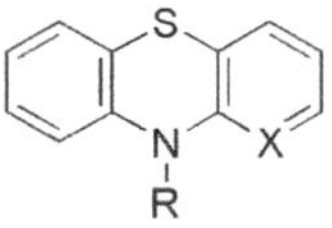

(**5.40**) R = CH$_2$CH(CH$_3$)N(CH$_3$)$_2$, X = N; isothipendyl
(**5.41**) R = CH$_2$CH(CH$_3$)N(CH$_3$)$_2$, X = CH; promethazine
(**5.42**) R = CH$_2$CH$_2$CH$_2$N(CH$_3$)$_2$, X = N; prothipendyl
(**5.43**) R = CH$_2$CH$_2$CH$_2$N(CH$_3$)$_2$, X = CH; promazine

Earlier examples of –N=/–CH= substitutions are to be found in the sulfonamide antibacterials with the development of sulfapyridine (**5.44**) and sulfadiazine (**5.45**), in which the heterocyclic ring confers greater acidity on the sulfonamide group, leading to the required degree of ionization at physiological pH.

Aromatic ring substitution of –CH=CH– by –S– or –O–, previously referred to as "ring equivalents," has been profitable. These are nonspecific bonding moieties. Replacements in the pyridyl ring of sulfapyridine (**5.44**) has produced the five-membered ring structures sulfathiazole (**5.46**), sulfisoxazole (**5.47**), and sulfamethizole (**5.48**), which are more soluble in urine and less liable to crystallize in the renal tubules.

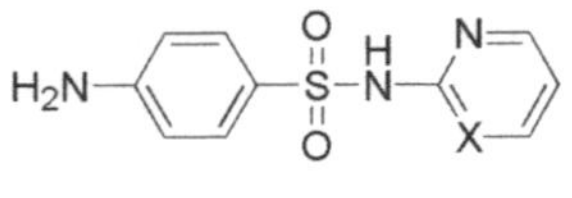

(**5.44**) X = CH; sulfapyridine
(**5.45**) X = N;   sulfadiazine

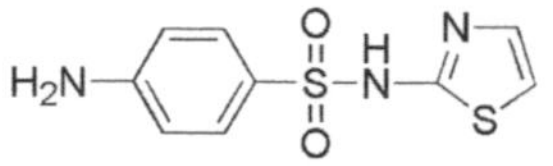

(**5.46**) sulfathiazole

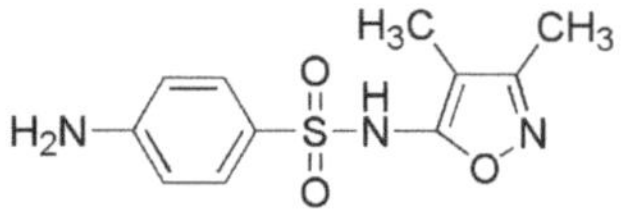

(**5.47**) sulfisoxazole

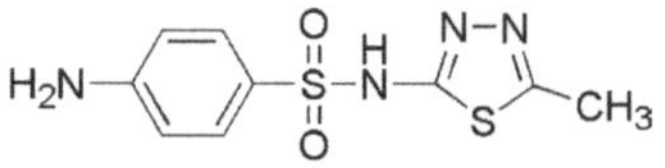

(**5.48**) sulfamethizole

The reason for increased solubility is that pyridine and pyrimidine analogs confer increased water solubility compared to the phenyl ring because the nitrogen lone pair does not participate in the heteroaromatic ring resonance (π-electron-deficient ring) and is able to hydrogen-bond with water. Alternatively, pyrrole, furan, and thiophene, ring equivalent analogs of phenyl, are almost insoluble in water because the N, O, and S lone pairs participate in ring resonance and are not available for hydrogen-bonding to water. However, replacement of –C= by –N= (thus converting pyrrole to imidazole, furan to oxazole, and thiophene to thiazole) increases water solubility as expected, due to introduction of a π-electron-deficient nitrogen into the ring system. The interchanging of phenyl with sulfur-, oxygen-, and nitrogen-containing heterocyclic rings has also been extensively exploited in the development of semisynthetic penicillins and cephalosporins with broader spectra of activity and greater stability towards β-lactamases.

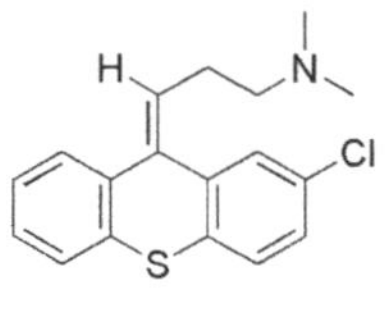

(**5.49**) chlorprothixene

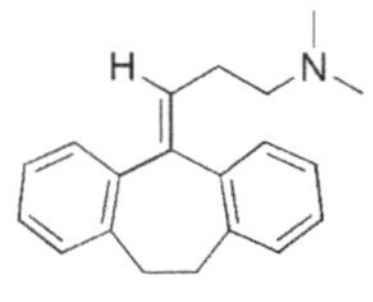

(**5.50**) amitriptyline

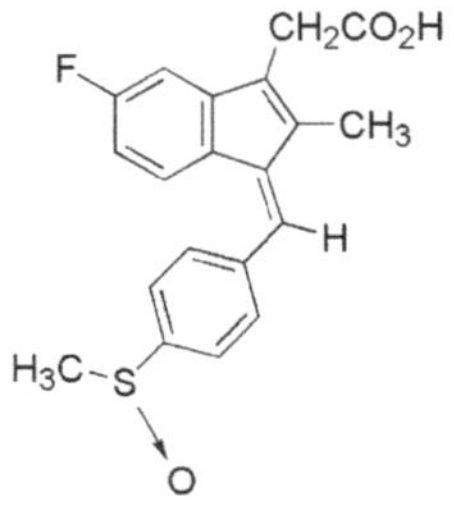

(**5.51**) sulindac

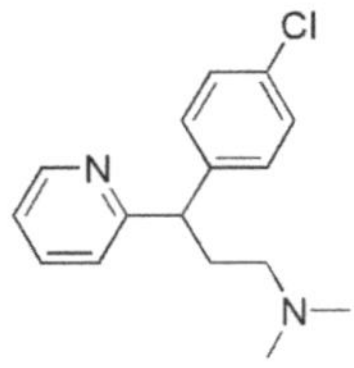

(**5.52**) chlorpheniramine 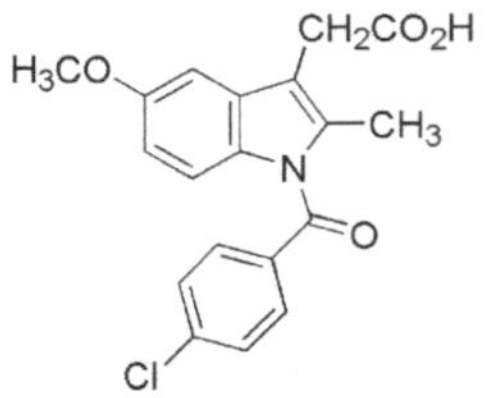   (**5.53**) indomethacin

Ring replacement of –N< by –HC< and its subsequent modification to >C= have resulted in a variety of useful drugs. This is seen in the development of psychotherapeutics such as chlorprothixene (**5.49**) and amitriptyline (**5.50**), and the antiinflammatory drug sulindac (clinoril) (**5.51**). Substitution of the pyridyl amino –N< by –HC< in the antihistamine mepyramine (**5.38**) (and 4-methoxybenzyl by 4-chlorophenyl) produces chlorpheniramine (**5.52**) valued for its short, powerful action and relative freedom from sedation. Similarly, the indole nucleus of the antiinflammatory drug indomethacin (**5.53**) (a prostaglandin synthetase (COX) inhibitor) has been modified (>N–C(=O)– → >C=CH–) in sulindac (**5.51**).

5.4.1.1.1.4   Other Isosteric Modifications
Several modifications are summarized here, in which the modified molecule bears a more general resemblance to the parent and are only termed isosteric for convenience.
   Reversal of groups — An ester may be reversed from –OC(=O)R to –C(=O)–OR, or an amide from –C(=O)NHR to –NHC(=O)R. The dipolar character and hydrogen-bonding capacity of the molecule is maintained.
   Ring opening and closure — Sulfonamide oral hypoglycemic agents arose directly from the clinical observation that a sulfathiazole derivative (**5.54**), which was being used specifically for treating typhoid, lowered the blood sugar almost to a fatal level. Modifications involving opening of the thiazole ring to give a thiourea unit in which =S was ultimately replaced by =O yielded carbutamide (**5.55a**) that was later replaced by the less toxic tolbutamide (**5.55b**).

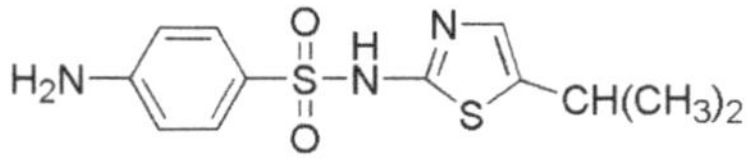

(**5.54**) 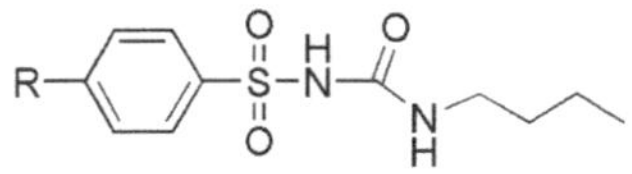   (**5.55a**) R = NH$_2$; carbutamide
                                    (**5.55b**) R = CH$_3$; tolbutamide

The antitubercular thiosemicarbazones were developed from the observation that sulfathiadiazole (**5.56**) possessed weak antitubercular properties and through subsequent testing of intermediates in the synthesis of aminothiadiazoles, i.e., the open chain analog (**5.57**). The thiosemicarbazone group is also associated with antiviral activity, and methisazone (**5.58**) was developed, which prevents smallpox infection among people who have previously been in contact with fatal cases.

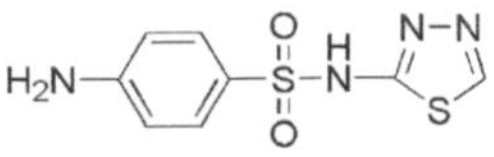

(5.56) sulfathiadiazole 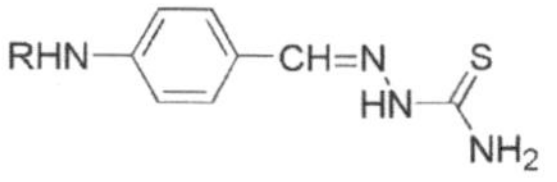   (5.57) aminothiadiazoles 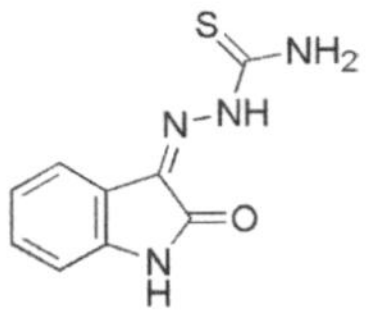   (5.58) methisazone

The closure of rings has frequently led to useful agents; conversion of the diphenyl ester (**5.59**) to a fluorene analog (**5.60**) increases anticholinergic spasmolytic activity markedly.

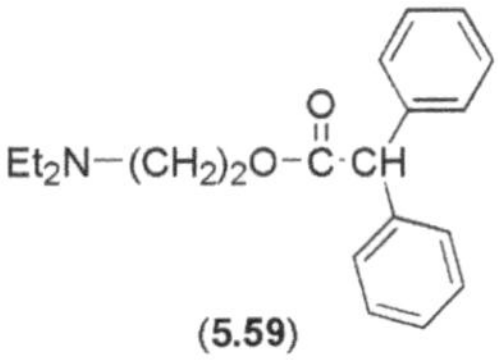

(5.59) 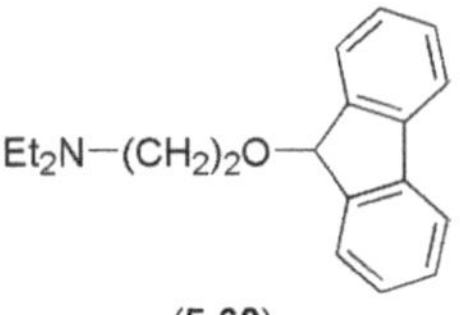   (5.60)

Terminal *N,N*-dialkylamino groups have been replaced by piperidino, leading to greater potency in tranquillizers, local anesthetics, antihistamines, and spasmolytics, the additional methylene groups increasing lipophilicity and membrane penetration.

Groups with similar polar effects — The COOH has been replaced isosterically with tetrazole, in which the ionized N-anion is stabilized by distribution over the other three nitrogen atoms (**5.61**). The α–COOH of carbenicillin (**5.62**) has been replaced by 5-tetrazolyl to improve oral stability because carbenicillin, being a β-amido (keto) acid, decarboxylates on oral administration at the acidic pH of the stomach. The unstable 3-acetoxymethyl side chain ($-CH_2-OC(=O)-CH_3$) of the cephalosporin cephalothin (**5.63**) has been replaced by the more stable thiotetrazole isostere (**5.64**). In both these cases the isosteric groups are probably binding non-specifically well away from the acylating β-lactam-ring-binding area on the enzyme targeted in the bacteria.

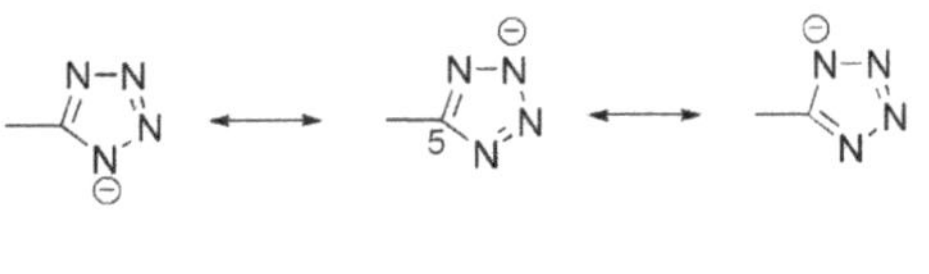

(5.61) 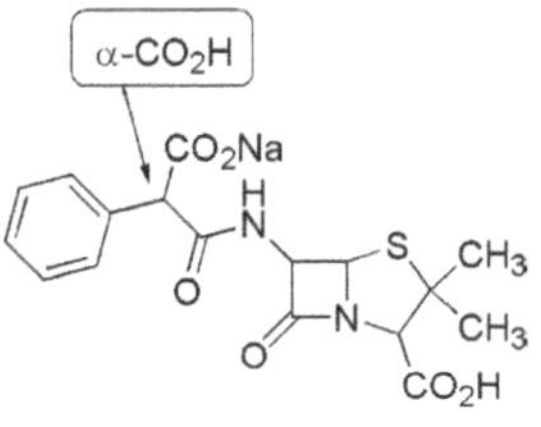   (5.62) carbenicillin

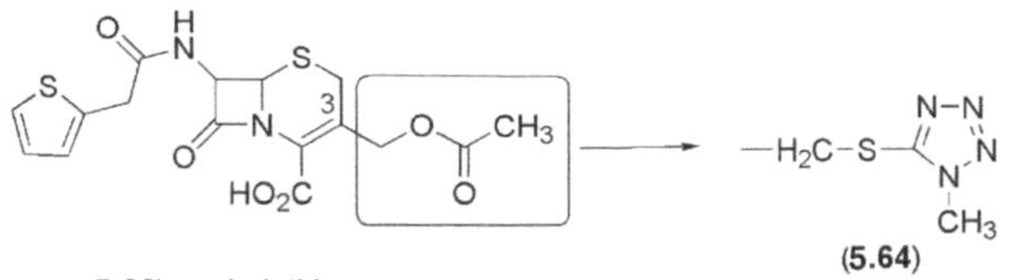

(5.63) cephalothin

(5.64)

5.4.2 DESIGN FROM A KNOWLEDGE OF THE CATALYTIC MECHANISM

A productive design approach if a model of the enzyme active site does not exist (as is not unusual for a new target enzyme) is based on a knowledge of the substrate, a lead inhibitor (perhaps from a related enzyme), and of the mechanism of the catalytic reaction. Molecular modeling (see Section 5.4.3) may enter into the design process at a later stage. A few selected examples are now given to illustrate this approach.

5.4.2.1 Examples

5.4.2.1.1 Angiotensin I-Converting Enzyme (ACE)

Angiotensin converting enzyme (ACE; peptidyldipeptide carboxy hydrolase) converts the inactive decapeptide angiotensin I (AI) to the vasoconstrictor octapeptide angiotensin II (AII) (see Equation 5.22) and is a key enzyme component of the renin–angiotensin system, which helps to maintain blood pressure following events such as hemorrhage that lower the blood volume.

$$\text{Asp-Arg-Val-Tyr-Ile-His-Pro-Phe-His-Leu} \xrightarrow{\text{ACE}} \text{Asp-Arg-Val-Tyr-Ile-His-Pro-Phe}$$

Angiotensin I   Angiotensin II   (5.22)

Several peptides, some from snake venom, were found to be inhibitors of the enzyme but were not orally active due to degradation in the alimentary tract.

Little was known at that time concerning the structure of the active site of the enzyme and the manner in which the natural substrate was bound; therefore insufficient information was available from this source for the design of more stable and efficient inhibitors. The observation that the enzyme was a metalloprotein and the fact that its substrate specificity resembled that of carboxypeptidase A, about which much was known, led to the view that the active sites of the two enzymes bore some similarity although there were obvious differences; cleavage of the carboxyl-terminal dipeptides occurred with ACE, whereas only single carboxyl-terminal amino acids were released by carboxypeptidase A. Guesses were made as to the structural features present in ACE that could account for these differences, and these provided a basis for the design of potential inhibitors of the enzyme.

The discovery that D-benzylsuccinic acid was a potent "biproduct inhibitor" (it resembled both parts of the products formed by substrate cleavage) of carboxypeptidase A was followed by incorporation into the molecule of the terminal Ala-Pro sequence of the snake venom peptides to yield the ACE inhibitor methylglutamyl-proline (**5.65**). Introduction of the thiol group for chelation to the active site zinc atom of ACE gave captopril (**5.66**), the first potent reversible ACE inhibitor, subsequently shown to be a nontoxic, orally active antihypertensive agent.

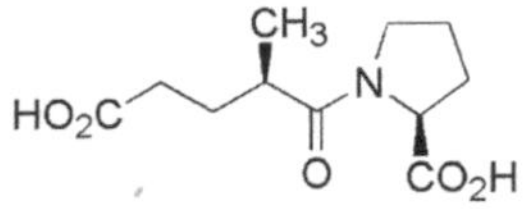

(**5.65**) methylglutamylproline 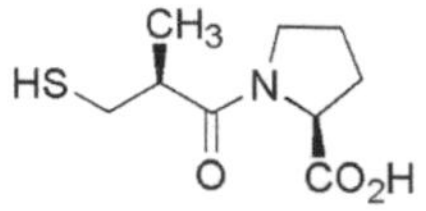   (**5.66**) captopril

Modification of (**5.65**) by increasing its resemblance to a peptide led to the discovery of the potent inhibitor enalaprilat (**5.67a**). Enalaprilat was not well absorbed after dosing, but this was improved by conversion to the ethyl ester enalapril (**5.67b**), which hydrolyzes to **5.67a** *in vivo*. Enalapril has advantages over captopril, as it is less likely to produce a rash and loss of taste, side effects that were reported especially for high doses of captopril. A further development along these lines is ramipril (**5.68**), which has a particularly long *in vivo* duration of action. A new class of bicyclic ACE inhibitors based on piperazic acid has been designed by computer graphic modeling to meet the requirements for binding of captopril to ACE. Cilazapril (**5.69b**), a prodrug of cilazaprilat (**5.69a**) to improve absorption, was designed in this manner.

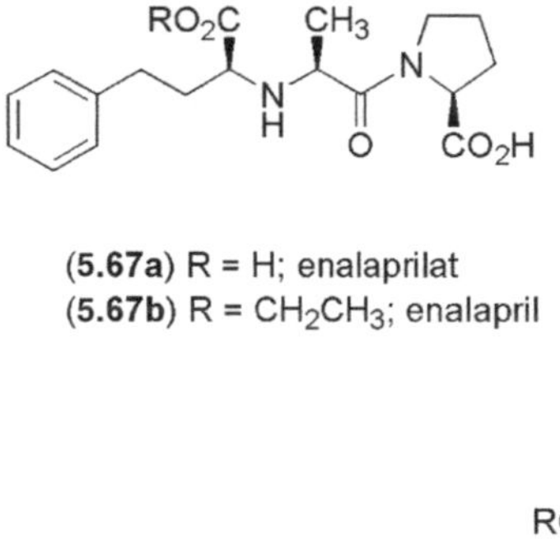

(**5.67a**) R = H; enalaprilat   (**5.68**) ramiprilat
(**5.67b**) R = CH₂CH₃; enalapril

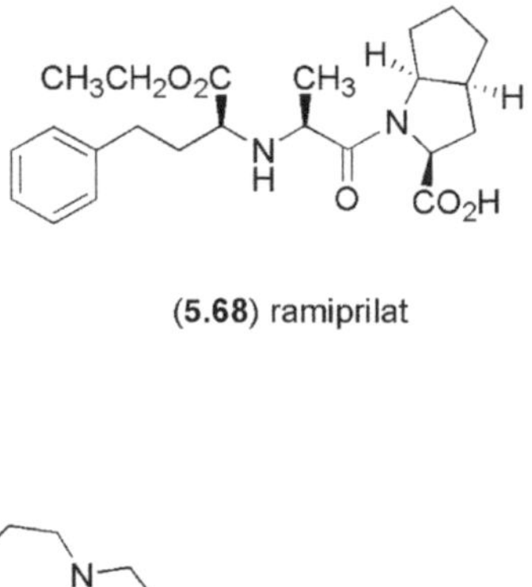

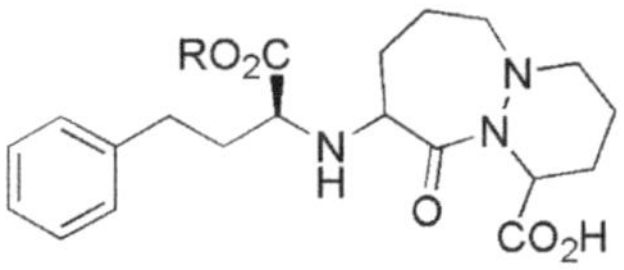

(**5.69a**) R = H; cilazaprilat
(**5.69b**) R = CH₂CH₃; cilazapril

Figure 5.3 shows the proposed manner in which the inhibitors fit into the active site of ACE. Other newly developed antihypertensives are quinaprilat, spiraprilat, lisinopril, benzeprilat, and perindoprilat.

More recently, following the x-ray crystallography of human testes ACE, the inhibitors have been designated as having N– or C–domain interactions or both with respect to substrate (angiotensin 1, C–), bradykinin (N– and C–) and the hemoregulatory hormone of AcSDKP, AcSDK (N–), opening up this field to the design of inhibitors as antiproliferative and antifibrotic agents.

5.4.2.1.2 Pyridoxal Phosphate–Dependent Enzymes

Many mechanism-based inactivators of pyridoxal phosphate–dependent enzymes are known, some of which were designed from a knowledge of the mechanism of action of their respective target enzymes. Inhibitors of AADC, histamine decarboxylase, ornithine decarboxylase, and GABA-transaminase, designed in this way, have proved to be useful drugs.

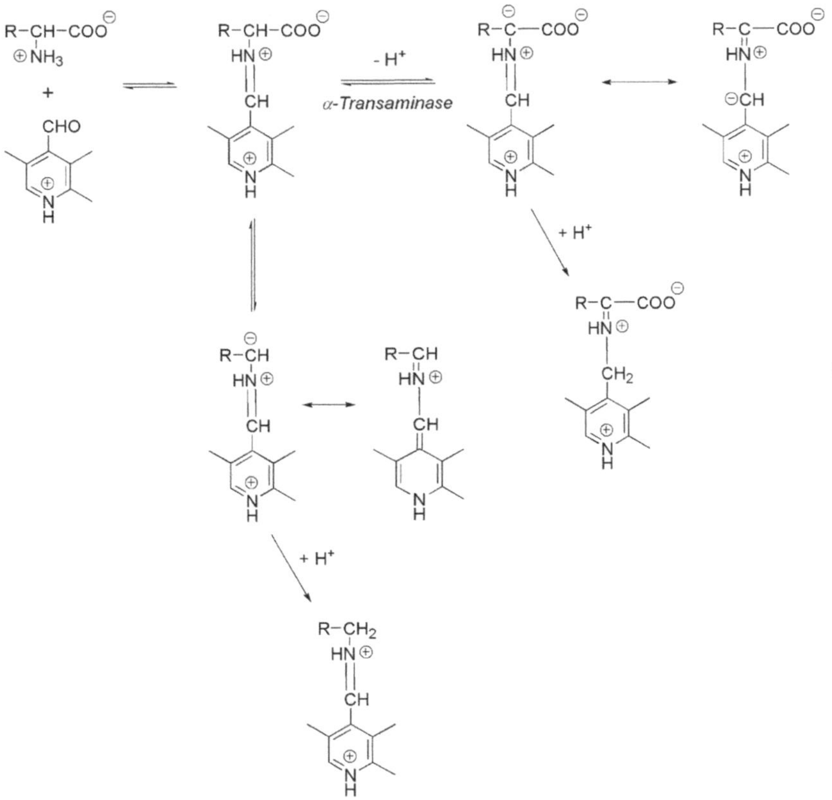

(5.23)

Using pyridoxal phosphate as coenzyme, these enzymes catalyze several types of reactions of amino acid substrates such as (1) transamination to the corresponding α-ketoacid, (2) racemization, (3) decarboxylation to an amine, (4) elimination of groups on the β- and γ-carbon atoms, and (5) oxidative deamination of ω-amino acids. The coenzyme is bound to the enzyme by the formation of an aldimine (Schiff base) with the ω-amino group of a lysine residue. The first step in the reaction with the amino acid substrate is an exchange reaction to form an aldimine with the α-amino group of the amino acid (see Equation 5.23). Either by hydrogen abstraction (transamination, racemization) or decarboxylation, a negative charge is developed

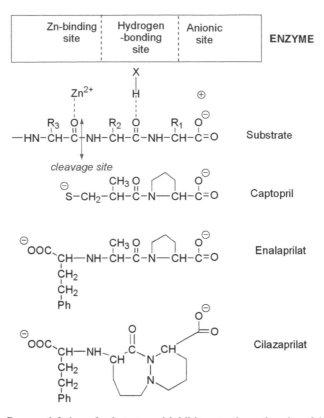

FIGURE 5.3 Proposed fitting of substrate and inhibitors to the active site of ACE.

on the α-carbon atom, and this is distributed over the whole conjugated cofactor system. Protonation then occurs on either the α-carbon atom (decarboxylation, racemization) or on the carbon atom adjacent to the pyridine ring (transamination) as shown in Equation 5.23. The direction of the fission that occurs is dictated by the nature of the protein at the active site so that a specific enzyme catalyzes a particular type of reaction. Inhibitors, as mechanism-based enzyme inactivators, act as substrates of the enzyme, but their structures vary from the natural substrate such that they either (1) divert the electron flux from the formed α-carbanion away from the coenzyme moiety or (2) using the normal electron flux give rise to reactive species. Both mechanisms lead to irreversible inhibition of the enzyme.

5.4.2.1.2.1   GABA Transaminase (GABA-T) Inhibitors
γ-Aminobutyric acid (GABA) is considered as the main inhibitory neurotransmitter in the mammalian central nervous system. There has been much interest recently in the design of inhibitors of the pyridoxal phosphate-dependent enzyme, α-ketoglut-arate-GABA transaminase. This enzyme governs the levels of GABA in the brain (see Equation 5.24). Inhibitors of the enzyme would allow a buildup of GABA that could be used as anticonvulsant drugs for the treatment of epilepsy.

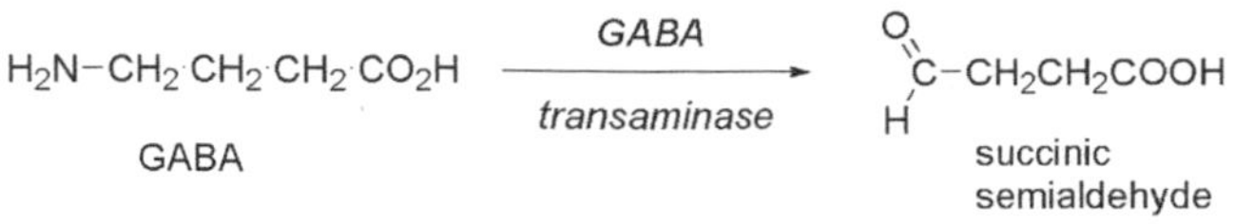

$$H_2N-CH_2\,CH_2\,CH_2\,CO_2H \xrightarrow[\text{transaminase}]{GABA} \overset{O}{\underset{H}{C}}-CH_2CH_2COOH$$

GABA   succinic
                                                            semialdehyde   (5.24)

γ-Acetylenic GABA (**5.70**) is a time-dependent inhibitor of GABA-T but also inhibits other pyridoxal phosphate-dependent enzymes. γ-Vinyl GABA (vigabatrin, **5.71**) acts in a similar manner through its (*S*)-enantiomer but has a more specific action. Vigabatrin (Sabril) has shown promise as a drug for the treatment of epilepsy. However, because about one third of patients using the drug have visual field defects (irreversible), there are restrictions on its prescribing, and new monitoring recommendations have been introduced.

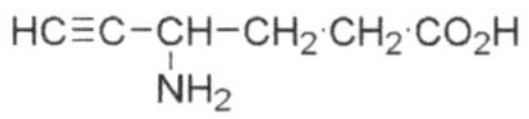

$$HC\equiv C-\underset{NH_2}{CH}-CH_2\,CH_2\,CO_2H \qquad H_2C=CH-\underset{NH_2}{CH}-CH_2\,CH_2\,CO_2H \qquad FH_2C\underset{NH_2}{\diagdown}CO_2H$$

(5.70) 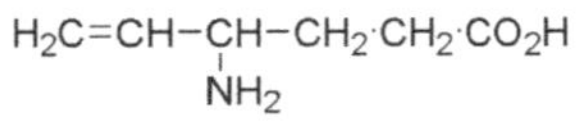   **(5.71)** vigabatrin 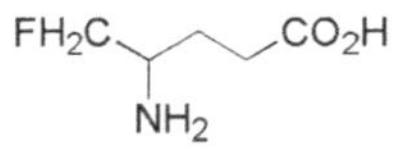   **(5.72)**

Halomethyl derivatives of GABA have also been described as inhibitors of GABA-T, e.g., the fluoromethyl derivative (**5.72**), in which the inactivation is accompanied by elimination of fluoride ion. However, *in vivo* (**5.72**) was almost 100-fold more active than vigabatrin but showed unexplained delayed toxicity after a single administration and was not further developed.

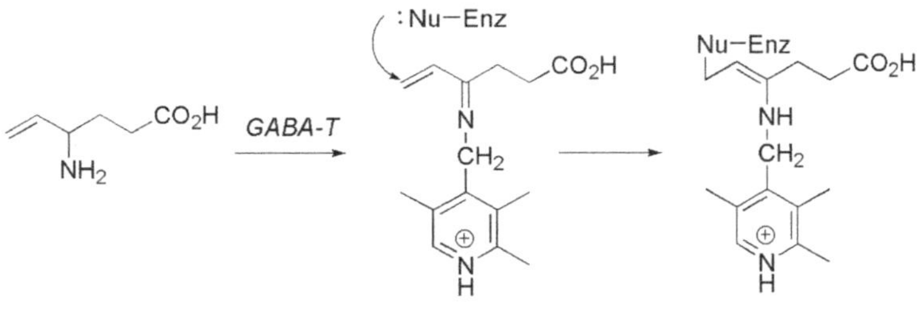

(5.73)   (5.25)

The initially postulated mechanism of action of the GABA-T inhibitors, based on GABA and bearing either an unsaturated function or a leaving group, was that they formed a Schiff-base with pyridoxal phosphate, followed by loss of the α-carbon proton; with the unsaturated derivatives the electron flow was towards the coenzyme moiety to give the vinylimine (**5.73**), whereas with the fluoromethyl derivatives the electron flow was away from the coenzyme and accompanied by loss of fluoride ion to give the enimine (**5.74**) (see Equation 5.25 and Equation 5.26).

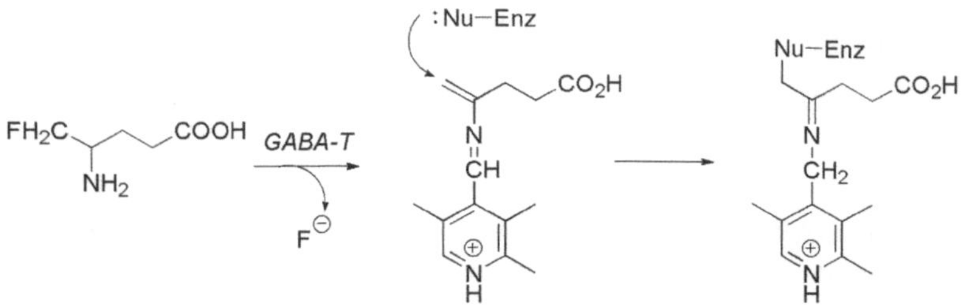

$$(5.26)$$

The electrophilic centers developed in the conjugated systems by the normal or abnormal electron flow react with a nucleophile at the active site of the enzyme. An alternative mechanism (see Equation 5.27) for the fluoromethyl derivatives is that the enimines dissociate from the pyridoxal phosphate to give an enamine that then recombines with the lysine of the active site. The electrophilic center of the enamine then attacks the cofactor to yield a stable complex at the active site, which leads to irreversible inhibition of the enzyme.

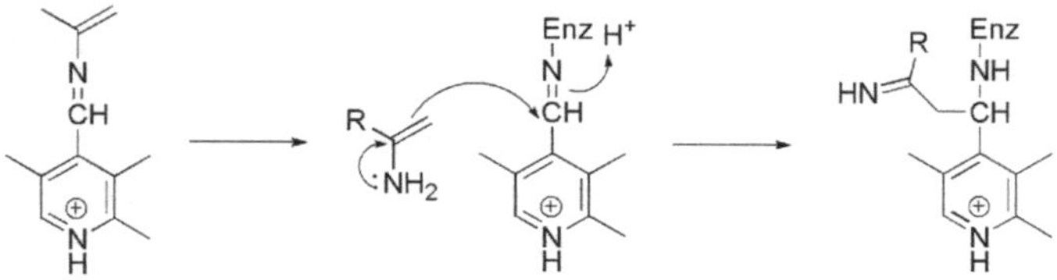

$$(5.27)$$

5.4.2.1.2.2 Ornithine Decarboxylase (ODC) Inhibitors

Naturally occurring polyamines such as putrescine, spermidine, and spermine are required for cellular growth and differentiation. Spermidine and spermine are derived in human-type cells from putrescine, which is synthesized by decarboxylation of ornithine catalyzed by the pyridoxal phosphate-dependent enzyme ornithine decarboxylase (ODC) (Equation 5.28). ODC has a very short biological half-life, and its synthesis is stimulated "on demand" by trophic agents and controlled by putrescine and spermidine levels. ODC has been considered a suitable target enzyme for the control of growth in tumors and diseases caused by parasitic protozoa.

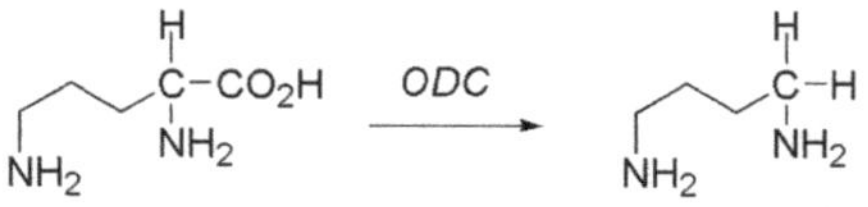

$$(5.28)$$

α-Difluoromethylornithine [eflornithine; (**5.75**)] is a mechanism-based inactivator of the enzyme, as is **5.76**, and irreversibly inhibits the enzyme by the general mechanism previously shown with elimination of a single fluoride ion to produce a conjugated electrophilic imine (c.f. **5.73**) that reacts with the nucleophilic thiol of

Cys-390. A further fluoride ion is then eliminated, which, following transaldimation with Lys-69, leads to the species Cys-390–S–CH=C(NH$_2$)–(CH$_2$)$_3$–NH$_2$ that loses ammonia and cyclizes to (2-(1-pyrroline)methyl) cysteine.

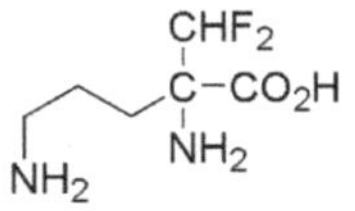

CHF$_2$
C–CO$_2$H
NH$_2$
NH$_2$

CHF$_2$
C–CO$_2$H
NH$_2$
NH$_2$

(5.75) eflornithine 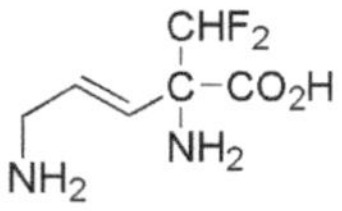 **(5.76)**

Eflornithine has low toxicity in animals and has shown antineoplastic and antiprotozoal actions in clinical trials especially as the methyl or ethyl esters that are effectively hydrolyzed at the higher cellular concentrations attained due to improved absorption. Eflornithine is used for the treatment of trypanosomiasis (sleeping sickness) as an alternative to the arsenic-containing drug melarsoprol with its dire side effects, but more recently it has been used as a topical prescription treatment (Vaniqua) for women with unwanted facial hair by slowing hair growth.

5.4.2.1.3 HIV Protease Inhibitors

Two genetically distinct subtypes, HIV-1 and HIV-2, of human immunodeficiency virus (HIV) have been identified. Reverse transcriptase (RT) inhibitors such as zidovudine (AZT) have had limited success because of emergence of viral resistance and drug toxicity. Blockade of the virally encoded protease, which is critical for viral replication, has become a major target in the search for an effective antiviral agent, and several inhibitors have been approved for use in patients in combination with an RT inhibitor (see Section 5.9).

Aspartate proteases, such as renin and HIV-protease, catalyze the hydrolysis of their substrates by aspartate-ion-catalyzed activation of water, the weak nucleophile, effectively to hydroxyl ion, the strong nucleophile. The hydroxyl ion attacks the carbonyl of the scissile amide bond in the substrate to give a tetrahedral intermediate that collapses to the products of the reaction (Equation 5.29).

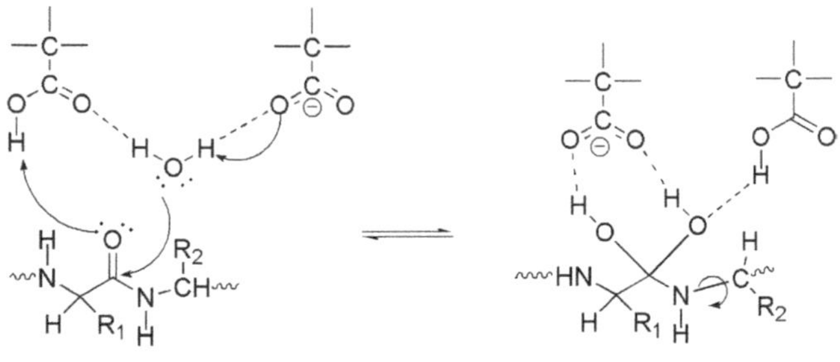

(5.29)

HIV-1 protease catalyzes the conversion of a polyprotein precursor (encoded by *gag* and *pol* genes) to mature proteins needed for the production of an infectious HIV particle and cleaves the polyprotein precursor at different sites, of which Tyr-Pro and Phe-Pro residues (occurring as P$_1$-P$_1'$ at three of the cleavage sites of HIV-

1) are of particular interest in relation to the development of inhibitors, other sites being Leu-Ala, Leu-Phe, Met-Met, and Phe-Leu. The amide bonds N-terminal to proline are not hydrolyzed by mammalian aspartic proteases and therefore offer selectivity for the viral enzyme.

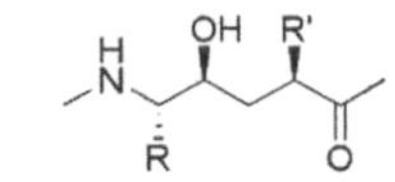

(5.77) hydroxyethylene isostere 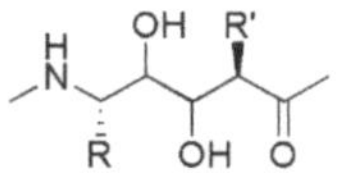   **(5.78)** dihydroxyethylene isostere 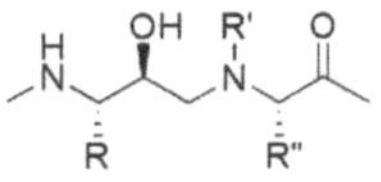   **(5.79)** hydroxyethylamine isostere

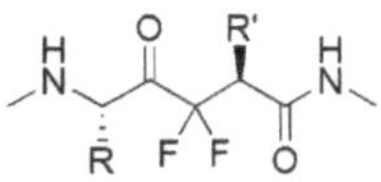

(5.80) difluoroketone 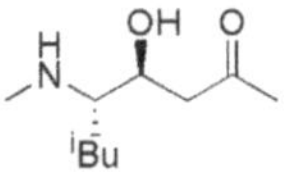   **(5.81)** statine 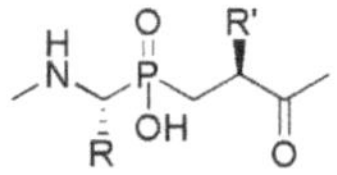   **(5.82)** phosphinate 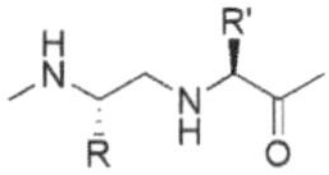   **(5.83)** reduced amide

Before the crystallography of enzyme and enzyme–inhibitor complexes was established, drugs that incorporated a transition-state mimic into substrate analogs were designed. The transition-state mimetics hydroxyethylene (**5.77**), dihydroxy-ethylene (**5.78**), and hydroxyethylamine (**5.79**) isosteres provide the greatest intrinsic affinity for the enzyme, followed by difluoroketones (**5.80**) = statine (**5.81**) > phosphinate (**5.82**) > reduced amide isostere (**5.83**). The principal structural feature in most transition-state analogs designed to inhibit HIV protease is the critical hydroxyl group that has subsequently been shown by x-ray analysis to bind both aspartic acid groups.

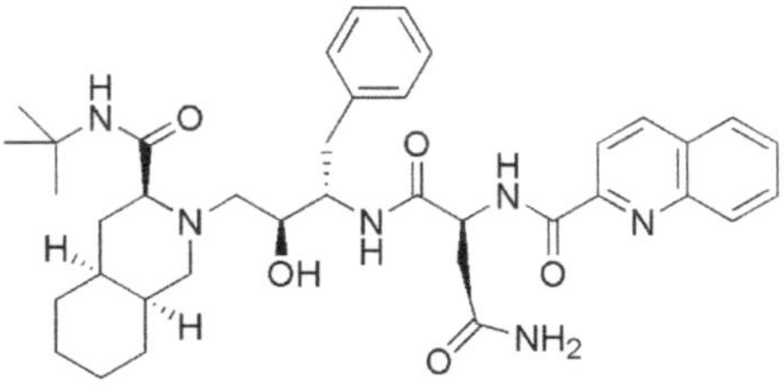

(5.84) saquinavir

The first clinically used inhibitor, designed using the substrate sequence 165-9 (Leu-Asp-Phe-Pro-ILe) for a particular polyprotein, was a stable tripeptide analog possessing a hydroxyethylamine moiety ($-CH(OH)-CH_2$) to resemble the tetrahedral intermediate ($-C(OH)_2-$) (see Equation 5.29). This compound saquinavir (**5.84**), has $IC_{50} = 0.4$ nM and is now in clinical use in drug combinations as an agent to prevent the spread of viral infection. Stable amino- and carboxy-terminal-blocking groups are present, and the hydrophobicity of the proline in the substrate has been increased in the perhydroisoquinoline residue. Other HIV protease inhibitors now in the clinic have been designed based on the amino acid sequence around a scissile bond of the polyprotein substrate and the structure of the tetrahedral intermediate. The chemical manipulations of the structure of the lead compounds discovered so as to produce the required characteristics of compounds for the clinic are described in detail in Section 5.9.

5.4.3 MOLECULAR MODELING

Computerized molecular modeling is now an essential part of the design process, but its relative importance in this process is determined by the state of knowledge concerning the target enzyme.

5.4.3.1 Crystal Structure of Enzyme or Enzyme–Inhibitor Complex Available

Ideally, a high-resolution crystal structure of the target enzyme with the active site identified by cocrystallization with an inhibitor provides a knowledge of binding sites on the inhibitor and enzyme and their relative disposition; from these parameters chemical libraries can be searched for suitable lead compounds of novel structure. Furthermore, an additional binding site may be identified on the enzyme so that a modified inhibitor using this additional site may be more potent or selective.

5.4.3.1.1 Practicalities

The ability in recent years to obtain the 3-D disposition of atoms in inhibitors, enzymes, and enzyme–inhibitor complexes by crystallographic techniques or of enzymes less efficiently by nuclear magnetic resonance (NMR) has led to a technique known as computer-assisted molecular modeling. Using suitable software, inhibitor structures lodged in the Cambridge Crystallographic Database at Daresbury, U.K., and enzyme and enzyme–inhibitor complexes from the Protein Data Bank (PDB) can be downloaded and the interaction of inhibitors with the active sites of enzymes studied (see Chapter 2 for examples).

If an inhibitor structure is not available in the database a 3-D model can be generated using a range of chemical drawing and *ab initio* software. The 3-D structure is subjected to manipulations to check bond lengths and bond angles and, where appropriate, with manual manipulation of groups or chains into positions that are usually expected by the operator. Finally, the model is minimized to obtain the lowest energy conformations. The inhibitor 3-D model is then "docked" into the active site of the enzyme structure obtained from the PDB. The inhibitor is initially docked manually and positioned to maximize favorable interaction with the active site amino acid residues; the enzyme–inhibitor complex is then minimized. Using a range of sophisticated docking software, all of which run on UNIX, techniques can be performed to optimize the theoretical enzyme–inhibitor complex and maximize the useful data obtained. One such technique is annealing, which vibrates the atoms (kinetic impulse) followed by a "cooling" period so that they reposition themselves in the complex as a low-energy-binding form.

Interaction in the complex between functions of the inhibitor and the enzyme, as well as further potential sites on the enzyme for a structurally modified inhibitor, can be noted. Such sites could include (1) a hydrophobic area composed of alkyl side chains of amino acid residues for binding similar areas introduced into the inhibitor although excess bulk would not give the required fit, (2) hydrogen-bonding donors such as –OH (threonine, serine, bound H_2O) and acceptors (C=O of peptide backbone, the imidazole nitrogen), (3) charged groups for electrostatic interaction (lysine and guanidine R_3NH^+; aspartate and glutamate CO_2^-) with groups of opposite charge introduced into the inhibitor.

Because the disposition of the binding functions of the inhibitor can be calculated on the basis of distances and angles, these parameters can be fed into the database or industrial chemical libraries using suitable software, and structures with similar disposition will be identified. This approach can lead to novel inhibitor lead structures being developed. An alternative *de novo* design approach is to join up the coordinates on the enzyme for the binding positions of a parent inhibitor using "filler" groups (rings, chains, atoms) to provide an appropriate carbon skeleton that again can generate novel inhibitor structures.

5.4.3.2 Prediction of 3-D Structure of Enzyme by Other Means

Usually, for a newly discovered target, the enzyme crystal structure is not known, and the 3-D structure of the protein has to be less satisfactorily predicted from either high-resolution NMR studies or by homology modeling from a related protein of known 3-D structure. For homology modeling, the sequence similarity between the two proteins should be at least 30%. Either of these techniques can lead to the identification of prospective binding sites at the enzyme active site and counter positions on a lead inhibitor by "docking" the inhibitor at the active site, using standard computer modeling software in the manner previously described. The design outcomes for this approach are the same as for those where crystallography is known (see Chapter 2 for examples).

In the absence of a model of the enzyme active site, modeling with a series of inhibitors by superimposition (matching) of key functional groups, similar areas of electrostatic potential, and common volumes may identify areas, i.e., the pharmacophore (that part of the molecule responsible for activity), with similar physical and electronic properties in the more active members of a series. Whereas this approach is suitable for rigid structures, it is less applicable to flexible molecules because the conformation (relative arrangement of atoms or groups in space) in solution may be different from that required to efficiently bind to the enzyme active site. Alternatively, the common conformational space available to a range of active inhibitors can be used to distinguish this from the space available to less active or inactive analogs that may lead to a defined model for the pharmacophore.

5.5 DEVELOPMENT OF A DRUG CANDIDATE FROM THE BENCH TO THE MARKETPLACE

The development of an inhibitor from its conception through to clinical trials and then on to the market is fraught with difficulties, and very few are successful. After satisfactory *in vitro* screening of a potent inhibitor for selectivity towards its target enzyme (i.e., little effect on related enzymes), *in vivo* studies in animals are undertaken to show that it has the required biological effect ("proof of concept") and establish that the candidate drug is well absorbed when administered orally (the usual route of administration), has a low rate of metabolism (long biological half life, $t_{1/2}$), and is free from toxic side effects. The *in vivo* studies present a formidable barrier to the development process, and many candidate drugs fail at this stage.

5.5.1 ORAL ABSORPTION

A drug needs to be absorbed through the gastrointestinal membrane and then carried by the plasma to its site of action. In the case of an inhibitor, the site of action is the target enzyme, and access may require further membrane penetration, e.g., cell membrane, bacterial cell, viral particle, cell nucleus, and blood–brain barrier. Successful passage through the body's membranes requires the correct balance between the hydrophilic and hydrophobic properties of the drug, that is, its water: oil solubility ratio or partition coefficient (distribution between water and lipid). Membranes are mainly lipid in nature, and if the drug is too hydrophobic, it will penetrate and remain in the membrane because it will have little tendency to pass through into the hydrophilic media of the plasma. Alternatively, if the drug is too hydrophilic, it will have little tendency to penetrate into the gastrointestinal membrane from the aqueous media of the gut; either situation amounts to poor drug absorption.

Many potent inhibitors of thrombin are known from *in vitro* studies, but few have the required hydrophilic: hydrophobic ratio to become useful clinical agents, and this feature has dogged development of antithrombotic agents for a long time (see Section 5.8 for a full account).

Aqueous solubility of a compound is increased by $-OH$, R_2NH, $-CONH-$, SO_2NH, R_3NH^+, and CO_2H groups, whereas lipid solubility is increased by alkyl chains and aryl or heteroaryl rings as well as halogen atoms, especially fluorine. Ionization of amine, carboxyl, or sulfonamide functions in a drug affect its absorption through lipid membranes (Equation 5.30, Equation 5.31, and Equation 5.32) because the unionized form of a drug is absorbed; this then equilibrates on the other side of the membrane with the ionized form.

$$R_2R_1-\overset{+}{N}H_3 \xrightleftharpoons{K_a} R_2R_1-NH_2 + \overset{+}{H} \tag{5.30}$$

$$RCOOH \xrightleftharpoons{K_a} RCOO^- + \overset{+}{H} \tag{5.31}$$

$$R_1SO_2NHR \xrightleftharpoons{K_a} RSO_2\overset{-}{N}R + \overset{+}{H} \tag{5.32}$$

Consequently, acid drugs, as the unionized CO_2H, may be absorbed from the acid environment of the stomach, and basic drugs (as unionized R^1R^2NH) from the small intestine, the pH of stomach and intestine being pH1 (acid) and pH6–7 (neutral), respectively.

The extent of ionization of a carboxylic acid in media of differing pH is given by the Henderson–Hasselbach equation:

$$pH = pK_a + \log_{10}[\text{Base A}^-]/[\text{Acid HA}] \qquad (5.33)$$

where pK_a is the $-\log_{10}$ of the ionization constant K_a, and HA and A$^-$ are the unionized and ionized species. In a similar manner, the pK_a of the conjugate acid ($R^1R^2NH_2^+$) is considered so that all pK_a values for acids and bases are on a single 0–14 pH scale.

The proportion of ionized and unionized species can be derived using Equation 5.33 for an acid or base at different pH values from their pK_a (which is the pH where HA = A$^-$ and R^1R^2NH = $R^1R^2NH_2^+$ = 50% ionization). Thus at physiological pH = 7, a base with pK_a = 8 (typical tertiary amine drug $R^1R^2R^3N$, such as morphine) will be ionized to the extent of 90% and exist as 10% free base in the intestine, whereas at pH 1 in the stomach, very little will exist as the free base. Similarly, a typical acidic drug such as the nonsteroidal antiinflammatory drug (NSAID) ibuprofen with pK_a = 4 will be unionized to the extent of 99% in the stomach but only 1% in the intestine. However, it can be adequately absorbed in the intestine even at this small concentration of unionized drug owing to the huge surface area of the absorbing villi making up the surface of the small intestine. Dicarboxylic acids are less likely to be absorbed.

Oral absorption of a drug may be improved by chemical manipulation to a biologically inert but more absorbable form of a drug, i.e., prodrug, which after absorption is converted by the body's enzymes to the active parent drug. This approach has proved particularly useful for drugs possessing a carboxylic acid group that, being in the ionized form at pH7, may not be well absorbed from the small intestine.

The ACE inhibitor enalaprilat (**5.67a**) (see Section 5.4.2.1.1) is well absorbed as its monoethyl ester, enalapril, and the enkephalinase A (MEP) inhibitor SCH 32615 (**5.85**), a dicarboxylic acid, is well absorbed as the acetonide of the glycerol ester, SCH 34826 (**5.86**).

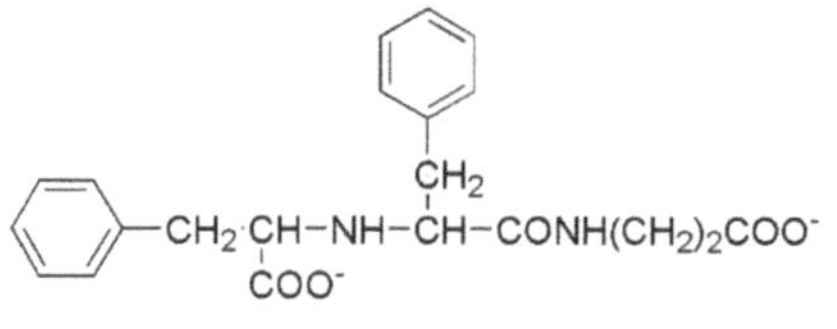

(5.85) SCH 32615

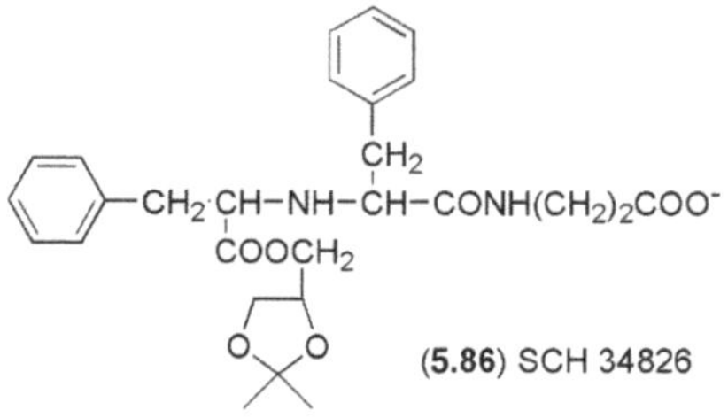

(5.86) SCH 34826

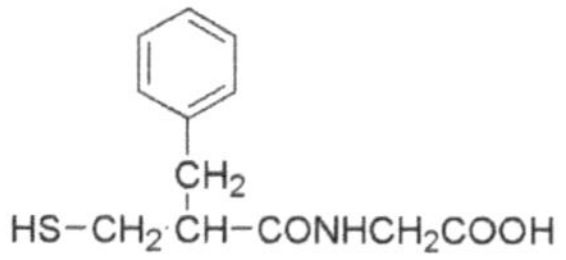

(5.87) thiorphan

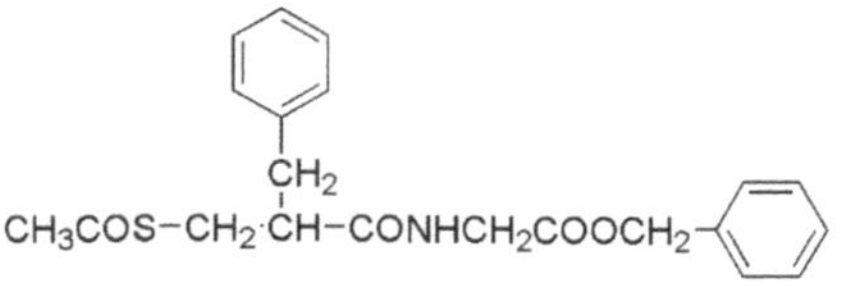

(5.88) acetorphan

The potent enkephalinase inhibitor thiorphan (**5.87**) is not active parenterally (by injection), but the protected prodrug acetorphan (**5.88**) is absorbed through the blood–brain barrier and subsequently converted by brain enzymes to the active drug. The absorption of the antiviral acyclovir (**5.89**), an inhibitor of HSV DNA polymerase, has been improved as the valine ester valaciclovir (**5.90**), and other analogs penciclovir (**5.91**) and famciclovir (**5.92**) are further improvements.

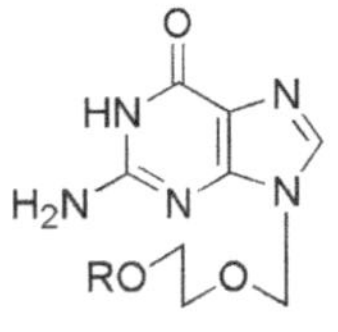

(**5.89**) acyclovir: R = H

(**5.90**) valaciclovir: R = $-\overset{O}{\overset{\|}{C}}-\overset{CH(CH_3)}{\overset{|}{CH}}-NH_2$

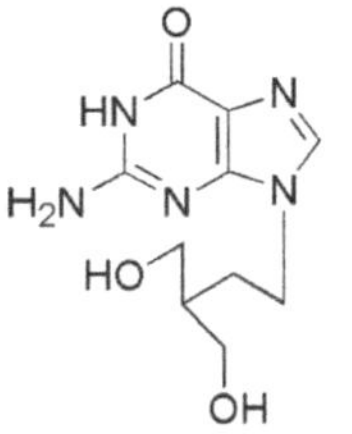

(**5.91**) penciclovir 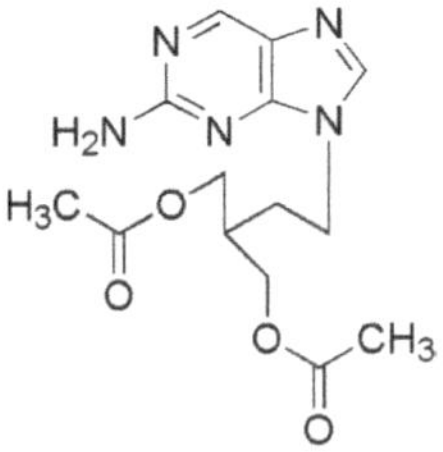   (**5.92**) famciclovir

Many penicillins are not absorbed efficiently when administered orally, and their lipophilic esters have been used to improve absorption. Simple aliphatic esters of penicillins are not active *in vivo,* and activated esters are necessary for the release of active penicillin from the inactive prodrug ester.

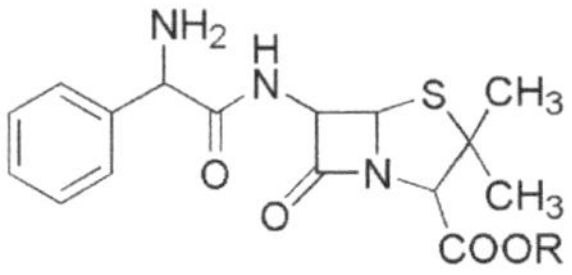

Penicillin	R
(**5.93**) ampicillin	H
(**5.94**) pivampicillin	$-CH_2O-\overset{O}{\overset{\|}{C}}-C(CH_3)_3$
(**5.95**) bacampicillin	$-\overset{CH_3}{\overset{\|}{CH}}-O-\overset{O}{\overset{\|}{C}}-O-CH_2CH_3$
(**5.96**) talampicillin	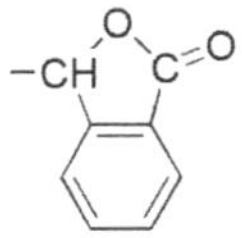

Ampicillin (**5.93**), a wide-spectrum antibiotic, is readily absorbed orally as the inactive prodrugs pivampicillin (**5.94**), bacampicillin (**5.95**), and talampicillin (**5.96**), which are then converted by enzymatic hydrolysis to ampicillin. Pivampicillin, the pivaloyloxymethyl ester, contains an acyloxymethyl function that is rapidly hydrolyzed by enzymes to the hydroxymethyl ester. This ester, being a hemiacetal of formaldehyde, spontaneously cleaves with the release of ampicillin and formaldehyde (Equation 5.34). In a similar manner, bacampicillin and talampicillin are cleaved and decompose to give ampicillin together with acetaldehyde and 2-carboxybenzaldehyde, respectively.

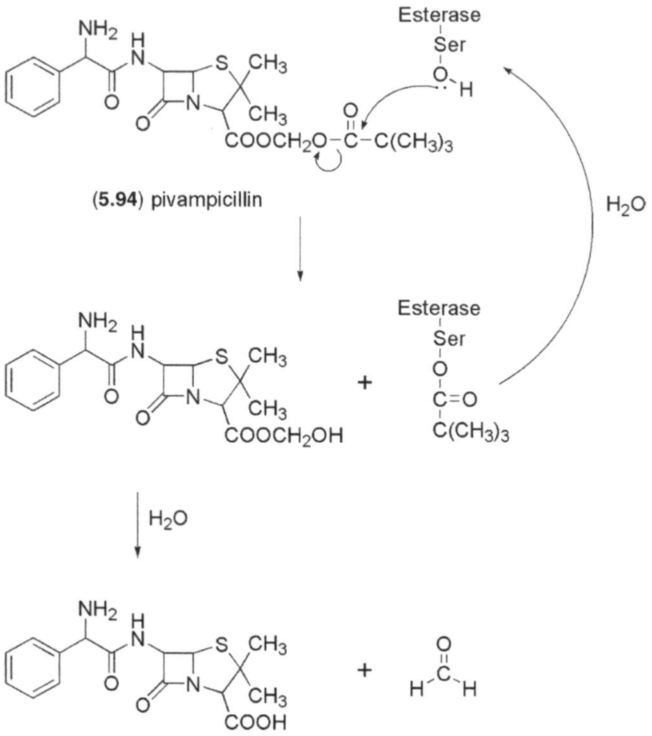

(5.94) pivampicillin

$$H_2O$$

$$H_2O$$

(5.34)

    Peptides as substrates of a peptide-degrading enzyme or agonists at a receptor site bind to the respective protein through a network of (>NH...0=C<) hydrogen bonds and consequently become hydrophilic molecules. Despite the introduction of terminal N and C hydrophobic residues or nonbinding (>N–CH$_3$) modifications, the oral administration of peptide-like enzyme inhibitors may lead to poor absorption due not only to the polar nature of the peptide backbone but also to degradation losses by intestinal proteases. Consequently, high potency with an IC$_{50}$ value in the low-nanomolar range is required for such drugs. Saquinavir (**5.84**), an HIV protease inhibitor, has a low oral absorption (c.2%), but this is offset by a low IC$_{50}$ of 0.4 nM.

    Large-molecule drugs with molecular weights greater than 350 Da, after absorption, may be excreted in the bile through the hepatobiliary system back into the intestine. This makes the plasma levels of the drug less predictable and a design aspect to be avoided.

5.5.2 METABOLISM

Most drugs, prior to removal from the body, undergo biotransformation (metabolism). The enzyme-catalyzed reactions leading to these changes are classified as Phase I (oxidation, reduction, and hydrolysis) and Phase II reactions (conjugation:

glucuronide or sulfate formation, methylation, acylation, and glutathione conjugation). Some Phase I reactions catalyzed by liver cytochrome P450s have been previously described (see Chapter 2). Modification of the original drug in these ways is an elimination mechanism because the metabolites formed usually have a greater hydrophilic nature and are excreted in the urine because they are not readily reabsorbed via renal tubules back into the system.

For a reversible inhibitor to be a useful drug, it must exist sufficiently long at the site of its target enzyme to exert its therapeutic effect. Because the level of the inhibitor at the site is a function of its plasma level, liver metabolism of the drug in the plasma to biologically inert products, followed by excretion, leads to progressive dissociation from the active site and, in time, reversal of the inhibition and its clinical effect. Plasma levels of drugs fall with time even with nonmetabolized drugs due to excretion, and drug metabolism enhances this fall. Irreversible inhibition of an enzyme once achieved is unaffected by an excess of the drug in the plasma and, thus, drug metabolism aspects; any increase in enzyme levels is due to chemical reactivation of the inhibited enzyme or its resynthesis by the body.

Metabolic processes other than prodrug activation lead to a shortening of the half-life ($t_{1/2}$) of the drug in the body. The t_f of an inhibitor in man is not directly related to that obtained from animal experiments, although it is usually longer than that observed in rats. A half-life of approximately 8 h for a drug is an acceptable value in man, although for cancer chemotherapy, a longer half-life of 12 to 36 h is required to provide adequate drug coverage in the event of patient noncompliance with the dose regimen.

5.5.2.1 Examples

Several aromatase inhibitors used for the treatment of breast cancer (see Section 5.6.1) are imidazoles (fadrozole, **5.97**) or triazoles [(+)-vorozole (**5.98**), letrozole (**5.99**), and anastrazole (**5.100**)]. In general, within this group of inhibitors, replacement of imidazole by triazole may lead to a decrease in *in vitro* potency, but *in vivo*, due to hydroxylation of the imidazole ring and the subsequently decreased potency, this is reversed as the triazole nucleus is more stable under metabolic hydroxylation.

(**5.97**) fadrozole   (**5.98**) vorozole   (**5.99**) letrozole   (**5.100**) anastrazole

In general, an electron-withdrawing substituent attached to a C–H function vulnerable to P450 oxidation will deter oxidation due to the increasing electropositivity of the hydrogen atom and the less likelihood of its bonding with the iron complex (see Equation 5.35).

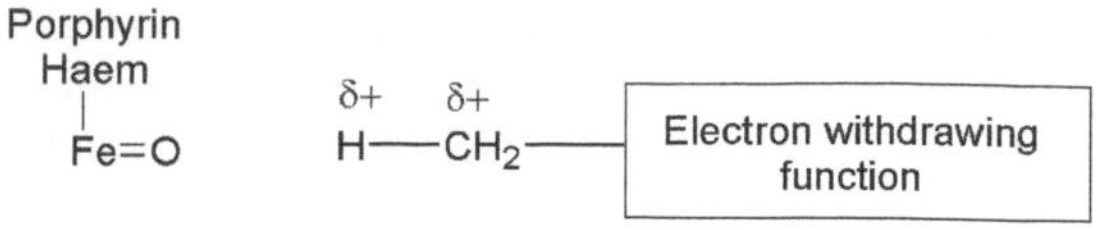

$$
\begin{array}{c}
\text{Porphyrin} \\
\text{Haem} \\
| \\
\text{Fe=O}
\end{array}
\qquad
\begin{array}{cc}
\delta + & \delta + \\
\text{H} & \text{CH}_2
\end{array}
\boxed{
\begin{array}{c}
\text{Electron withdrawing} \\
\text{function}
\end{array}
}
$$

(5.35)

This approach is also illustrated in the development of fluconazole (**5.104**). Ketoconazole (**5.101**), an antifungal agent, has a short $t_{1/2}$ and requires frequent dosing when administered orally at high dose for prostate cancer (since withdrawn because of liver complications). It is highly protein bound due to its lipophilic nature; hence less than 1% of the unbound form exists at the site of action. Modification led to UK-46,245 (**5.102**), which had twice the potency in a mouse candidosis model. Further manipulation to improve metabolic stability and decrease lipophilicity led to UK-47,265 (**5.103**) which, although 100 times more potent than ketoconazole on oral dosing, was hepatotoxic to mice and dogs and teratogenic to rats. Alteration of the aryl substituent to 2,4-difluorophenyl gave fluconazole (**5.104**) which, with >90% oral absorption and infrequent dosing ($t_{1/2}$ = 30 h), is used for the treatment of candida infections and as a broad-spectrum antifungal. The stability of fluconazole under metabolism could be attributed to the possession of the stable triazole nucleus (which is not hydroxylated, unlike imidazole), as well as to the protection of the $-CH_2-$ groups from hydroxylation by flanking electron-withdrawing groups (hydroxyl, triazole, and difluorophenyl).

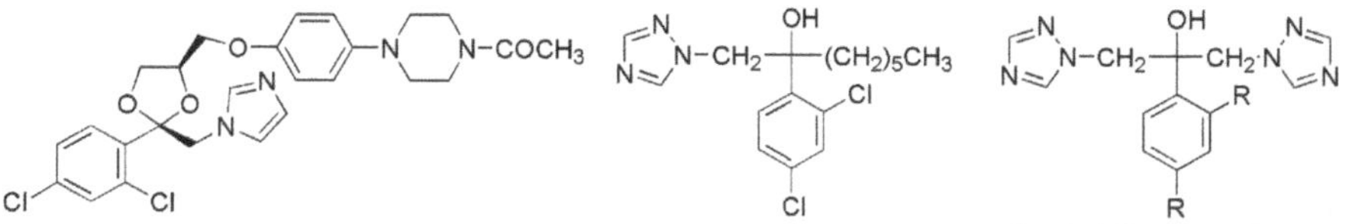

(**5.101**) ketoconazole   (**5.102**) UK-46,245   (**5.103**) R = Cl: UK-47,265
                                                             (**5.104**) R = F: fluconazole

Drug metabolism may be put to a beneficial use where, as described previously (see Section 5.5.1), a poorly absorbable active drug may be chemically manipulated to a readily absorbable inert prodrug in which the active drug is released on metabolism after passage through the gastrointestinal tract; esterases are the main class of enzymes for such conversions.

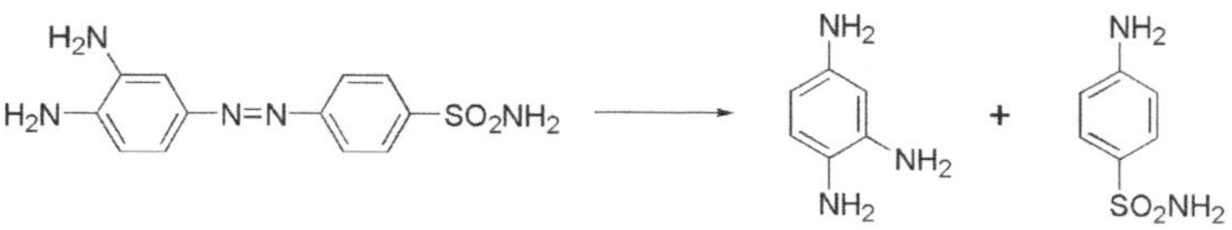

(**5.105**) prontosil   (**5.106**)
                                     p-aminobenzene sulfonamide   (5.36)

The development of the first antibacterial agents, the sulfonamides, stemmed from a discovery by Domagk in 1935 that the azo dye prontosil (**5.105**) has antibacterial activity, subsequently shown to be due to a metabolite, p-aminobenzene sulfonamide (**5.106**) (Equation 5.36). This directed research led to the subsequent development of a wide range of clinically superior sulfonamides through modification of the aminobenzene sulfonamide molecule.

5.5.3 Toxicity

Toxic effects may become apparent on chronic dosing during animal pharmacology studies and clinical trials, or even after marketing, leading in the latter situation to withdrawal of the drug with huge financial losses for the drug house concerned. The toxic side effects may merely be a matter of inconvenience or may be more severe.

5.5.3.1 Examples

A well-known example, but with a satisfactory outcome, is aminoglutethimide (**5.112**), which was introduced as an antiepileptic and subsequently withdrawn due to effects on steroidogenesis enzymes, leading to a "medical adrenalectomy." It was later reintroduced as an anticancer agent for the treatment of breast cancer. Its action on the last step of the steroidogenic pathway that forms estrogen (which stimulates tumor growth), governed by aromatase, capitalizes on this toxic effect. However, because it also much more weakly inhibits an enzyme involved in the first step of the pathway, the cholesterol side-chain cleavage enzyme, leading to a deficiency of adrenal corticol steroids, it is supplemented with hydrocortisone when used clinically.

Carbonic anhydrase consists of 14 isoenzyme forms with different subcellular localization and tissue distribution (see Section 2.6). They catalyze interconversion between carbon dioxide and a bicarbonate ion. As bicarbonate leaves the tissue with a sodium counterion and its water sheath, a diuretic (removal of water) effect is observed; this is useful in reducing the intraocular pressure observed in glaucoma. A number of systemic carbonic anhydrase inhibitors (orally administered) are available for this condition, e.g., acetazolamide (**5.18**), but because they are not selective to the isozymes concerned (CAII and IV), they could show side effects in other processes. These inhibitors are not suitable for topical application (eye drops) in which their actions would be localized due to nonpenetration of the general circulation. However, by adding lipophilic residues (tails) to existing systemic inhibitors, the hydrophilic/hydrophobic balance is altered in favor of lipophilicity, and the resulting agents can be administered topically as aqueous solutions because they enable better penetration of the lipid membranes of the cornea.

Ace inhibitors, e.g., captopril (**5.66**), are used as blood-pressure-reducing agents in hypertensive patients. Drug-induced dry cough affects about 40% of patients and could be due to the buildup of bradykinin, a substrate of ACE that increases NO generation through NO synthase with inflammatory effects on bronchial epithelial cells. The cough is said to be reduced by iron supplements ($FeSO_4$) because the activity of the NO synthase is reduced.

Many drugs, e.g., cimetidine, erythromycin, choramphenicol, isoniazid, verapamil, and enzyme inhibitors such as ketoconazole (**5.101**), are nonspecific inhibitors of liver cytochrome P450 enzymes, i.e., they inhibit many isoenzyme forms. Because cytochrome P450 enzymes metabolize many drugs and the dose given for effective

$$
\begin{array}{cc}
& CO_2H \\
H & \overset{|}{\underset{OH}{\diagup}}\cdots CH_3 \\
\end{array}
\qquad
\begin{array}{cc}
& CH_3 \\
H & \overset{|}{\underset{OH}{\diagup}}\cdots CO_2H \\
\end{array}
$$

FIGURE 5.4 Optical isomers of lactic acid.

drug plasma levels reflects this, they consequently affect the metabolism of other drugs given concurrently in a multidrug regimen, leading to enhanced levels of these drugs and the appearance of toxic effects (see Chapter 3). This interaction is particularly significant where the enhanced levels are for a drug (such as phenytoin) with a narrow therapeutic range (i.e., the range between the therapeutically effective dose and the toxic dose). Further, the situation is complicated by the fact that due to genetic variations in the normal adult population, there are "slow" and "fast" drug metabolizers and, so, the effect of one drug on the metabolism of another is not always predictable between patients (see Chapter 3).

Sildenafil (Viagra), a phosphodiesterase type 5 inhibitor, is metabolized by the cytochrome P450 enzymes CYP 3A4 and CYP 2C9, and should not be administered with drugs that inhibit CYP 3A4, such as erythromycin and HIV protease inhibitors (indinavir, nelfinavir, and saquinavir; rotonavir also inhibits CYP 2C9), which may increase sildenafil plasma levels.

5.5.4 STEREOCHEMISTRY

5.5.4.1 Optical Stereoisomerism

Optical isomerism is the most frequent situation that occurs in a compound when four different substituents are attached to the same carbon atom. The carbon center is referred to as chiral and asymmetric. The two different tetrahedral arrangements (configurations) of the substituents around the chiral center (Figure 5.4) lead to forms of a compound (enantiomers) that have identical physical and chemical properties (solubility, melting point and NMR, infrared spectra, etc.) but differ in their ability to rotate the plane of polarized light in a spectropolarimeter; one has a laevo (−) and the other a dextro (+) rotation. A compound produced by a standard synthetic technique is a racemate (±) which consists of 50% of each enantiomer and thus lacks a net rotation and is optically inert. The optical isomers of a racemate can be separated or synthesized by special methods.

The relative configuration between enantiomers of different compounds is made readily observable by the use of Fischer plane-projection formulae. Thus, the two optical isomers of lactic acid (**5.107** and **5.108**) may be distinguished as follows:

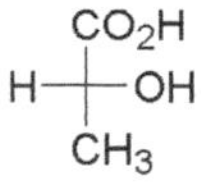

$$
\begin{array}{c}
CO_2H \\
H-\!\!\!+\!\!\!-OH \\
CH_3
\end{array}
\quad \text{and} \quad
\begin{array}{c}
CO_2H \\
HO-\!\!\!+\!\!\!-H \\
CH_3
\end{array}
\qquad
-\!\!\!\!+\!\!\!\!-
$$

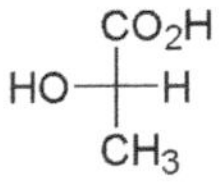

          (5.107)   **(5.108)**

The atoms to the right and left (−H and −OH) are attached to the central carbon by bonds that project towards the reader, whereas the bonds attaching the −CH₃ and

–COOH groups project below the plane of the paper (**5.108**). However, prior to modern techniques of x-ray crystallography, there was no means of telling which of the two structures represented the (+)- or (–)-enantiomorph. Nevertheless, their relative configurations may be correlated with arbitrarily assigned configurations for (+)- or (–)-glyceraldehyde, designated as D and L, respectively. This kind of correlation requires a particular enantiomorph to be chemically related, via a series of reactions precluding racemisation, to either D(+)- or L(–)-glyceraldehyde. Thus, (–)-lactic acid has been shown to have the D configuration (**5.109**) and is designated as D(–).

$$
\begin{array}{ccc}
\text{CHO} & \text{CHO} & \text{CO}_2\text{H} \\
\text{H}\!-\!\!\!-\!\text{OH} & \text{HO}\!-\!\!\!-\!\text{H} & \text{H}\!-\!\!\!-\!\text{OH} \\
\text{CH}_2\text{OH} & \text{CH}_2\text{OH} & \text{CH}_3
\end{array}
$$

D(+)-glyceraldehyde   L(-)-glyceraldehyde   (5.109)

X-ray crystallography later revealed the absolute configuration of salts of (+)-tartaric acid that had previously been related to L(–)-glyceraldehyde. It also revealed that D(+)-glyceraldehyde had the configuration that was previously arbitrarily assigned to it. This enabled the absolute configuration of molecules whose stereochemistry had been related to the glyceraldehyde system to be determined.

Proteins are formed by polymerization of L-amino acids and are related to L(–)-serine (**5.110**).

$$
\begin{array}{cc}
 & \text{CH}_2\text{NHCH}_3 \\
 & \text{H}\!-\!\!\!-\!\text{OH} \\
\text{CO}_2\text{H} & \\
\text{H}_2\text{N}\!-\!\!\!-\!\text{H} & \\
\text{CH}_2\text{OH} & \text{OH} \\
 & \text{OH}
\end{array}
$$

(5.110) L(-)-serine   (5.111) D(-)-adrenaline

Biological activity usually resides mainly in one enantiomer of a receptor agonist, antagonist, or enzyme inhibitor (see Section 5.5.4.1.1), the other enantiomer exhibiting low or no activity. The biological target (receptor, enzyme) has a 3-D-structure and, consequently can distinguish between prospective 3-D-ligand enantiomers. For example, D(–)-adrenaline (**5.111**) is 800 times more potent than the L(+)-isomer because the former can establish a three-point contact with its receptor involving (1) the basic group, (2) the benzene ring with its two phenolic groups, and (3) the alcoholic hydroxyl group. L(+)-adrenaline can only attain a two-point contact with relevant complementary groups (A, B, and C) on the receptor (see Figure 5.5).

The D and L convention in assigning absolute configurations presents difficulties when there is more than one center of asymmetry in the compound, and to overcome these difficulties, Cahn, Ingold, and Prelog evolved their Sequence Rule. This requires the four substituents around each chiral center to be considered in the order of descending atomic numbers of the connecting atoms — a, b, c, and d (Figure

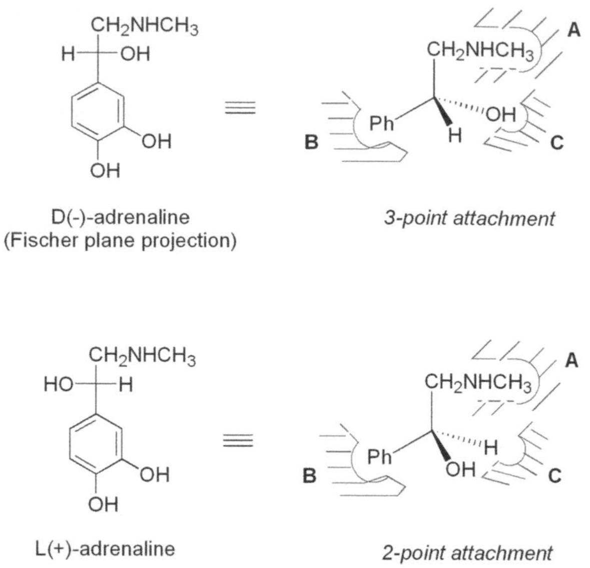

FIGURE 5.5 Stereoselective binding of D(−)-adrenaline.

5.6). The asymmetric center is then viewed from the side opposite to that of the substituent with the lowest priority, d. If the sequence of the remaining three substituents in descending order of priority appears clockwise, then the configuration is designated as R (rectus). Conversely, if the substituents in descending order of priority appear anticlockwise in sequence, the configuration is referred to as S (from "sinister," meaning "left" in Latin). Where the connecting atoms of two or more substituents are identical, priority is given to the substituent whose second atom has the highest atomic number. For D(−)-adrenaline, $OH > CH_2N > C–C=C$ (3Cs) > H and it has an R-configuration (Figure 5.6).

The stereochemistry of enzyme inhibitors possessing chiral centers is usually important in determining their potency toward a specific enzyme. This is a problem to be addressed in the early stages of drug design because it can sometimes be avoided by limiting the design to achiral (nonchiral) compounds.

Drug registration authorities worldwide have moved toward a requirement that for all new drugs, the most active enantiomeric form must be marketed unless the activity of the separate enantiomers is available for the racemate, and enantioselective methods of chemical and biological analysis have been used in both animal and human studies. These requirements take into account the pharmacological consequences of the use of racemic drugs as inhibitors or receptor antagonists, which are conveniently summarized elsewhere (Hutt, 1998).

Whereas the literature abounds in examples of activity residing mainly in one enantiomer of an inhibitor following in vitro studies, relatively few of these compounds have, as yet, reached the clinic or been subjected to registration requirements because in vivo information is not available from animal studies.

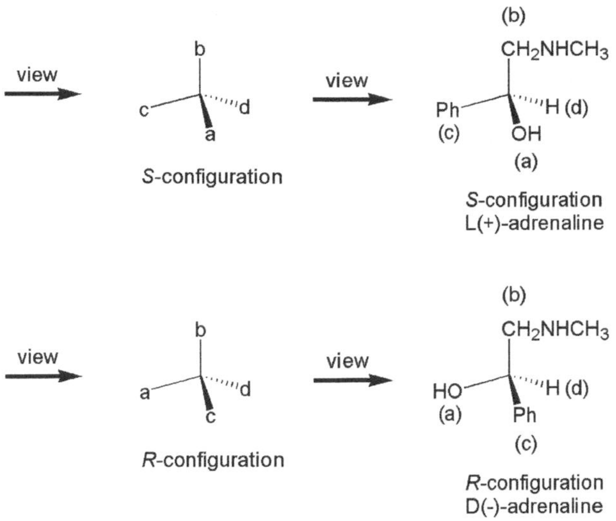

FIGURE 5.6 (*R*)- and (*S*)-configuration of the enantiomers of adrenaline.

5.5.4.1.1 Examples

Aminoglutethimide (AG) (**5.112**), a long-established aromatase inhibitor, is used clinically (after surgery) as the racemate in the treatment of breast cancer in post-menopausal women to decrease their tumor estrogen levels and to prevent the growth of metastases (from the spread of clumps of tumor cells) in other parts of the body. The (+) (*R*)-form is about 38 times more potent as an inhibitor of aromatase than the (−) (*S*)-form. AG is also a weaker inhibitor of the side-chain cleavage enzyme (CSCC), which converts cholesterol to pregnenolone in the adrenal steroidogenic pathway, accounting for its side effect of depleting corticosteroids. Depletion of corticosteroids in this manner requires adjuvant hydrocortisone administration with the drug. Here the (+) (*R*)-form is about 2.5 times more potent than the (−) (*S*)-form.

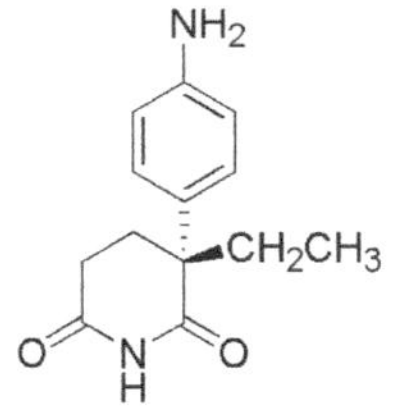

(5.112) (*R*)-aminoglutethimide 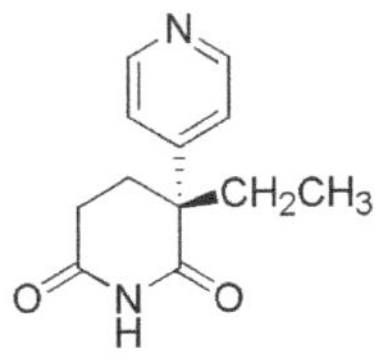   **(5.113)** (*R*)-pyridoglutethimide

For pyridoglutethimide (rogletimide) (**5.113**), an analog of AG, in which the phenylamine has been converted to pyridine (without the undesirable depressant effect), the inhibitor potency resides mainly in the (+) (*R*)-form (20 times that of the (−) (*S*)-form). 1-Alkylation improves potency *in vitro* but the activity for the most potent inhibitor in the series, the 1-octyl, resides in the (−) (*S*)-form due to a change in the mode of binding of the inhibitor to the enzyme.

A more selective inhibitor of aromatase than AG is the triazole vorozole (**5.98**), which is about 1000 times more potent as an inhibitor. The (+) (*S*)-form is 32 times more active than the (−) (*R*)-form, but the very small inhibitory activity of the racemate toward the other steroidogenic pathway enzymes, 11β-hydroxylase and 17,20-lyase, originates in the (−)- and (+)- forms, respectively.

It is of interest that in the benzofuranyl methyl imidazoles (**5.114**), some of which are 1000 times more potent as aromatase inhibitors in the racemic form than AG, comparable activity lies in both enantiomers. Construction of the aromatase active site by homology modeling and docking of inhibitors at the site showed that the two aryl ring structures fit equally well into the site for either stereoisomer.

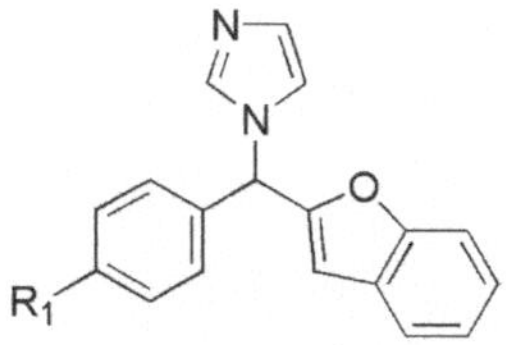

(**5.114**)

MAO occurs in two forms, MAO-A and MAO-B. The use of MAO unselective inhibitors as antidepressants is complicated by a dangerous hypertensive reaction, termed the "cheese-effect," with tyramine-containing foods (cheese, chianti); this reaction is due to the inhibition of MAO-A located in the gastrointestinal tract, which would otherwise remove the tyramine. L-(−) deprenyl (selegiline) (**5.24**), a selective inhibitor of MAO-B and, thus not associated with the cheese-effect, is widely employed to limit dopamine breakdown in Parkinson's disease with a selective inhibitory dosage. The (−)-isomer is much more potent than the (+)-isomer, and because the products of metabolism are (−)-metamphetamine and (+)-metamphetamine, respectively, the more potent stimulatory (+)-metamphetamine side effects are removed from the racemate by the use of L-(−) deprenyl.

γ-Aminobutyric acid (GABA) transaminase inhibitors allow a buildup of the inhibitory neurotransmitter GABA and are potential drugs in the treatment of epilepsy. The inhibitory action of γ-vinyl GABA [vigabatrin (**5.71**)], a drug now restricted to the treatment of this disease, resides mainly in the (*S*)-enantiomer [see Section 5.4.2.1 (2)].

The HIV protease inhibitor saquinavir (**5.84**), used clinically in combination with other inhibitors directed at different targets in the virus multiplication machinery to decrease resistance to the infection, has an (*R*)-configuration for the hydroxyl group. This is important for binding because activity lies in the form (*R*)-enantiomer $IC_{50} = 0.4$ n*M*, (*S*)-enantiomer = > 100 n*M*.

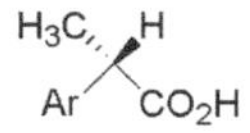

(**5.115**)

$(CH_3)_2CHCH_2$—[aryl ring]—$\overset{CH_3}{\underset{CO_2H}{\overset{|}{C}}}$H

(5.116) (S)-ibuprofen

H_3CO—[naphthalene ring]—$\overset{CH_3}{\underset{CO_2H}{\overset{|}{C}}}$H

(5.117) (S)-naproxen

[diphenyl ether]—$\overset{CH_3}{\underset{}{\overset{|}{C}}}$$CO_2H$

(5.118) fenoprofen

The 2-arylpropionic acids **(5.115)** are a group of nonsteroidal antiinflammatory drugs (NSAIDs), including ibuprofen **(5.116)**, naproxen **(5.117)**, and fenoprofen **(5.118)**, which exert their effect by inhibition of the cycloxygenase (COX) responsible for prostaglandin synthesis. Their activity mainly resides in the (S)-form, but they are marketed, with a few exceptions, as the racemates. The large differences in *in vitro* activity between the isomers decrease, and many disappear *in vivo* due to chiral inversion in mammals of the inactive (R)-forms to the active (S)-forms. Consequently, there would be no gain and much expense in marketing the resolved active (S)-forms of these drugs.

5.5.5 DRUG RESISTANCE

A major setback to the clinical use of an established inhibitor for bacterial, parasitic, and viral diseases is that the inhibitor becomes less effective against its target with the passage of time, a phenomenon known as *drug resistance*. Replacement of one inhibitor with another aimed at the same target, once resistance has occurred, does not lead to improvement; this is known as "cross-resistance" to drugs. To ensure that resistance resulting from the development of resistant forms is minimized, patient compliance in taking the antibiotic therapy for the full scheduled time is essential. Drug resistance may occur in many ways, some of which undoubtedly complement one another:

- The inhibitor sensitive step in a pathway may be bypassed by duplication of the target enzyme in parasitic diseases, the second version being less sensitive to drug action (e.g., resistance to methicillin, trimethoprim, and sulfonamides).
- In viral and parasitic diseases, development of mutants occurs under drug pressure when an amino acid residue of the natural (wild type) enzyme is changed in the transcription process. Suboptimal drug therapy selects for mutants that have a growth advantage over the wild (natural) type. Subsequent transmission of resistant variants to uninfected individuals may lead to infections that are drug resistant from the outset and require a new structural type of inhibitor for their suppression. Related drugs with a similar action and structure will show cross-resistance in that none will be superior in tackling the developed resistance. Resistance to drugs targeted at the reverse transcriptase (RT) of HIV-1 is due to mutational changes in the enzyme caused by careless transcription, leading to failure, with time, of a drug to clear the virus. In effect, the binding surfaces on the enzyme in the mutant no longer efficiently match those of the inhibitor, leading to less efficient drug binding and to escape of the viral process

from inhibition. The nonnucleoside RT inhibitors (NNRTIs) nevirapine (**5.119**) and efavirenz (**5.120**) develop mutants that are resistant, with point changes in the enzyme–amino acid sequence being observed at 103 (lysine → asparagine), 101 (lysine → glutamic acid), 188 (tyrosine → leucine), and 190 (glycine → serine) in different mutants. In patients failing therapy with these drugs, cross-resistance to all available NNRTIs followed. Combination with other drugs that attack different targets is an approach to overcoming resistance to individual drugs, i.e., an RT inhibitor with an HIV aspartate proteinase. Resistance to HIV aspartate proteinase inhibitors due to mutation is discussed in detail in Section 5.9.

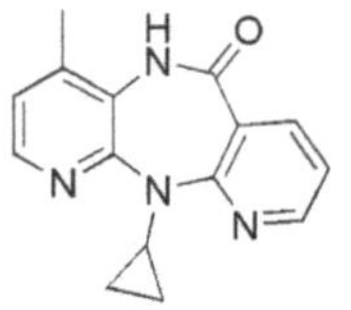

(5.119) nevirapine 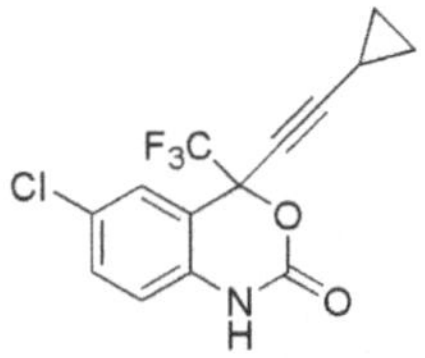   **(5.120)** efavirenz

- General phenomena leading to drug resistance are overproduction of target enzyme (because of which higher inhibitor concentrations are needed to inhibit growth, e.g., resistance to trimethoprim), as well as development of pathways that bypass the inhibited process, and reduced uptake of the drug or increased efflux (pushing out of the cell) of the drug by altering the number of transmembrane pumps in the cell membrane.

FURTHER READING

Acharya, K.R., Sturrock, E.D., Riordan, J.F., and Ehlers, M.R.W. (2003). Ace revisited: a new target for structure-based drug design. *Nature Reviews (Drug Discovery)*, 2, 891–902.

Ala, P.J. and Chang, C.-H. (2002). HIV aspartate proteinase: resistance to inhibitors. In *Proteinease and Peptidase Inhibition: Recent Potential Targets for Drug Development*, H.J. Smith and C. Simons, Eds., pp. 367–382. London: Taylor & Francis.

Basuray, I. (2003). Neutral peptidase inhibitors: new drugs for heart failure. *Indian Journal of Pharmacology*, 35, 139–145.

Davies, R.H. and Williams, H. (1988). Structural modifications in drug design. In *Introduction to the Principles of Drug Design*, H.J. Smith and H. Williams, Eds., 2nd ed., pp. 82–95. London: Wright.

Edwards, P.D. (2002). Human neutrophil elastase inhibitors. In *Proteinease and Peptidase Inhibition: Recent Potential Targets for Drug Development*, H.J. Smith and C. Simons, Eds., pp. 154–177. London: Taylor & Francis.

(a) Frick, L. and Wolfenden, R. (1989). Substrate and transition-state analogue INHIB. (b) Shaw, E. (1989). Active-site-directed irreversible inhibitors. (c) Tipton, K. (1989). Mechanism-based inhibitors. In *Design of Enzyme Inhibitors as Drugs*, M. Sandler and H.J. Smith, Eds., Vol. 1, (a) pp. 19–48; (b) pp. 49–69; (c) pp. 70–93. Oxford: Oxford University Press.

Hooper, N.M. (2002). Zinc metallopeptidases. In *Proteinase and Peptidase Inhibition: Recent Potential Targets for Drug Development*, H.J. Smith and C. Simons, Eds., pp. 352–366. London: Taylor & Francis.

Hutt, A.J. (1988). Drug chirality and its pharmacological consequences. In *Introduction to the Principles of Drug Design and Action*, H.J. Smith, Eds., 3rd ed., pp. 51–166. London: Harwood Academic.

Luscombe, D.K., Tucker, M., Pepper, C.M., Nicholls, P.J., Sandler, M., and Smith, H.J. (1994). Enzyme inhibitors as drugs: from design to the clinic. In *Design of Enzyme inhibitors as Drugs*, M. Sandler and H.J. Smith, Eds., Vol. 2, (a) pp. 42–64; (b) pp. 1–41. Oxford: Oxford University Press.

Silverman, R.C. (1988). *Mechanism-based Enzyme Inactivation: Chemistry and Enzymology*, Vol. 1, pp. 3–23. Boca Raton: CRC Press.

Smith, H.J., Nicholls, P.J., Simons, C., and Le Lain, R. (2001). Inhibitors of steroidogenesis as agents for the treatment of hormone-dependent cancers. *Expert Opinion on Therapeutic Patents*, 11, 789–824.

The Royal College of Ophthalmology. (2001). Annual Report. The ocular side-effects of vigabatrin (Sabril). Information and guidelines for screening.

Vanden Bossche, H. (1992). Inhibitors of P450-dependent steroid biosynthesis: from research to medical treatment. *Journal of Steroid Biochemistry and Molecular Biology*, 43, 1003–1021.

5.6 ENZYME INHIBITOR EXAMPLES FOR THE TREATMENT OF BREAST CANCER

L.W. Lawrence Woo

5.6.1 INTRODUCTION

Breast carcinoma is the most common form of female cancer. Each year, there are about 38,000 new cases of breast cancer in the U.K., and the lifetime risk of breast cancer in women is one in nine. Many breast tumors are initially hormone dependent, with estrogens playing a pivotal role in supporting the growth and development of such tumors. The highest incidence of breast cancer occurs in postmenopausal women at a time when their ovaries have become nonfunctional and estrogens are synthesized exclusively in peripheral tissues such as the adipose tissue, brain, hair follicles, and muscle. In women who are suffering from hormone-dependent breast cancer (HDBC), it has been established that estrogens are produced not only in normal but also in malignant breast tissue. These estrogens are no longer an endocrine factor but act locally at the sites of production in a paracrine or even intracrine manner. This, in effect, ensures that the tumor will develop in an environment where only small amounts of estrogen are produced.

After breast cancer has been diagnosed, the primary treatment for patients is invariably the removal of the lumps or lesions by surgery, with or without adjuvant radiotherapy. This is then often followed by drug intervention that targets the micrometastases that could be present even in the very early stages of the disease. The choice of such follow-up treatment after surgery is much dependent on the hormone dependency of the tumors. If the tumor tissues express the estrogen receptor

(ER+) and progesterone receptor (PR+), they are deemed to be sensitive to estrogens, and endocrine therapy will be the first-line treatment for the patient.

Like most carcinomas, breast cancer becomes life threatening when metastases have taken place and colonies of tumor cells have been established in the vital organs of the body. Although breast cancer in an advanced state is incurable, it remains sensitive to drug treatment. Both endocrine therapy and chemotherapy are shown to be effective in controlling the progression of the disease, and in improving the patient's quality of life and chances of survival.

5.6.2 ENDOCRINE THERAPY

The principle of endocrine therapy is to block the effects of estrogen on cell proliferation. The strategy involves one or more of the following: (1) blocking the hormone receptor itself, (2) suppressing estrogen production, and (3) destroying the ovaries. Although progestins (e.g., medroxyprogesterone acetate), estrogens (e.g., diethylstilbestrol), and gonadotrophin-releasing hormone agonists (e.g., buserelin) have a role in the management of breast cancer, the most widely used forms of chemical endocrine therapy for postmenopausal women with HDBC include, primarily, an early nonselective antiestrogen tamoxifen, and, more recently, selective estrogen receptor modulators (SERMs) and aromatase inhibitors. SERMs are drugs with mixed agonist/antagonist action on ERs in different tissues. They are designed to block the binding of estradiol to the ERs in the tumor tissue of the breast and endometrium, and hence to block the transcriptional activation of genes required for cellular proliferation without interfering with the desirable agonist effect of estrogen on the bone and in the lowering of serum LDL cholesterol levels. Aromatase inhibitors inhibit the biosynthesis of estrogens and hence decrease the amount of estrogen available for binding to the receptors.

5.6.3 AROMATASE

The aromatase complex consists of two proteins, aromatase (EC 1.14.14.1, cytochrome P450 19, estrogen synthetase, $P450_{arom}$, CYP 19) and NADPH-cytochrome P450 reductase. The latter is common to most cell types and functions to donate electrons to the cytochrome P450. Aromatase, a member of the cytochrome P450 superfamily of enzymes, is responsible for the final step of the biosynthesis of estrogen. The heme protein of the aromatase enzyme catalyzes the conversion of C19 androgens, androstenedione and testosterone, to C18 phenolic estrogens, estrone and estradiol, respectively (Figure 5.7 and Figure 5.8), in three main steps, consuming 3 mol of oxygen and 3 mol of NADPH.

In the first step (Figure 5.7), molecular oxygen is bound by the reduced heme iron to form a peroxide. Upon splitting of the oxygen–oxygen bond with the formation of a molecule of water, the resulting iron–oxo intermediate abstracts a proton from the angular methyl group, leading to the first hydroxylation at C19. There is evidence to suggest that two amino acid residues (aspartic acid 309 and glutamic acid 302) in the enzyme active site close to C19 of the substrate catalyze the first step of the reaction. The second step involves another hydroxylation at C19

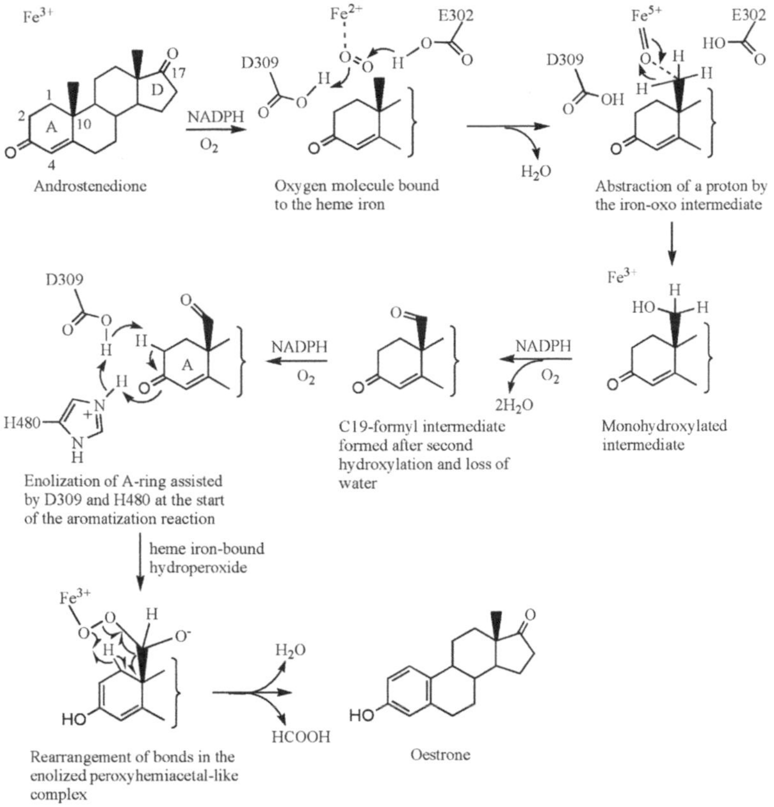

FIGURE 5.7 Proposed mechanism for the conversion of androstenedione to estrone by human aromatase and the possible role of amino acid residues: E302, glutamic acid 302; D309, aspartic acid 309; and H480, histidine 480. (Adapted from Kao, Y.C., Korzekwa, K.R., Laughton, C.A. et al. (2001). Evaluation of the mechanism of aromatase cytochrome P450. A site-directed mutagenesis study. *European Journal of Biochemistry*, 268, 243–251.)

that proceeds in a similar manner to the first to yield a *gem*-diol. This unstable structure then collapses nonenzymatically to form a 19-formyl androgen with a stereospecific loss of the C19 pro-*R* hydrogen and water. The mechanism for the final aromatization step is still debatable. One proposed theory involves an attack of the electrophilic 19-formyl group by the heme-iron-bound peroxide to form an enzyme-bound peroxyhemiacetal-like intermediate. It is thought that an abstraction of a hydrogen atom from C2 by the neighboring aspartic acid 309 occurs simultaneously, which facilitates the enolization of the C3-keto group catalyzed by histidine 480 in the vicinity. Rearrangement of bonds in this enolized peroxyhemiacetal-like complex then follows, with the introduction of a double bond between C1 and C10 and the formation of formic acid to give phenolic estrone as product. The estrone produced is then reduced by the enzyme 17β-hydroxysteroid dehydrogenase Type 1 (17β-HSD I) to 17β-estradiol, which is the most biologically active estrogen of natural origin.

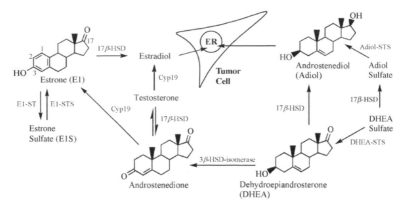

FIGURE 5.8 The origin of estrogenic steroids in postmenopausal women: CYP 19, aromatase; ST, sulfotransferase; STS, sulfatase; 17β-HSD, 17β-hydroxysteroid dehydrogenase; 3β-HSD-isomerase, 3β-hydroxysteroid dehydrogenase Δ^5, Δ^4-isomerase; ER, estrogen receptor. (Adapted from Woo, L.W.L., Purohit, A., Malini, B. et al. (2000). Potent active site-directed inhibition of steroid sulfatase by tricyclic coumarin-based sulfamates. *Chemistry & Biology*, 7, 773–791.)

5.6.4 Inhibition of Aromatase as Endocrine Therapy

Aromatization is the final and rate-limiting step in the cascade of estrogen biosynthesis. Any interference in this unique process should not affect the production of other steroids further up the cascade. For this reason, aromatase is a desirable target for inhibition. Once the HDBC cells are deprived of estrogenic stimulation, they are effectively being starved, and their growth and development are impeded, leading to regression of the disease.

5.6.4.1 Nonsteroidal Aromatase Inhibitors (NSAIs)

The design and development of aromatase inhibitors for the treatment of hormone-dependent cancers began with the discovery of the aromatase inhibitory activity of aminoglutethimide (AG). AG is a simple chemical derivative of the sedative glutethimide. It was originally introduced as an anticonvulsant, but was found to cause adrenal insufficiency. This unwanted property was due to the ability of AG to block adrenal steroidogenesis by inhibiting the enzymatic conversion of cholesterol to pregnenolone. More interestingly, this agent also blocked the peripheral conversion (aromatization) of androgenic precursors to estrogens, which is the source of estrogen in postmenopausal women. AG binds to aromatase, producing a Type II optical difference spectrum with an absorption peak occurring between 425 and 435 nm and a trough at 390 nm showing competitive inhibition. The commercial product is a mixture of D- and L-isomers, and they have different pharmacological potencies. The D-isomer is 30 times more potent at inhibiting aromatase activity, whereas the L-isomer is more potent at inhibiting cholesterol side-chain cleavage (steroidogenesis).

The discovery of the aromatase inhibitory property of AG in the 1970s led to massive research programs in the following decades that aimed at the design and development of more potent and selective NSAIs. The initial design strategies adopted by several research groups focused on modifying AG structurally, and the modifications included: (1) replacing the *p*-aminophenyl group with a pyridyl ring, (2) contracting the piperidinedione to a five-membered pyrrolidinedione ring, (3) bridging C3 and C5 of the AG to give a bicyclo derivative, and (4) replacing the C3 ethyl group or substituting at the *N*-atom with longer alkyl chains or both. Most of these modifications produced inhibitors that were more potent against and selective for aromatase.

With aromatase being a cytochrome P450 enzyme, it had been reasoned that the reversible inhibition observed for AG could involve interaction of its nitrogen atom with the heme iron of the enzyme, in accordance with its Type II spectrum as a competitive inhibitor. This reasoning prompted investigation into compounds that contain aza-heterocycles and led to the discovery that a group of imidazole antifungal agents (miconazole, clotrimazole, and ketoconazole) were better aromatase inhibitors than AG. The research efforts that followed produced a series of second-generation NSAIs with *S*-fadrozole (**5.121**), liarozole (**5.122**), and CGS18320B (**5.123**) showing the most promising biological activities. The replacement of the imidazole ring with a triazole also produced compounds that were highly effective aromatase inhibitors. Substituted triazoles such as vorozole (**5.124**), letrozole (**5.125**), and anastrozole (**5.126**) were prime examples of third-generation NSAIs.

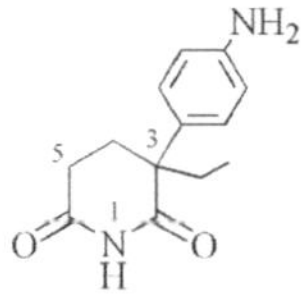

(AG) 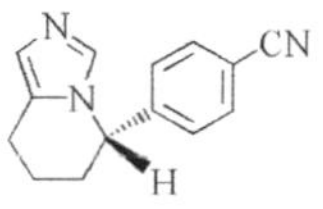   (5.121) 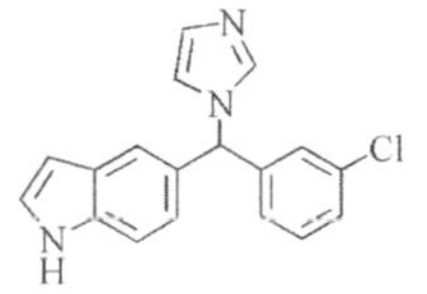   (5.122) 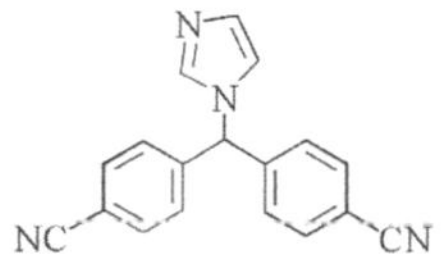   (5.123)

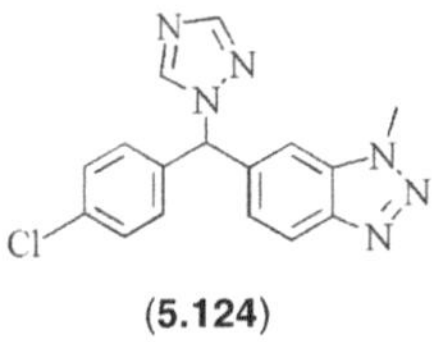

(5.124) 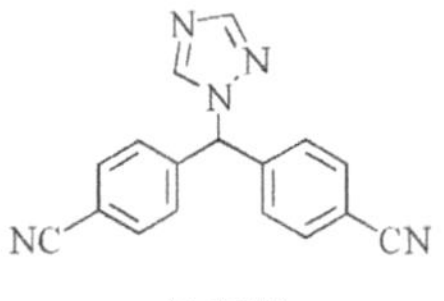   (5.125) 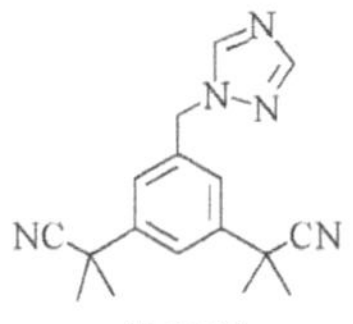   (5.126)

However, not every potent NSAI discovered has successfully reached the clinical development stage. For the second generation of inhibitors, only fadrozole is used clinically, primarily in Japan where a license has been granted for the treatment of postmenopausal women with advanced, hormone-responsive mammary carcinoma. Fadrozole has been shown to be less effective than the triazole derivatives, and may cause a degree of inhibition of cortisol and aldosterone production at higher doses than normally used clinically.

Being chiral, vorozole (Rivizor®) was developed as its (+)-*R* isomer. This agent was found to be more potent than anastrozole and letrozole *in vitro,* but this apparent superiority was not evident *in vivo.* Vorozole was withdrawn from clinical development in the U.S. Phase III trials several years ago, and it is uncertain if this agent will be developed further.

The most successful and highly developed third generation of NSAIs are letrozole and anastrozole, which are highly potent and selective against the aromatase enzyme. Letrozole (Femara®) and anastrozole (Arimidex®) are currently available in the U.K. for the treatment of advanced breast cancer in postmenopausal women, and both agents are the gold standard NSAIs in the clinics. Recent Phase III trials of letrozole or anastrozole vs. tamoxifen as the first-line therapy for advanced breast cancer in postmenopausal women have shown significant advantages of either NSAI over tamoxifen in prolonging the survival of treated patients. The results of a large head-to-head, multicenter, and randomized trial of letrozole vs. anastrozole as the second-line treatment in postmenopausal women with advanced breast cancer were presented at the 2002 annual meeting of the American Society of Clinical Oncology, which showed an overall response rate of 19.1% for women taking letrozole at a dose of 2.5 mg/d compared with 12.3% for those taking anastrozole at a dose of 1 mg/d. Clinical benefit was seen in 27% of the women treated with letrozole compared with 23% with anastrazole. This preliminary data of a single trial suggests that postmenopausal women with advanced HDBC might respond better to letrozole.

Despite the success in the clinical development of several third-generation NSAIs, new compounds of considerable structural diversity are still being reported in scientific journals as novel inhibitors of aromatase. Like their predecessors, these fourth-generation NSAIs are Type II competitive inhibitors and possess the pharmacophore for aromatase inhibition, which is a heme-ligating nitrogen heterocycle with substituents that explore favorable interactions with the amino acid residues lining the enzyme active site. It is beyond the scope of this chapter to discuss in detail all the new NSAIs reported to date. Those compounds of particular significance and interest (**5.127** to **5.143**) are depicted here.

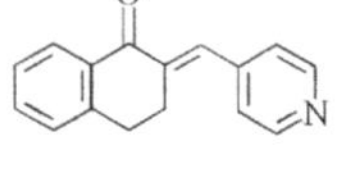

(**5.127**) 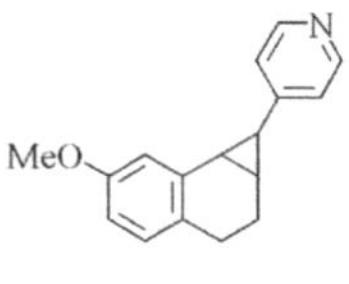 (**5.128**) 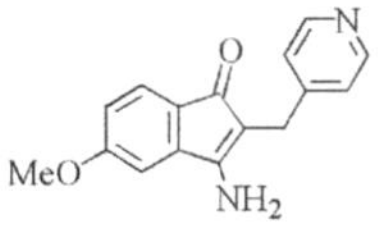 (**5.129**) 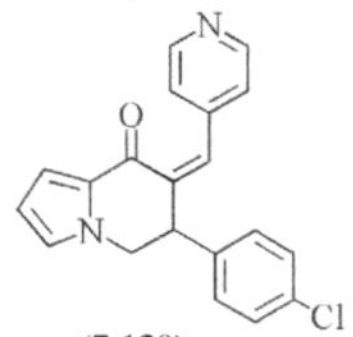 (**5.130**)

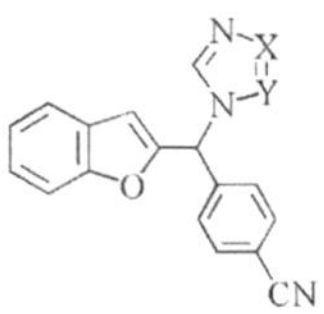

(**5.131**) X=CH, Y=N
(**5.132**) X=N, Y=CH

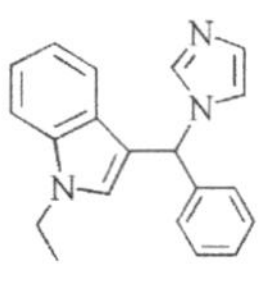

(**5.133**) 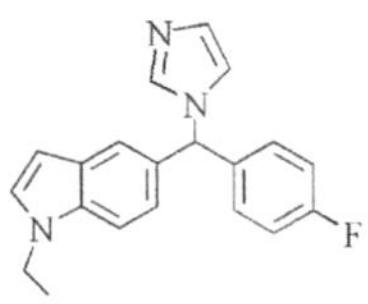 (**5.134**) 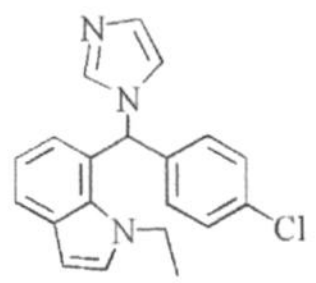 (**5.135**)

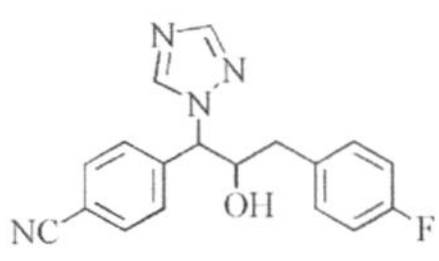

(5.136) 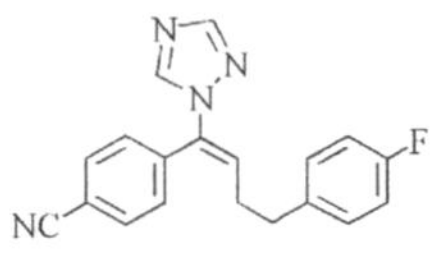   **(5.137)**

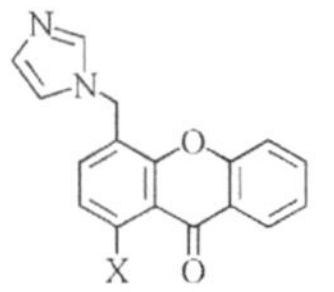

Xanthone derivative

(5.138) X=CN
(5.139) X=NO$_2$
(5.140) X=Br

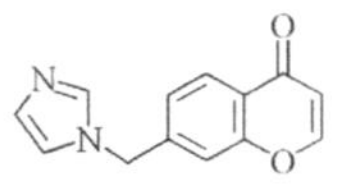

Chromone derivative

(5.141)

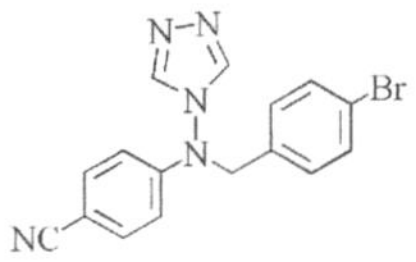

(5.142) 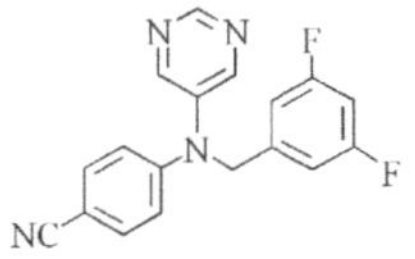   **(5.143)** 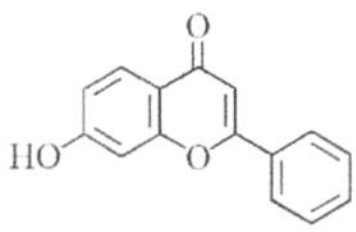   **(5.144)** 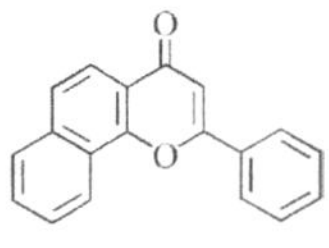   **(5.145)**

Phytoestrogens of natural origin such as 7-hydroxyflavone (**5.144**) and α-naph-thoflavone (**5.145**) are moderate to good inhibitors of aromatase. Because these agents do not contain a nitrogen heterocycle, their inhibitory activities have been attributed to the heme-ligating potential of their carbonyl moiety.

The most advanced and evaluated fourth-generation NSAIs were compounds **5.142** and **5.143**, developed by Yamanouchi Pharmaceuticals in Japan. Both compounds are tertiary amines possessing three different substituents (i.e., a p-cyanophenyl group, a halogenated benzyl group, and a heterocyclic group) on a central N-atom, thus avoiding the presence of a tetrahedral center and chirality complications for drug registration purposes. The designing out of chirality is of particular pharmaceutical significance because it has been demonstrated that enantiomers of NSAIs might exhibit different biological activities — both in terms of potency and selectivity — against the aromatase enzyme (e.g., AG). Compounds (**5.142**) and (**5.143**) were among the most active NSAIs reported to date with respective IC$_{50}$ values of 0.12 nM and 0.04 nM. Both agents were shown to display a high selectivity against the aromatase enzyme and to inhibit insignificantly other enzymes that are responsible for steroidogenesis. However, despite their promising biological profiles, Yamanouchi Pharmaceuticals has recently discontinued the clinical development of this class of compounds, probably as a result of competition and the market dominance of letrozole and anastrozole.

5.6.4.2 Steroidal Aromatase Inhibitors

An alternative strategy for designing aromatase inhibitors is to use the natural substrate of the enzyme as the initial template and to modify it structurally to produce derivatives that impede the binding of androstenedione to the enzyme active site. By definition, steroidal inhibitors should be selective against the aromatase enzyme, although they themselves and their metabolites might have undesirable endocrinological effects.

Steroidal aromatase inhibitors can be classified into three main types: (1) reversible (competitive), (2) tight binding, and (3) enzyme generated (mechanism-based inactivators and suicide substrates).

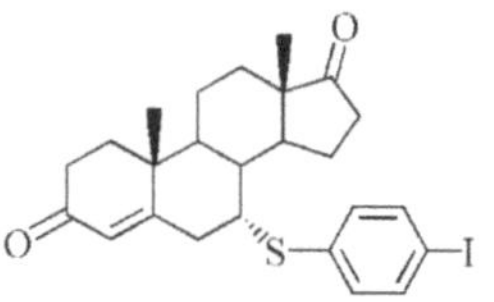

(5.146) 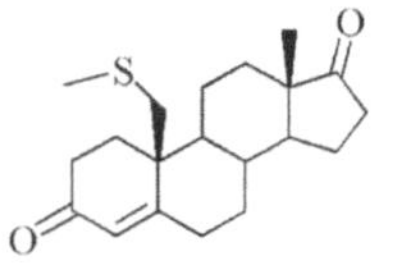   (5.147) 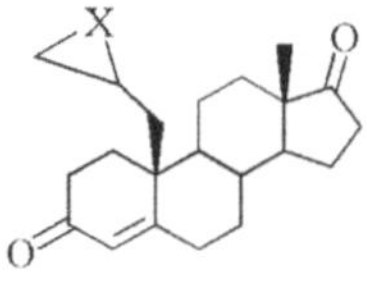   (5.148) X=O
                     (5.149) X=S

Reversible inhibitors are compounds that exclude the access of substrate to the catalytic site of the enzyme by direct competition. Such agents are effective as therapeutic agents as long as appropriate concentrations can be maintained. 7α-[(4′-Iodophenyl)thio]androstenedione (**5.146**) was a competitive inhibitor of aromatase with a K_i value of 12 nM (cf K_m value for androstenedione is 30 nM). The inhibition observed for **5.146** might be due to its p-iodophenylthio moiety interacting with a hydrophobic pocket of the enzyme that has been suggested to be available in the vicinity of C7 of androstenedione. Unlike AG, **5.146** did not interact with the heme iron. For this reason, the changes in the UV-absorption spectrum of the enzyme-bound heme upon binding of **5.146** resembled that of androstenedione, which is characterized by a shift in the maximum from about 420 to about 390 nm (Type I spectrum).

Tight-binding compounds are inhibitors of aromatase that display very high affinity for the target enzyme. They are, by definition, reversible inhibitors, but they can produce stoichiometric pseudo-irreversible inhibition because of their slow rate of dissociation from the enzyme. Examples of this type of inhibitor include the 10β-MeSCH$_2$– (**5.147**), (19R)-10β-oxirane (**5.148**), and the (19R)-10β-thiirane (**5.149**) derivatives of androstenedione, whose K_i values are 1, 7, and 1 nM, respectively. These compounds, on binding to aromatase, showed UV-absorption spectra similar to that observed for AG, suggesting that their potent inhibiton of aromatase was due to coordination of their 10β-heteroatom to the heme iron through lone-pair electrons, in addition to the binding of their steroidal skeleton to the enzyme active site.

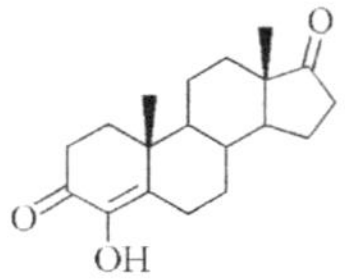

(5.150) 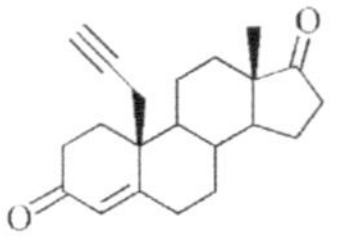   (5.151) 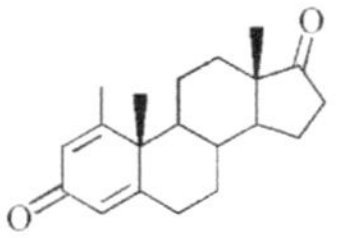   (5.152) 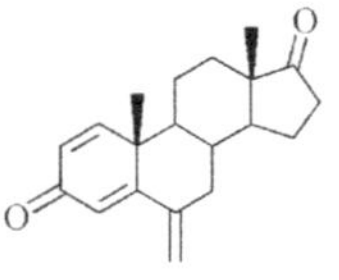   (5.153) 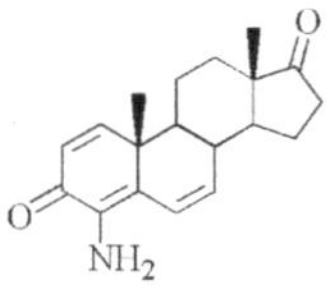   (5.154)

Enzyme-generated inhibitors are those agents that bind in a latent form to the target enzyme, and are then activated by the enzyme's usual catalytic process to form reactive species that attack the active site covalently prior to their dissociation. This type of compound is also described as a suicide inhibitor or mechanism-based inactivator. Because the inhibition is irreversible, these agents, in principle, should have long-lasting effects *in vivo* until new enzyme is synthesized, and their administration to patients can be reduced, thus minimizing potential drug-related side effects. A number of enzyme-generated aromatase inhibitors have been developed over the years, and the more prominent ones were: 4-hydroxyandrostenedione (4-OHA, formestane, Lentaron®, **5.150**), 10β-(2-propynyl)estr-4-ene-3,17-dione (**5.151**), 1-methylandrosta-1,4-diene-3,17-dione (**5.152**), 6-methylen-androsta-1,4-diene-3,17-dione (**5.153**, exemestane, Aromasin®), and 4-aminoandrosta-1,4,6-triene-3,17-dione (**5.154**). Formestane and exemestane were the two most developed steroidal aromatase inhibitors, and are now in clinical use in the U.K. for the treatment of advanced breast cancer in postmenopausal women. Formestane is administered by deep intramuscular injection every 2 weeks. The response rate in the third- and fourth-line treatment settings in metastatic disease was 22%, with a median response duration of 10 months. Exemestane, the newest steroidal aromatase inhibitor to be developed, is orally active and is given as a tablet on a daily basis. The overall response rate in the third-line treatment setting in metastatic disease has been shown to be similar to that of formestane. However, it is expected that exemestane will be more widely prescribed than formestane because of its oral bioavailability and hence better patient compliance.

In recent years, there has been a decline in the volume of papers published on the design and discovery of new steroidal aromatase inhibitors compared to that on NSAIs. This, in part, might be due to the remarkable success of NSAIs in the clinic. In addition, the potential endocrinological side effects and the poor drug delivery of steroidal compounds via the oral route, in general, might also be contributory factors.

5.6.4.3 Computer-Aided Drug Design of Aromatase Inhibitors

Molecular modeling and computational techniques have essentially become an integral part of modern drug design and discovery programs. Although computers with powerful computing power are still needed for providing some of the state-of-the-art computational support, modeling software suites are increasingly being developed for PCs, rendering them more affordable and accessible to the medicinal-chemical community. The ability to rationalize the interactions between a ligand and a biological target such as the active site of an enzyme, and to predict structures that will interact favorably with the biological target, has revolutionized the whole approach to lead identification and optimization in the drug design and discovery process. With the advent of genomics and proteomics, it is anticipated that the *in silico* approach will be pivotal to the identification of reasonable lead candidates for the prosecution of a lead optimization program with medicinal chemistry.

The impact of computational support on the design of aromatase inhibitors has, to some extent, been limited by the lack of a crystal structure of the human enzyme. Because it is a membrane-bound protein, so far it has not been successfully isolated and crystallized. At a time when no reliable protein structure of human aromatase was available, the initial attempts to model the binding mode of aromatase inhibitors to the enzyme active site relied on the overlaying of ligands onto the natural substrate of the enzyme, i.e., androstenedione. For NSAIs, the heme-ligating moiety of the inhibitors was first positioned in the region below the heme iron in close proximity to the angular C19-methyl group of androstenedione. The rest of the inhibitor molecule was then modeled around the substrate for potential geometric and electronic anchoring points, with much emphasis on overlaying hydrogen-bond-accepting substituents of the inhibitor onto the C3- and C17-carbonyl groups of androstenedione. However, when the x-ray structures of bacterial enzymes P450cam, P450bm3, and P450terp were published, the aromatase enzyme-inhibitor interactions were studied and predicted on the basis of a plausible model of the protein 3-D structure. Using a variety of techniques, including aligning the amino acid sequence of the human aromatase active site with those of bacterial origin because of a high degree of homology, secondary structure prediction, molecular mechanics, and molecular dynamic stimulations, several 3-D homology models of human aromatase have been constructed. These computer-generated structures of human aromatase in conjunction with site-directed mutagenesis studies, which identify the essential amino acid residues of the protein for biological activity, have provided a qualitative picture of how aromatase inhibitors may interact with the enzyme active site.

In recent years, attempts have been made to rationalize the structure–activity relationships of azole NSAIs in terms of their 3-D properties. By using the 3D-QSAR (Quantitative Structure–Activity Relationship) technique called CoMFA (Comparative Molecular Field Analysis), analogs of S-fadrozole (**5.121**) and pyridyl tetralone (**5.127**) were aligned onto S-fadrozole, the reference template, in accordance with the pharmacophoric hypothesis for the inhibitor. The model developed relates the steric and electrostatic characteristics of inhibitors to their inhibitory activities, and provides a quantitative description of favorable and unfavorable regions around the inhibitors studied. Compounds **5.138** to **5.141** were indeed designed as a result of a CoMFA model. The chromone and xanthone templates were selected in order to exploit the steric zone around the phenyl ring of fadrozole for favorable interactions as identified by the CoMFA model. The imidazole ring linked to the templates by a methylene unit and an H-bond-accepting nitro group located on the aromatic ring at a suitable distance from the azole nitrogen atom carrying the lone pair electrons were engineered to provide the pharmacophore for aromatase inhibition. Compounds (**5.138**, IC_{50} = 43 nM and **5.139**, IC_{50} = 40 nM) that showed good inhibitory potency against, and selectivity for, the human aromatase enzyme were found to be more potent than fadrozole.

5.6.5 STEROID SULFATASE

5.6.5.1 The Enzyme and Breast Cancer

The enzyme steroid sulfatase (E.C. 3.1.6.2, STS) catalyzes the hydrolysis of sulfate esters of 3-hydroxy steroids, which are the biologically inactive forms, to their active 3-hydroxy steroids (Figure 5.8). Major substrates for STS include estrone sulfate (E1S), dehydroepiandrosterone (DHEA) sulfate, pregnenolone sulfate, and cholesterol sulfate, indicating that the A-ring of the steroid can be aromatic or alicyclic. STS is a 65-kDa membrane-bound protein predominantly associated with the endoplasmatic reticulum and present in almost all mammalian tissues. It is one of the eleven mammalian sulfatases that have been identified, and its amino acid sequence has been resolved. In the past decade, STS has been the subject of intensive research due to its potential involvement in the pathogenesis of a number of diseases such as breast cancer, androgen-dependent skin disorders, cognitive dysfunction, and immune function. Among these potential indications, the inhibition of STS as endocrine therapy for HDBC has been the most investigated.

Much of the estrone (E1) synthesized from androstenedione via the aromatase pathway can be converted to estrone sulfate (E1S) by estrone sulfotransferase. Plasma and tissue concentrations of E1S are considerably higher than those of unconjugated estrogens. The half-life of E1S (10 to 12 h) is significantly longer than that (20 to 30 min) of unconjugated E1 and estradiol (E2). Hence it has been postulated that the high tissue and circulating levels of E1S could act as a reservoir for the formation of biologically active estrogens via the action of STS. Furthermore, the STS activity in breast tumors has been shown to be much higher than that of the aromatase enzyme. There is evidence to show that as much as a 10-fold greater amount of E1 may originate from the sulfatase pathway than via the aromatase route. More recently, studies at the molecular level have revealed that in 87% of breast cancer patients investigated, STS mRNA levels were higher in malignant than in nonmalignant tissues.

In addition, the production of androstenediol (Adiol), and to a lesser extent E2, via DHEA (Figure 5.8) could also significantly contribute to the estrogenic stimulation of hormone-dependent breast tumors. Adiol, an androgen that binds to estrogen receptors and has estrogenic properties, originates from DHEA sulfate once it has been hydrolyzed to DHEA (Figure 5.8) by DHEA-sulfatase. Therefore, STS inhibitors, when used alone or in conjuction with an aromatase inhibitor, may enhance the response of hormone-dependent breast tumors to this type of endocrine therapy by reducing not only the formation of E1 from E1S but also the synthesis of other estrogenic steroids such as Adiol from DHEA-S.

5.6.6 Inhibition of Steroid Sulfatase as Endocrine Therapy

5.6.6.1 Steroidal STS Inhibitors

An initial attempt to design STS inhibitors had resulted in the synthesis of several steroidal 3-O-sulfates, of which Adiol 3-sulfate (**5.155**, K_i = 2 μM) was shown to be the most potent, although this inhibitor would be of little value clinically because it was a substrate for, and hence hydrolyzed by, STS with the release of the estrogenic Adiol. Continual search for STS inhibitors led to the discovery of danazol (**5.156**), a drug used for treating endometriosis, as a weak STS inhibitor. However, the main strategy employed by various groups for generating a reasonable lead STS inhibitor was to replace the sulfate group (OSO_3^-) of E1S with surrogates or mimics such as methylthiophosphonates [$-OP(=S)(OH)Me$], phosphonates [$-OP(=O)(OH)R$], sulfonates ($-OSO_2R$), sulfonyl halides ($-SO_2Cl$ and $-SO_2F$), sulfonamide ($-SO_2NH_2$), and methylsulfonyl ($-SO_2CH_3$). These E1 derivatives were designed to compete with E1S for the enzyme active site without acting as substrates for the enzyme. Despite the efforts involved, these agents were found to be only weak inhibitors of STS. However, the breakthrough came when the sulfate group of E1S was replaced by a sulfamate moiety (OSO_2NH_2). The resulting analog, estrone 3-O-sulfamate (EMATE, IC_{50} = 18 nM, human placental microsomes), was found to inhibit STS not only potently but also in a time- and concentration-dependent manner, indicating that EMATE is an irreversible active site-directed inhibitor. Unexpectedly, despite its high potency, EMATE is not suitable for use in the treatment of HDBC because it was shown subsequently to be more estrogenic than ethinylestradiol when administered orally in rats. The reason for the estrogenicity of EMATE is most likely due to release of E1 as a consequence of its inactivation of STS. In order to overcome this undesirable property of EMATE, nearly all steroidal inhibitors that followed were designed to be less estrogenic than EMATE but with similar or superior inhibitory activities to that of EMATE.

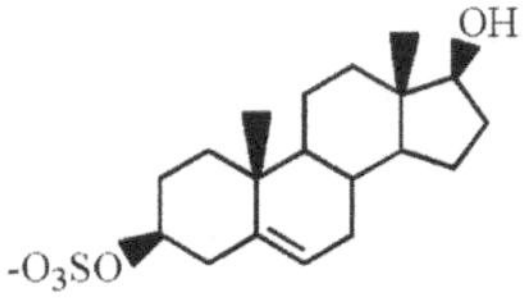

(5.155) 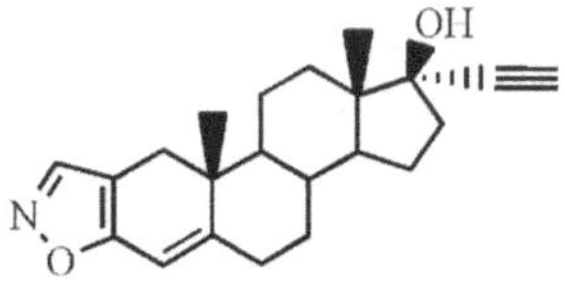   (5.156)

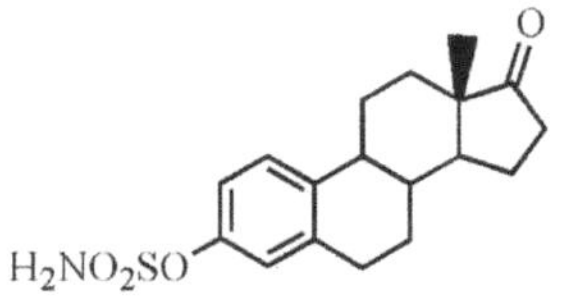

EMATE 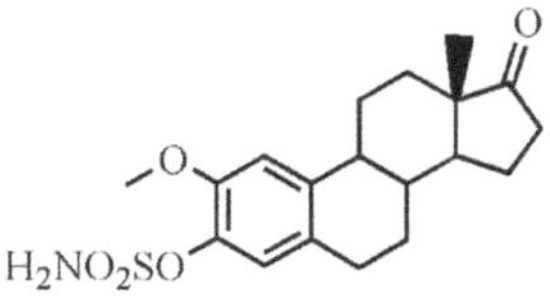   (5.157)

The initial strategy was to introduce substituents such as allyl, *n*-propyl, nitro, cyano, and halogens (F, Cl, Br, I) to the A-ring of EMATE at the 2- and/or 4-positions. Analogs of EMATE with electron-withdrawing substituents on the A-ring showed higher potency than EMATE *in vitro*. In comparison, those analogs with bulkier aliphatic substituents were found to be weaker STS inhibitors. The most successful A-ring modified analog of EMATE was 2-methoxyestrone 3-O-sulfamate (2-MeOEMATE, **5.157**). This inhibitor was as potent as EMATE in inhibiting the activity of STS *in vitro* but, in contrast, was completely devoid of estrogenicity. More significantly, 2-methoxyEMATE was found to induce apoptosis (programmed cell death) in a few cancer cell lines and inhibit tumor angiogenesis. 2-MethoxyE-MATE is therefore potentially useful for treating both HDBC and some hormone-independent cancers.

The D-ring of EMATE has been targeted for reducing the estrogenicity of the inhibitor. Early work had seen the reduction of the 17-carbonyl of EMATE to the methylene derivative NOMATE (**5.158**), which was as potent as EMATE but less estrogenic. Enlargement of the D-ring from a cyclopentanone to a lactone gave estralactone 3-*O*-sulfamate (**5.159**), which almost completely inhibited STS in MCF-7 (human breast cancer) cells at 10 μ*M*.

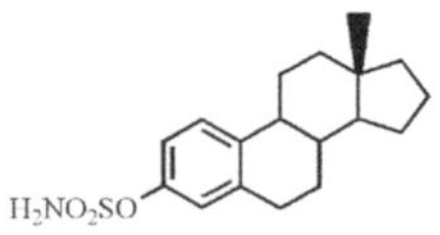

H_2NO_2SO

(**5.158**)

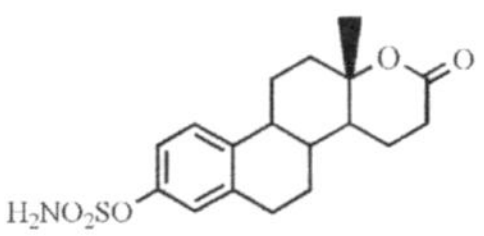

H_2NO_2SO

(**5.159**)

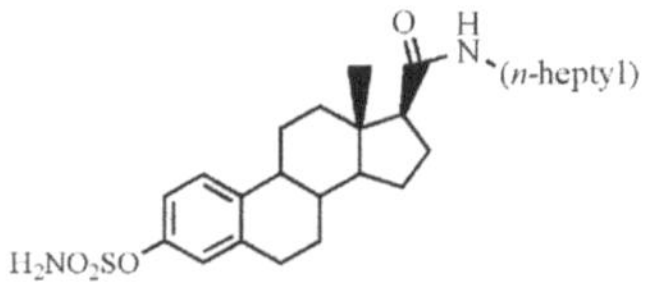

H_2NO_2SO

(**5.160**)

H_2NO_2SO

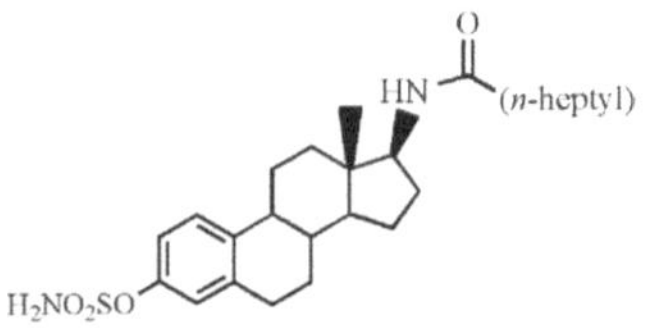

(**5.161**)

    Introduction of hydrophobic substituents into the D-ring of EMATE increased its potency and significantly reduced its estrogenicity. 17β-(*N*-Alkylcarbamoyl) estradiol-3-*O*-sulfamates and 17β-(*N*-alkanoyl) estradiol-3-*O*-sulfamates are highly potent STS inhibitors with optimal inhibitory activities shown by their *n*-heptyl derivatives **5.160** and **5.161**, respectively (both IC_{50} values = 0.4 n*M* in MDA-MB-231 cells). Using the estrogen-sensitive MCF-7 cell line that proliferates upon stimulation by estrogens, no significant estrogenic potential was observed in either inhibitor at a concentration of 1 μ*M*, a dose that is about 2000 times higher than their IC_{50} values against STS. It is evident that the hydrophobic side-chains of these inhibitors have the dual effect of increasing binding to the enzyme active site around the D-ring of EMATE and abolishing estrogenicity of the parent phenolic steroid.
    A stereoselective addition of Grignard reagents to the C17-carbonyl group of E1 gave a series of 17α-derivatives of E2 of which the 4′-*t*-butylbenzyl **5.162** and the 3′-bromobenzyl **5.163** derivatives were among the most potent inhibitors of this series, showing IC_{50} values (JEG-3 cells) of 28 n*M* and 24 n*M*, respectively. Because of the absence of a sulfamate group, both **5.162** and **5.163** are reversible inhibitors, but they are only seven times less potent than EMATE. In common with **5.160** and **5.161**, the reversible inhibitors **5.162** and **5.163** exploit the hydrophobic binding areas in the vicinity of the D-ring of EMATE. Sulfamoylation of **5.162** gave the corresponding sulfamate **5.164**, which was found to be even more potent as an STS inhibitor than EMATE (IC_{50} = 0.15 n*M* vs. 2.1 n*M* for EMATE, recombinant STS in cell extract with E1S as substrate). Because of the potency of **5.162**, it was reasoned that this phenolic steroid, released after the inactivation of enzyme by the irreversible STS inhibitor **5.164**, would still be exhibiting reversible inhibition against any unreacted STS.

	X	Y
(5.162)	OH	4'-^{t}Bu
(5.163)	OH	3'-Br
(5.164)	OSO$_2$NH$_2$	4'-^{t}Bu

(5.165) R = H

(5.166) R = n-Propyl

(5.167) R = CH$_2$ —

The most recent D-ring modification of EMATE was a series of N-substituted piperidinedione derivatives. This ring structure was designed to serve as a versatile anchor for introducing a variety of side-chains that can exploit the hydrophobic pockets and regions known to be present in the vicinity of the D-ring of EMATE. Two compounds, the N-(propyl)- (5.166) and N-(1-pyridin-3-ylmethyl)- (5.167) derivatives, showed exceptionally high potency, with both sharing the same IC$_{50}$ value of 1 nM in a human placental microsomes preparation. EMATE in the same assay gave an IC$_{50}$ value of 18 nM. The N-unsubstituted derivative 5.165 showed potency similar (IC$_{50}$ = 20 nM) to EMATE, indicating that the six-membered piperidinedione ring is a good mimic of the D-ring of EMATE. After an oral dose of 10 mg/kg per day for 5 d, 5.166 and 5.167 were found to inhibit rat liver STS by 99%. Both compounds were devoid of estrogenic activity in the rat uterine weight-gain assay.

5.6.6.2 Nonsteroidal STS Inhibitors

It has long been recognized that nonsteroidal agents themselves and their metabolites are less likely to exhibit unwanted endocrinological activities *in vivo* than their steroidal counterparts. This includes estrogenicity, which has been the major problem with EMATE. The desire to develop nonsteroidal STS inhibitors has also been driven by the finding that EMATE has a memory-enhancing effect in rats, that STS may regulate part of the immune response in humans, and that STS may be involved in the development of androgen-dependent skin disorders such as acne. It is clearly undesirable to treat some of these nonmalignant conditions with steroidal drugs.

The first attempt in developing nonsteroidal STS inhibitors was made with the synthesis of sulfate derivatives of 2-phenylindoles, of which 5.168 had an IC$_{50}$ value of 80 μM, albeit as a competitive inhibitor and substrate for STS. Several single-ring phenyl phosphates substituted with hydrophobic chains were synthesized as tools to probe the structural requirement of STS for substrate and inhibitor recognition. The best inhibitor was the monoanionic form of n-lauroyl tyramine phosphate (5.169), which had a K_i value of 520 nM at pH 7. This work led to the development of a series of (p-O-sulfamoyl)-N-alkanoyl tyramines, of which the N-tetradecanoyl derivative (5.170) was found to have an IC$_{50}$ value of 55.8 nM against STS derived from human placental microsomes. It was postulated that the phenyl ring and the amido functionality of these tyramine sulfamates mimic the A-ring and C17-carbonyl group of EMATE, respectively.

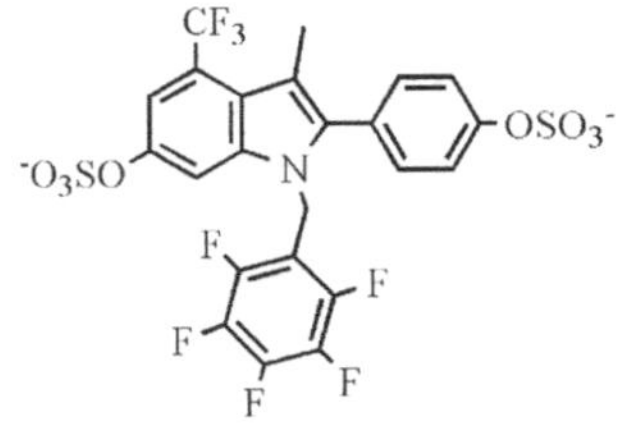

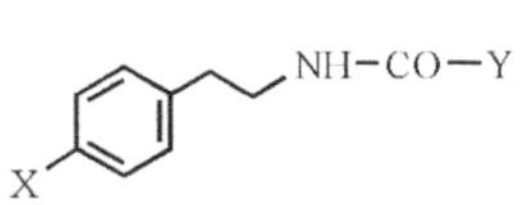

	X	Y
(5.169)	$OP(OH)O_2^-$	$(CH_2)_{10}CH_3$
(5.170)	OSO_2NH_2	$(CH_2)_{12}CH_3$

(5.168)

To explore the structure–activity relationships for EMATE and to reduce its estrogenicity, the A/B ring mimic of EMATE, tetrahydronaphthalene 2-O-sulfamate (5.171), was prepared, and it was found to exhibit STS inhibitory activity, but to a much weaker extent than EMATE. In pursuit of alternative nonsteroidal mimics of EMATE, a series of bicyclic coumarin sulfamates was synthesized, of which 4-methylcoumarin 7-O-sulfamate (5.172) showed an IC_{50} of 380 nM in an MCF-7 cells preparation, a potency about 3 times that of 5.171 but was still much weaker than EMATE (IC_{50} = 65 pM from the same assay).

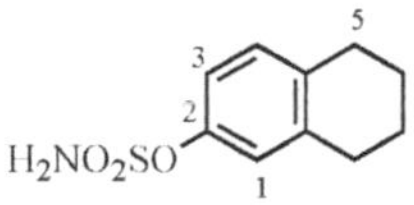

(5.171) 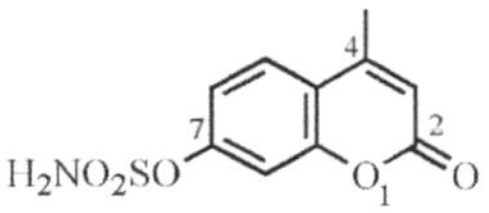   (5.172)

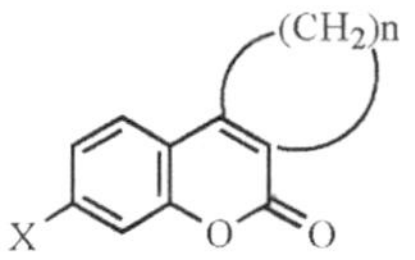

	X	n
(5.173)	OH	5
(5.174)	OSO_2NH_2	5
(5.175)	OSO_2NH_2	8

Further extension of the coumarin sulfamate series has established that derivatives with hydrophobic substituents introduced at the 3- or 4-position or both of the bicyclic ring system are more potent STS inhibitors. A series of tricyclic coumarin sulfamates was developed by allowing resorcinol to react with cyclic β-keto esters under Pechmann synthesis conditions. One compound, 667COUMATE (5.174), was found to inhibit STS activity with an IC_{50} value of 8 nM (placental microsomes), a potency about three times that of EMATE. The apparent K_i value for 667COUMATE was found to be 40 nM, which was significantly lower than that for EMATE (670 nM), suggesting that the lower IC_{50} value observed for this nonsteroidal inhibitor in comparison with EMATE could be attributed to a higher affinity of 667COUMATE for the enzyme active site. In addition, it was postulated that because the phenolic coumarin precursor (5.173) of 667COUMATE is a better leaving group than E1 [pK_a ca. 8.5 for (5.173) vs. ca. 10 for E1], the "sulfamoylation potential," which relates to the ability to inactivate the enzyme, is higher for 667COUMATE, making this agent a more potent STS inhibitor than EMATE. *In vivo*, a single dose (10 mg/kg, p.o.) of 667COUMATE inhibited rat liver STS activity by 93%. When administered to rats bearing nitrosomethylurea-induced mammary tumor, 667COUMATE caused regression of E1S-stimulated tumor growth in a dose-dependent manner. Like EMATE, 667COUMATE is an irreversible inhibitor of STS and is active by the oral route, but in contrast to EMATE, is devoid of estrogenicity. 667COUMATE is now in formal clinical development and has just entered Phase I trial. With the extension of the tricyclic coumarin sulfamates series, it has been shown that the *in vitro* inhibitory activity was the highest with 6610COUMATE (**5.175**) (IC$_{50}$ = 1 n*M*, placental microsomes). *In vivo*, this compound was found to be marginally more potent than 667COUMATE.

Because of the structural resemblance of the naturally occurring flavonoids to estrogens, sulfamates of flavones, isoflavones, and flavanones were prepared, and compounds such as **5.176** to **5.178** were shown to be moderate inhibitors of STS. Further exploitation of this class of compounds led to a series of chromenone- and thiochromenone-based sulfamates, of which **5.179** showed an IC$_{50}$ value of 0.34 n*M* in purified STS, making this compound about 170 times superior to EMATE (IC$_{50}$ = 56 n*M* in the same assay) and the most potent STS inhibitor discovered to date. The most likely explanation for the potency observed for **5.179** is that its aryl sulfamate moiety mimics the A-ring of EMATE, whereas the 1-adamantyl group exploits hydrophobic interactions that have been demonstrated from previous work to exist around the D-ring of EMATE.

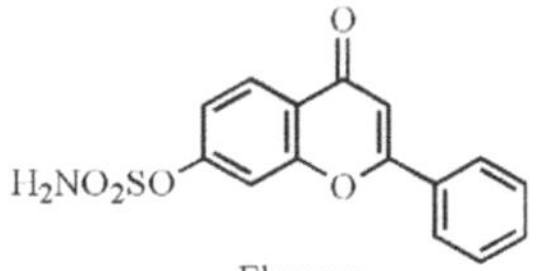

Flavone

(**5.176**)

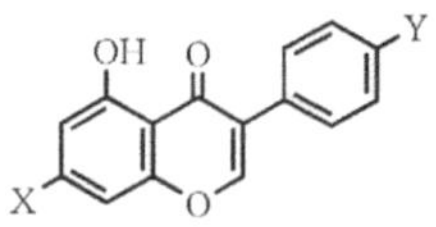

Isoflavone

	X	Y
(**5.177**)	OH	OSO$_2$NH$_2$
(**5.178**)	OSO$_2$NH$_2$	OSO$_2$NH$_2$

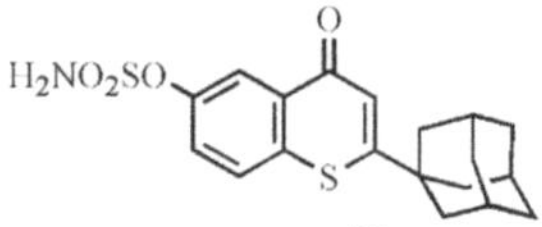

59

(**5.179**)

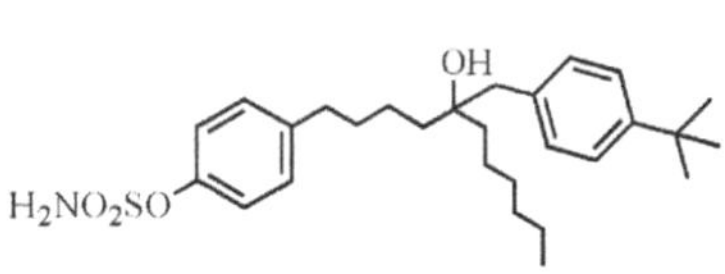

(**5.180**)

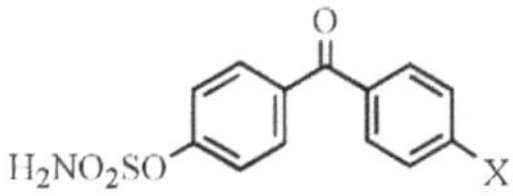

(**5.181**) X = OSO$_2$NH$_2$
(**5.182**) X = H

It had been shown earlier that the steroidal inhibitor **5.164** was a highly potent STS inhibitor. A series of structurally related 4-substituted monoaryl sulfamates were designed as nonsteroidal mimics of **5.164**. The optimal inhibitor in the series, **5.180**, was more potent than EMATE with an IC$_{50}$ value of 0.4 n*M* as compared with 0.9 n*M* for EMATE in the homogenates of HEK-293 cells transfected with STS.

A more recent publication has shown that benzophenone 4,4'-bissulfamate (**5.181**) was, surprisingly, a potent inhibitor that showed an IC$_{50}$ value of 190 n*M* (cf. 56 nM for EMATE) against recombinant human STS. Structure–activity relationship studies have shown that the bis-sulfamate is crucial to the high potency because the monosulfamate derivative, i.e., benzophenone 4-sulfamate (**5.182**), was about 25 times less active.

It is beyond the scope of this section to include all the steroidal and nonsteroidal STS inhibitors reported in the literature and patents to date. The more interesting and significant examples have been discussed here. Readers can refer to more recent publications and reviews in this field for a more comprehensive list of STS inhibitors.

5.6.6.3 Mechanism of Action for STS and STS Inhibitors

All potent irreversible STS inhibitors reported to date share a common pharmacophore, that is, a phenol sulfamate ester with substituents that exploit favorable hydrophobic interactions with the enzyme active site. Although nonsulfamoylated phenolic compounds such as **5.162** and **5.163** were potent reversible inhibitors of STS, the sulfamate derivative **5.164** was an irreversible inhibitor with a potency superior to its phenolic counterpart (see steroidal inhibitors in Section 5.6.6.1). The N-monoalkyl (e.g., **5.183**) and N,N-dialkyl derivatives of EMATE (e.g., **5.184**) were weak reversible inhibitors of STS, although N-acetyl-EMATE (**5.185**), but not the benzoyl derivative, inhibited the enzyme irreversibly, albeit less potently than EMATE. Analogs of EMATE in which the 3-O-atom was replaced by other heteroatoms (S, **5.186** and N, **5.187**) were only weak reversible inhibitors of STS. All these findings have demonstrated that a free sulfamate group (i.e., $H_2NSO_2O–$) with no substitutions at the N-atom is the prerequisite for highly potent irreversible STS inhibition.

(**5.183**) X = OSO_2NHMe
(**5.184**) X = OSO_2NMe_2
(**5.185**) X = $OSO_2NHCOCH_3$
(**5.186**) X = SSO_2NH_2
(**5.187**) X = $NHSO_2NH_2$

Despite the fact that STS is bound to the membrane of the endoplasmic reticulum, it was successfully isolated and purified. The crystal structure of human STS has recently been reported in the literature. The overall shape of the protein is mushroom-like, with the crown protruding toward the lumen side of, and the stalk traversing through the lipid bilayer of, the endoplasmic reticulum. STS is similar to its closely related soluble enzymes arylsulfatase A (ASA) and arylsulfatase B (ASB), whose crystal structures were published a few years ago, and shares with them a similar catalytic site topology and a unique characteristic universal to all sulfatases, i.e., posttranslational modification of a conserved cysteine residue to a formylglycine (FGly, •–CHO) residue. As observed for ASB, the resting state of human STS at the catalytic site consists of a sulfated gem-diol form of FGly [i.e., ~$CH(OH)OSO_{3-}$], which is coordinated to a bivalent Ca^{2+} cation. The steroid scaffold recognition site of STS is composed of mainly hydrophobic residues that interact favorably with

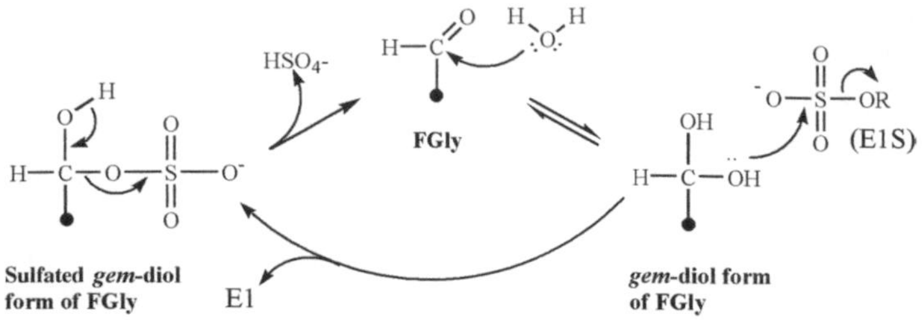

FIGURE 5.9 Proposed reaction scheme for cleavage of estrone sulfate by steroid sulfatase. (Adapted from Woo, L.W.L., Purohit, A., Malini, B. et al. (2000). *Chemistry & Biology*, 7, 773–791.)

E1S via hydrophobic contacts. It has been demonstrated that $[\sim CH(OH)OSO_{3-}]$ is crucial to the hydrolysis of sulfate ester by ASB. The putative mechanism of STS is depicted in Figure 5.9. The first step involves the regeneration of the *gem*-diol form of FGly via the attack of a molecule of water on the FGly intermediate. One of the hydroxyl groups of the *gem*-diol form of FGly then attacks the sulfur atom of E1S, releasing E1 and regenerating the sulfated *gem*-diol form of FGly.

Because the sulfamate group is designed to mimic the sulfate group, it is reasonable to expect that the mechanism of action of sulfamate-based STS inhibitors such as EMATE would also involve the $[\sim CH(OH)OSO_3{}^-]$ residue. One of the proposed mechanisms of STS inhibition by EMATE is shown in Figure 5.10. It is not clear if structure I is a dead-end product or will undergo further modifications to yield a species that irreversibly inactivates the enzyme.

However, it is also possible that irreversibly inhibiting sulfamate esters (e.g., EMATE) could inhibit STS in a more random manner by a specific or nonspecific sulfamoylation of amino acid residues in the active site. Such proposed mechanisms are shown in Figure 5.11. Path A involves an attack by a nucleophilic amino acid residue in the active site. Path B involves the generation of a sulfonylamine species via an E1*cB* process, possibly initiated by an enzyme-catalyzed *N*-proton abstraction and stimulated by hydrogen bonding to the bridging *O*-atom. The *N*-proton of a sulfamate group is fairly acidic, considering that EMATE has a pK_a value of 9.5

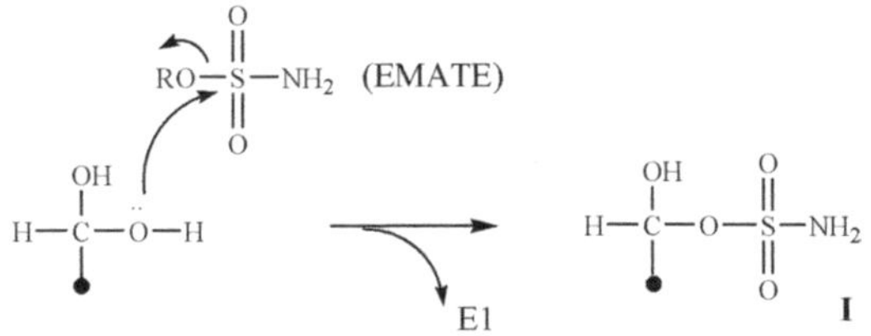

FIGURE 5.10 Proposed mechanism of steroid sulfatase inhibition by EMATE via a nucleophilic attack on the sulfamoyl group by the *gem*-diol form of formylglycine residue in the enzyme active site. Structure **I** is proposed to be a dead-end product. (Adapted from Woo, L.W.L., Purohit, A., Malini, B. et al. (2000). *Chemistry & Biology*, 7, 773–791.)

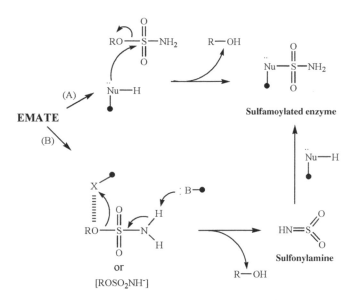

FIGURE 5.11 Proposed random specific or nonspecific sulfamoylation by EMATE of an essential amino acid residue in the steroid sulfatase active site. Path A: via an attack by a nucleophilic amino acid residue in the active site other than the *gem*-diol residue (:Nu-H). Path B: via the generation of a sulfonylamine species. No regeneration of the enzyme active form from the sulfamoylated intermediate is expected. :B, a proton-abstracting amino acid residue; X, a hydrogen-bond-donating amino acid residue or a coordinating metal ion. Dashed line: hydrogen-bonding. (Adapted from Woo, L.W.L., Purohit, A., Malini, B. et al. (2000). *Chemical Biology*, 7, 773–791.)

measured in 70% aqueous methanol. This value is expected to be lower, perhaps by one or two units, when EMATE is in an aqueous environment. With what is known about the topology of the active site of STS, the attacking nucleophile in Path A and the *N*-proton abstracting amino acid residue (:B) in Path B could well be either one of the conserved lysine or histidine residues lining the active site of STS. The end product of both paths is a sulfamoylated essential amino acid residue that would have led to irreversible inactivation of the enzyme.

5.6.7 FUTURE DIRECTIONS

It has clearly been established that inhibition of the aromatase enzyme is highly effective in the management of HDBC. Although letrozole and anastrazole were originally licensed as the second-line treatment after the failure of tamoxifen in postmenopausal women with advanced breast cancer, evidence has emerged from many recent trials to suggest that these agents are indeed superior to tamoxifen as the first-line treatment for HDBC. With these NSAIs showing better toxicity profiles than tamoxifen, it is envisaged that letrozole or anastrazole would eventually challenge the established role of tamoxifen in the treatment of breast cancer. Despite the reports of so many fourth-generation NSAIs, it is not known if any of these agents would successfully progress to the clinical development stage. Unless some of these newer agents are shown to possess a therapeutic profile superior to existing agents, it is anticipated that letrozole and anastrazole would remain the gold standard of NSAIs for many years to come.

The recent entry of the nonsteroidal STS inhibitor 667COUMATE into Phase I trial is a promising advance. The results of a follow-on Phase II trial would allow the role of STS inhibition in the management of HDBC to be assessed. Once a clinical proof-of-concept has been established, it is anticipated that a few more leading steroidal and nonsteroidal STS inhibitors will be scheduled for clinical development.

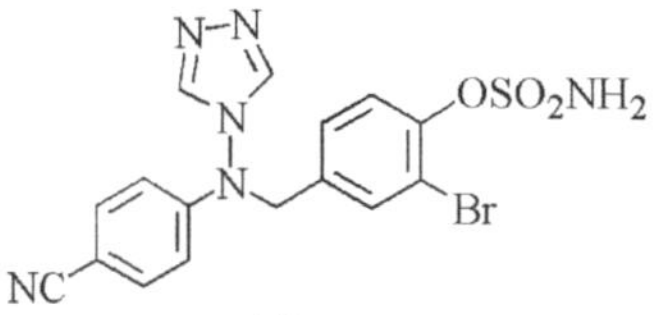

(A)

There is strong evidence to suggest that more effective estrogen deprivation in patients with breast cancer that is sensitive to estrogenic stimulation can only be achieved if both aromatase and STS are inhibited at the same time. Although administering an aromatase inhibitor in conjunction with an STS inhibitor as two separate agents might be an obvious choice in a combined endocrine therapy, an attractive alternative approach is to design a dual aromatase and sulfatase inhibitor (DASI) that will inhibit both enzymes as a single agent. This concept has recently been validated by the report on compound A (see Structure A) that showed dual inhibition of both enzymes. It is anticipated that DASIs such as compound A should allow the therapeutic potential of dual inhibition of estrogen formation in breast tumors to be evaluated.

ACKNOWLEDGMENT

The author thanks Barry V.L. Potter for comments on the manuscript.

FURTHER READING

Reviews on Aromatase Inhibitors

1. Recanatini, M., Cavalli, A., and Valenti, P. (2002). Nonsteroidal aromatase inhibitors: recent advances. *Medicinal Research Reviews*, 22, 282–304.
2. Brodie, A. (2002). Aromatase inhibitors in breast cancer. *Trends in Endocrinology and Metabolism*, 13, 61–65.

Aromatase Mechanisms

1. Cole, P.A. and Robinson, C.H. (1990). Mechanism and inhibition of cytochrome P450 aromatase. *Journal of Medicinal Chemistry*, 33, 2933–2942.

2. Akhtar, M., Njar, V.C., and Wright, J.N. (1993). Mechanistic studies on aromatase and related C–C bond cleavage P450 enzymes. *Journal of Steroid Biochemistry and Molecular Biology*, 44, 375–387.
3. Kao, Y.C., Korzekwa, K.R., Laughton, C.A. et al. (2001). Evaluation of the mechanism of aromatase cytochrome P450. A site-directed mutagenesis study. *European Journal of Biochemistry*, 268, 243–251.

Reviews on Steroid Sulfatase Inhibitors

1. Reed, M.J., Purohit, A., Woo, L.W.L., and Potter, B.V.L. (1996). The development of steroid sulfatase inhibitors. *Endocrine-Related Cancer*, 3, 9–23.
2. Poirier, D., Ciobanu, L.C. and Maltais, R. (1999). Steroid sulfatase inhibitors. *Expert Opinion on Therapeutic Patents*, 9, 1083–1099.
3. Nussbaumer, P. and Billich, A. (2003). Steroid sulfatase inhibitors. *Expert Opinion on Therapeutic Patents*, 13, 605–625.
4. Smith, H.J., Nichols, P.J., Simons, C., and Le Lain, R. (2001). Inhibitors of steroidogenesis as agents for the treatment of hormone-dependent cancers. *Expert Opinion on Therapeutic Patents*, 11, 789–824.

Crystal Structure of STS

1. Hernandez-Guzman, F.G., Higashiyama, T., Pangborn, W. et al. (2003). Structure of human estrone sulfatase suggests functional roles of membrane association. *Journal of Biological Chemistry*, 278, 22989–22997.

Proposed Mechanisms of Irreversible Steroid Sulfatase Inhibitors

1. Woo, L.W.L., Purohit, A., Malini, B. et al. (2000). Potent active site-directed inhibition of steroid sulfatase by tricyclic coumarin-based sulfamates. *Chemistry & Biology*, 7, 773–791.

Dual Aromatase and Sulfatase Inhibitors

1. Woo, L.W.L., Sutcliffe, O.B., Bubert, C. et al. (2003). First dual aromatase-steroid sulfatase inhibitors. *Journal of Medicinal Chemistry*, 46, 3193–3196.

5.7 ENZYME INHIBITOR EXAMPLES FOR THE TREATMENT OF PROSTATE TUMOR

Samer Haidar and Rolf W. Hartmann

5.7.1 5α-REDUCTASE AND ANDROGEN-DEPENDENT DISEASES

The nonmalignant enlargement of the prostate (benign prostatic hyperplasia [BPH]) is the most common benign tumor in men aged over 60 years. 5α-reductase (5α-R) is an NADPH-dependent, membrane-bound enzyme that catalyzes the conversion of the steroidal hormone testosterone (T) to the more potent androgen dihydrotestosterone (DHT, Scheme 5.1). Two isozymes, named Type 1 and Type 2, have been described with a different tissue-distribution pattern and with distinct biochemical and pharmacological properties.

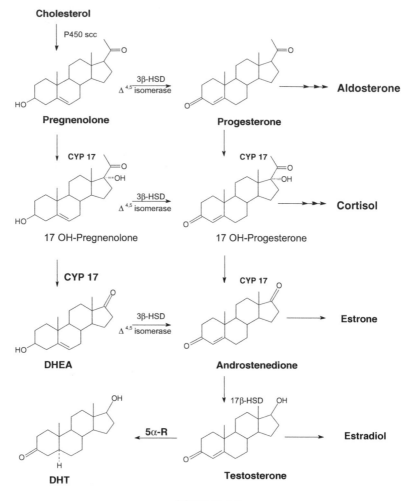

SCHEME 5.1

Scheme 5.2 shows a proposed mechanism of 5α-reductase reaction involving the direct donation of a hydride (H⁻) from NADPH to the C-5 position of T, leading to the formation of an enolate that presumably is stabilized by some electrophilic residue (E⁺) in the active site. However, this process may be viewed alternatively as activation of the enone by E⁺, leading to a positively polarized species that accepts a hydride from NADPH at C-5. Subsequently, the enzyme-mediated tautomerism leads to the product DHT, with the release of NADP⁺.

Androgens induce the differentiation and maturation of the male reproductive organs and the development of male secondary sexual characteristics. In men, androgens are formed mainly in the testes and adrenal glands. Androgens also stimulate the growth of some androgen-dependent diseases like prostate cancer (PC) or BPH.

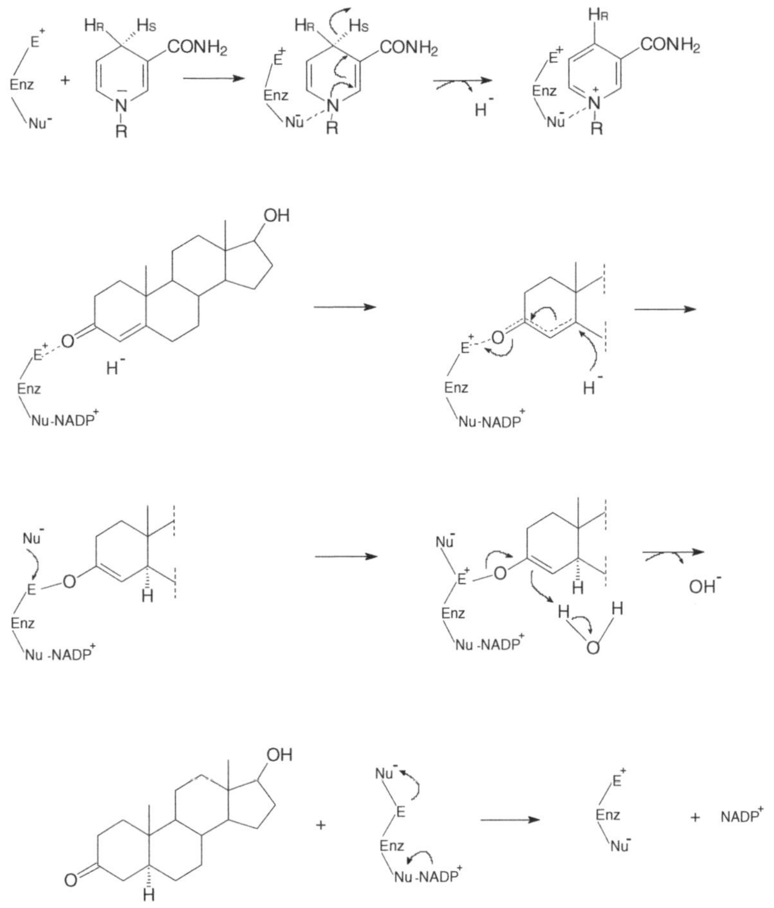

SCHEME 5.2

As a noninvasive treatment for BPH, 5α-reductase is a promising therapeutic target and preferable to other therapeutic agents such as antiandrogens. Tumor growth in BPH around the renal duct exerts pressure and decreases the diameter of the lumen, leading to decreased and subsequently sporadic urine flow. Selective and highly active 5α-reductase inhibitors reduce the formation of DHT and subsequently decrease the weight of the prostate and also increase the diameter of renal lumen.

It is important to mention that 5α-reductase inhibitors are also used in the treatment of male pattern baldness or androgenetic alopecia, and also as an alternative treatment for acne in adult patients.

There is an additional interesting application of 5α-reductase inhibitors, which is due to the fact that (1) androgens are cocarcinogens for the development of prostate cancer and (2) DHT is 10 times more potent as an androgen than T. Finasteride is currently being tested as a chemopreventive agent in the U.S. National Cancer Institute Clinical Cooperative Groups in the Prostate Cancer Prevention Trial (PCPT).

5.7.2 INHIBITORS OF 5α-REDUCTASE

An azasteroid series of inhibitors was first disclosed by Merck in the early 1980s and is the class most often studied. Structural modifications to exploit the 4-azasteroid series by Merck led to the discovery of 4-MA (Scheme 5.3), a potent dual inhibitor of both human 5α-R isozymes that was halted in clinical development due to hepatic toxicity. The unsaturated analog finasteride (Scheme 5.3) has been marketed for the treatment of BPH (Proscar®, 5 mg/d) and androgenetic alopecia (Propecia®, 1 mg/d), with minimal side effects. Finasteride is a potent irreversible inhibitor of 5α-R type 2. It suppresses the formation of intraprostatic DHT by up to 90% and reduces circulating DHT levels by 70%, resulting in the reduction of prostate volume by 20%. In fact, finasteride reduces the DHT levels incompletely, which might be due to the fact that at the clinical dosage, 5α-reductase type 1 is not inhibited sufficiently. This directed research toward development of dual inhibitors of 5α-R. Glaxo developed the dual inhibitor dutasteride based on the 4-azasteroid surrogate (Scheme 5.3). This compound is the most potent dual inhibitor and is currently being evaluated in clinical trial Phase 3 for the treatment of BPH, and appears to be more active than finasteride.

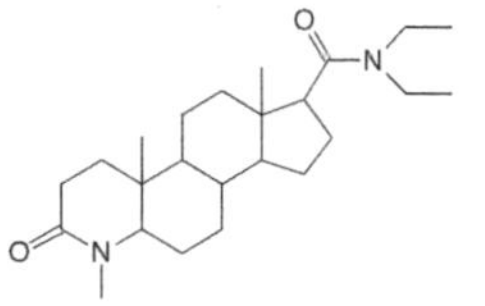

4-MA

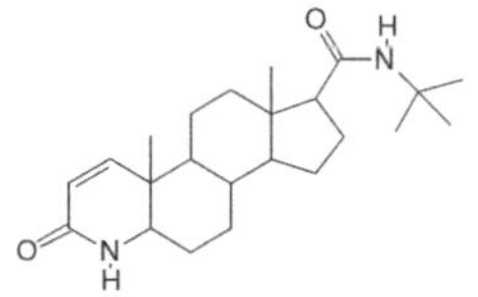

Finasteride

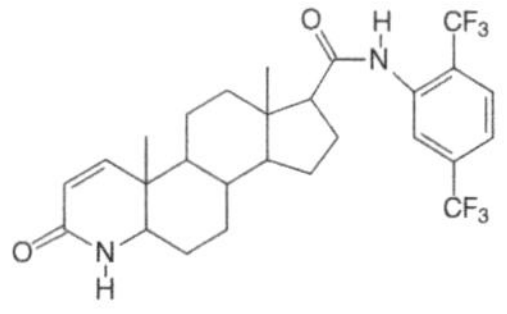

Dutasteride

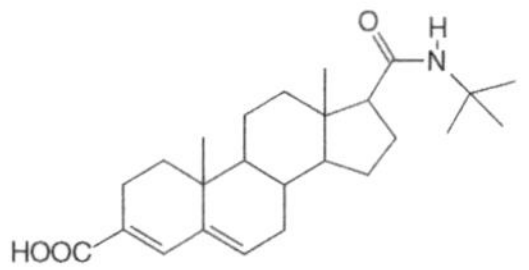

Epristeride

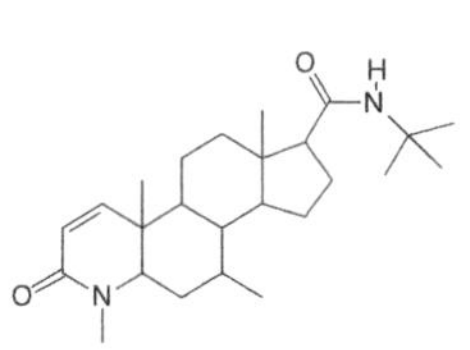

Turosteride

SCHEME 5.3

Another class of compounds designed as mimics of the enolate intermediate contain a 3-carboxylate 4-ene moiety. One example of this class is epristeride (Scheme 5.3), developed by SmithKline Beecham (SKB), which has entered clinical trial for the treatment of BPH. It is a selective inhibitor of 5α-R type 2 and reduces DHT clinically to a lesser extent than finasteride. Turosteride (Scheme 5.3) is another selective inhibitor of 5α-R type 2, which is in clinical development.

A number of classes of nonsteroidal inhibitors of 5α-R have been identified. Different strategies have been used for discovering active compounds. LY-320236 (Scheme 5.4) was designed as a mimic of the azasteroid inhibitors, and is a dual inhibitor of both isozymes. Recently, a series of potent nonsteroidal inhibitors with *in vivo* activity in rats was described. One of them is FP-58 (Scheme 5.4). Other compounds like FK-143 have been developed by optimizing the lead compound ONO-3805 (Scheme 5.4). FK-143 is a dual inhibitor of both human 5α-R isozymes. Other nonsteroidal inhibitors have been developed from high-throughput screening. An example of this strategy is compound SKB-24, which is a highly active 5α-R type 2 inhibitor. Further optimization of this compound recently led to the development of compound OR-12 (Scheme 5.4), which is far more active than SKB-24.

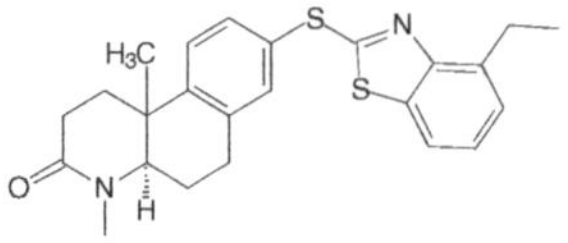

LY-320236

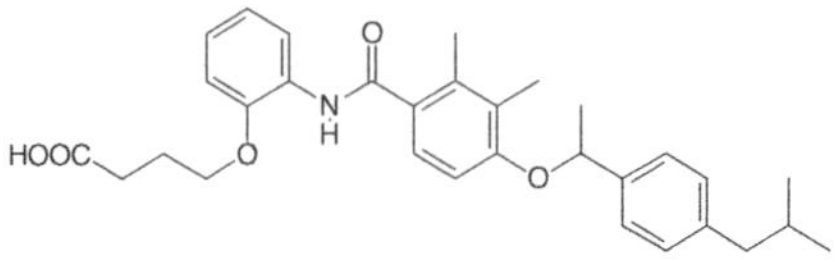

ONO-3805

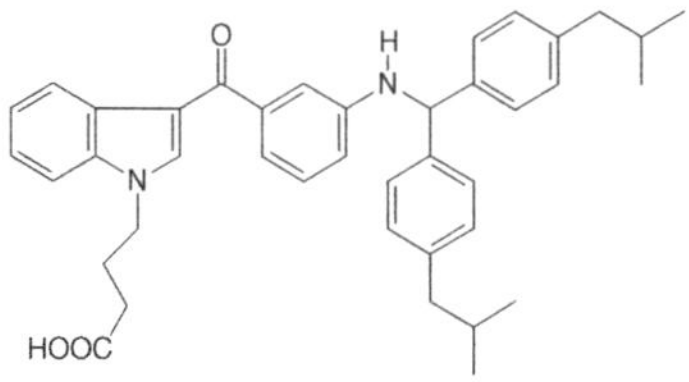

FK-143

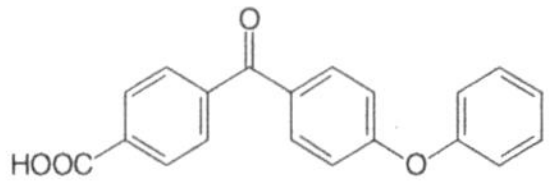

SKB-24

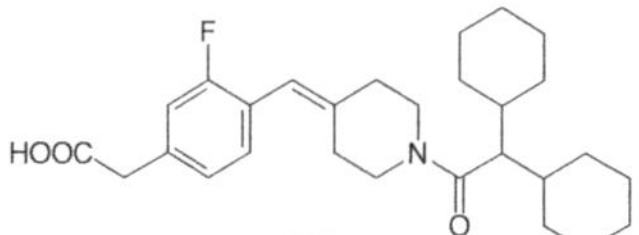

FP-58

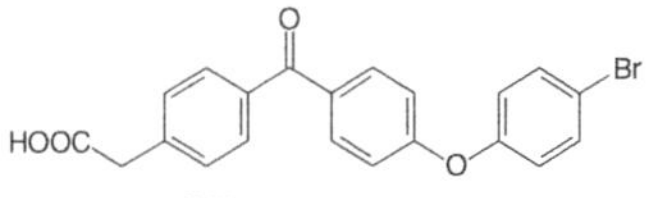

OR-12

SCHEME 5.4

5.7.3 PROSTATE CANCER AND CYP 17

Prostate cancer has surpassed lung cancer as the most often diagnosed male cancer in the West.

Following the report of Huggins and Hodges, who in the 1940s demonstrated that androgen ablation could benefit patients with prostate cancer, orchiectomy became the first-line treatment for androgen-dependent prostate cancer. In the 1970s, antiandrogen treatment made it possible to avoid surgical castration; later, "medical castration" using gonadotrophin-releasing hormone (GnRH) analogs was used as an alternative for the 80% of prostate cancers that are androgen-dependent.

Antiandrogens are intended to inhibit the interaction between an androgen and its receptor. However, antiandrogens fail in reducing androgen concentration and, furthermore, the receptors are likely to mutate and recognize the antiandrogen as a stimulator. Castration (surgical or medical) can eliminate the production of testicular androgens but does not affect androgen biosynthesis in the adrenals. A more recent novel approach involves inhibition of androgen production.

17α-Hydroxylase-C17, 20-lyase (P450 17, CYP 17), is a key enzyme in the biosynthesis of androgens. This enzyme converts the steroids progesterone and pregnenolone (Scheme 5.1) into the 19C androgens androstenedione and dehydroepiandrosterone, respectively. These are transformed to the more potent androgens testosterone and dihydrotestosterone.

During the catalytic cycle, a single oxygen from O_2 is inserted into the substrate, whereas the other oxygen atom is converted into a water molecule. The reaction requires two reduction equivalents (i.e., two electrons and two protons) that originate from NADPH/H$^+$ and are delivered by another enzyme, NADPH reductase. Scheme 5.5 describes both the mechanism of the oxygen activation and the hydroxylation of pregnenolone with the release of an acetic acid fragment, leading to the formation of DHEA.

Inhibition of CYP 17 blocks the synthesis of androgens in both tissues (testes and adrenals), decreasing the plasma levels of T and DHT. In fact, inhibition of CYP 17 might be a superior therapy to the combination of castration and antiandrogens.

5.7.4 INHIBITORS OF CYP 17

Ketoconazole (Scheme 5.6) has been used in the treatment of prostate cancer. This imidazole antifungal agent is an inhibitor of CYP 17 and accordingly an inhibitor of androgen synthesis. Although some benefits were obtained using this compound in the treatment of metastatic PC, the majority of clinical experience indicates that it has limited usefulness. This is mainly due to its relative lack of specificity, which explains some of its undesirable side effects, and its short plasma half-life, which makes frequent dosing necessary. The fact that ketoconazole showed some benefits encouraged the search for more active and selective inhibitors to overcome its shortcomings. Indeed, a large number of steroidal and nonsteroidal inhibitors have been subsequently described. A common feature of these compounds is a group (mainly, a nitrogen-bearing heterocycle) capable of complexing the heme iron located in the active site of the enzyme.

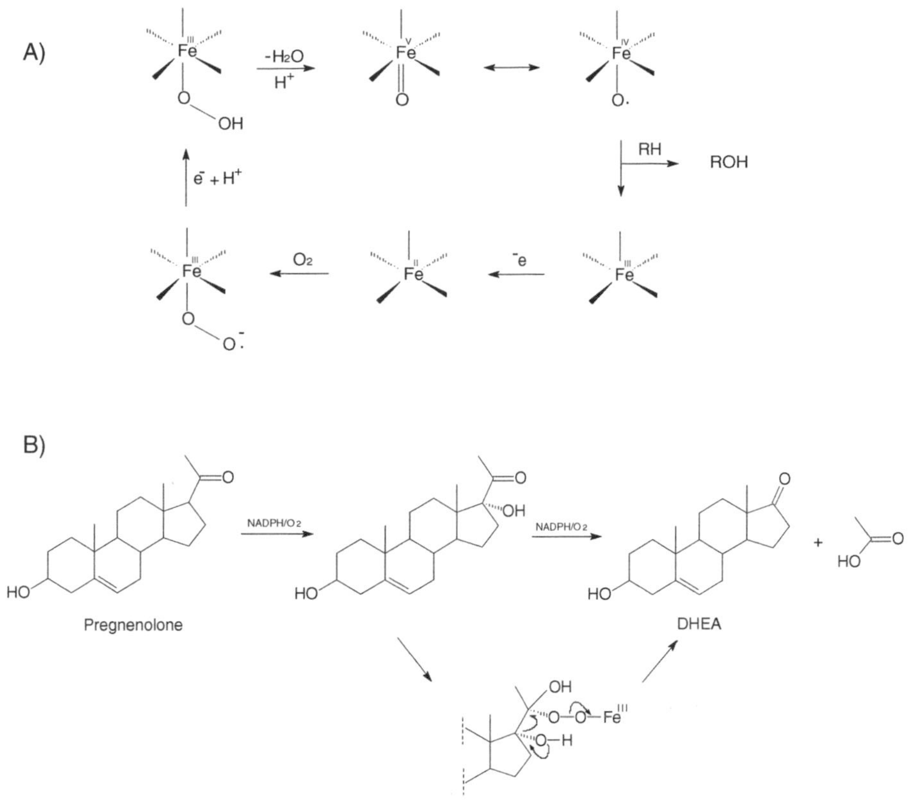

A)

B)

SCHEME 5.5

In the class of nonsteroidal inhibitors, several imidazole compounds were developed. Among them is liarozole (Scheme 5.6), which is a good inhibitor of rat CYP 17. In humans, liarozole reduced plasma testosterone concentration by 80%. Further studies indicated that its antitumor activity is due to its inhibition of retinoic acid metabolism. Liarozole is in Phase 3 clinical trials. Another very potent imidazole type CYP 17 inhibitor is YM-116 (Scheme 5.6), which is also in clinical trial. *In vivo*, it was very active in GnRH-treated rats and was reported to be more potent than the steroidal inhibitor abiraterone (Scheme 5.7). Our group developed several nonsteroidal imidazole, triazole, and pyrazole compounds as steroidal AB ring mimetics. One of them is BW-19 (Scheme 5.6), which is presently undergoing clinical evaluation. BW-19 was highly active *in vitro* and reduced testosterone levels and the weights of some related organs in rats compared with castration controls.

In the class of biphenyl compounds that were designated as steroidal AC ring mimetics, highly potent and selective compounds were discovered, one being YZ-5ay, which also showed high *in vivo* activity.

CB-7645 is a pyridyl inhibitor of CYP 17 (Scheme 5.6), which was described by Barrie et al. as having good inhibitory activity *in vitro*.

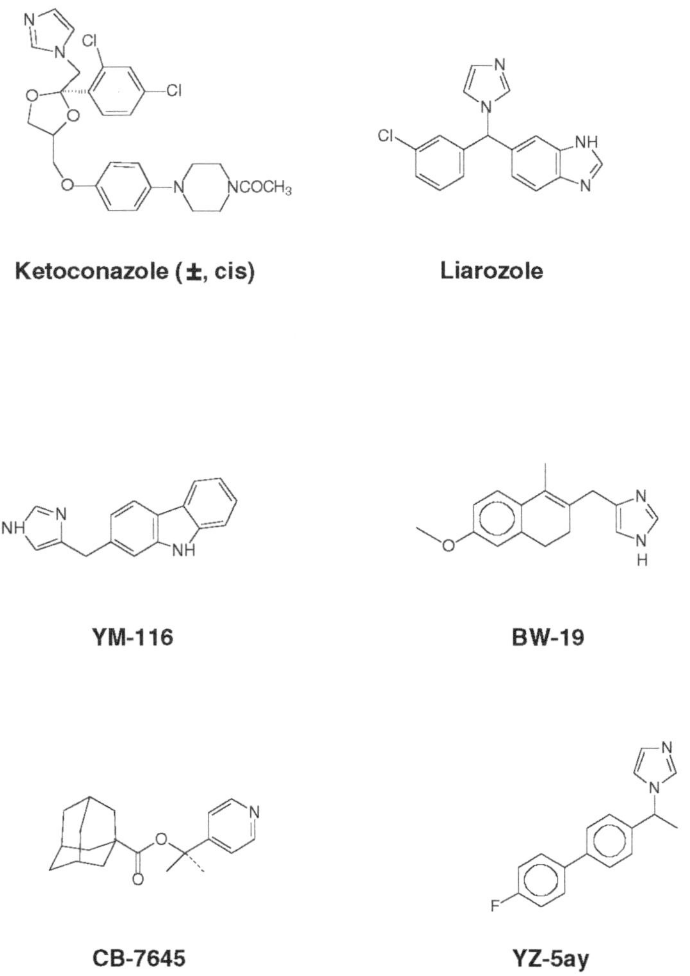

Ketoconazole (±, cis)

Liarozole

YM-116

BW-19

CB-7645

YZ-5ay

SCHEME 5.6

Several 17-pyridyl steroids have been synthesized and tested for CYP 17 inhibition. One of them, the 3-pyridyl compound abiraterone (Scheme 5.7), is a highly active and selective steroidal inhibitor. The profile of abiraterone in Phase 1 clinical trial was very promising because it was very active in lowering the T plasma level and did not generally affect the cortisol level. Recently, Sa-40, a compound derived from abiraterone, was described by our group. This pyrimidyl compound showed a threefold stronger inhibition of the human enzyme, and is equally active in rats.

Njar and Brodie described a number of steroidal CYP 17 inhibitors. Some of them were highly active *in vitro* and *in vivo* and were reviewed recently, e.g., the imidazole compounds VN-85-1 and I-49 (Scheme 5.7). The most potent compound of those is VN-85-1, which is 20 times more potent than ketoconazole *in vitro* and very effective *in vivo* as well.

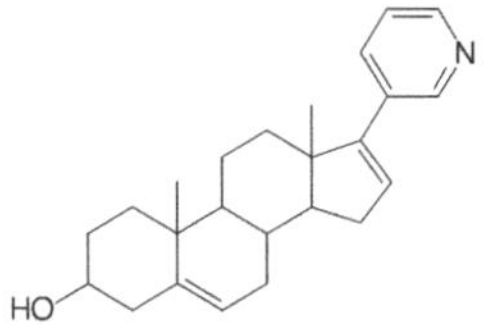

Abiraterone (CB-7598)

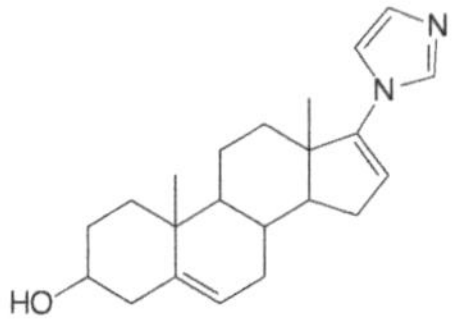

VN-85-1

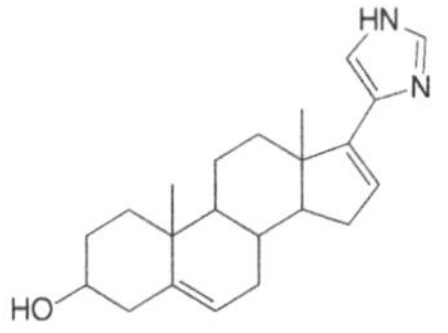

I-49

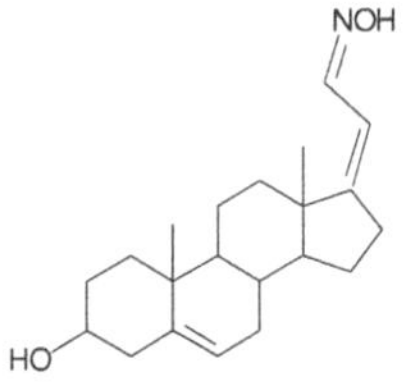

MH-47

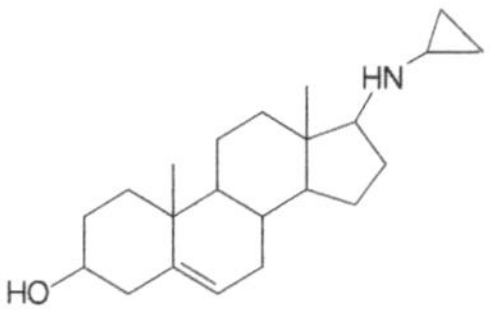

MDL-27,302

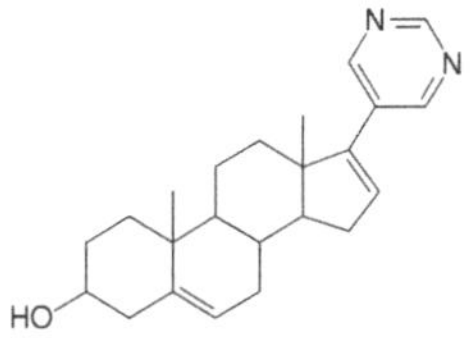

Sa-40

SCHEME 5.7

MDL-27,302 (Scheme 5.7) is one of the mechanism-based inhibitors that were described by Angelastro et al. Like reversible inhibitors, they compete with the natural substrate and interact with the active site of the enzyme. During the normal catalytic mechanism, a reactive intermediate is formed that covalently binds to the active site, thus inactivating the enzyme.

Interestingly, nonheterocyclic moieties were also found to be capable of complexing the heme iron. One example is the α, β unsaturated oxime compound MH-47 (Scheme 5.7), a highly active and selective *in vitro* and *in vivo* inhibitor.

Several researchers have focused recently on what is termed *dual inhibitors* of CYP 17 and 5α-reductase. Such a strategy might be more effective than inhibition of CYP 17 alone because a potent dual inhibitor would block the synthesis of both T and DHT. Indeed, several dual inhibitors have been reported up to now and some of them are highly active toward both enzymes. All of these inhibitors are steroids, in particular, progesterone-based compounds. LY-20720 was originally described as a 5α-reductase inhibitor and was later discovered to be active toward CYP 17 as well.

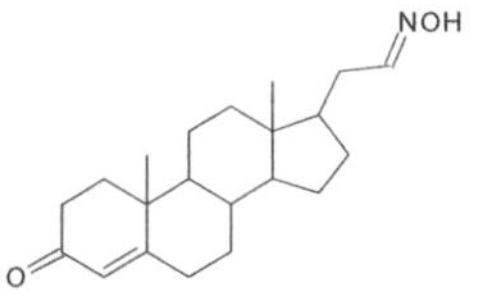

MH-54

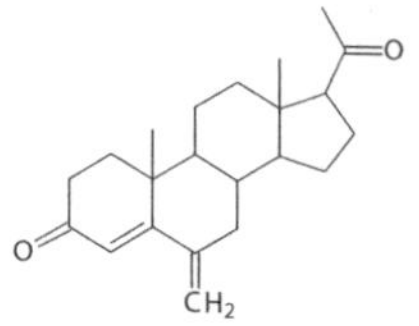

LY-20720

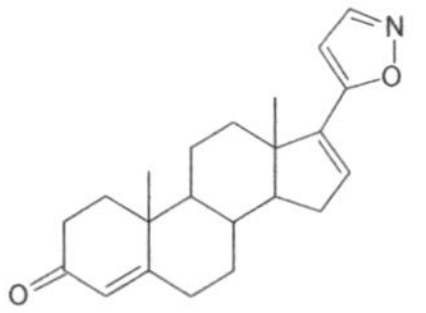

L-39

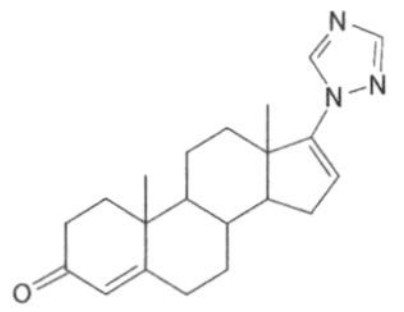

VN-107-1

SCHEME 5.8

More recently, compounds MH-54, L-39, and VN-107-1 (Scheme 5.8) have been described as dual inhibitors exhibiting strong inhibitory activity toward both enzymes.

FURTHER READING

5α-Reductase

Andersson, S. and Russell, D.W. (1990). Structural and biochemical properties of cloned and expressed human and rat steroid 5α-reductase. *Proceedings of the National Academy of Sciences U.S. A.*, 87, 3640–3644.

Bartsch, G., Rittmaster, R.S., and Klocker, H. (2000). Dihydrotestosterone and the concept of 5α-reductase inhibition in human benign hyperplasia. *European Urology*, 37, 367–380.

McConnell, J.D. (1990). Androgen ablation and blockade in the treatment of benign prostatic hyperplasia. *Urological Clinics of North America*, 17, 661–670.

Thiboutot, D.M., Gilliland, K.B., Light, J.L., and Lookingbill, D.M. (1999). Androgen metabolism in sebaceous glands from subjects with and without acne. *Archive of Dermatology*, 135, 1041–1045.

Inhibitors of 5α-Reductase

Audet, P., Nurcombe, H., Lamb, Y., Jorkasky, D., Loyd-Davies, K., and Morris, R. (1993). Effect of multiple doses of epristeride (E) a steroidal 5α-reductase inhibitor on serum dihydrotestosterone (DHT) in older male subjects. *Clinical Pharmacology and Therapeutics*, 53, 231.

Bramson, H.N., Hermann, D., Batchelor, K.W., Lee, F.W., James, M.K., and Frye, S.V. (1997). Unique preclinical characteristics of GG745, a potent dual inhibitor of 5α-reductase. *Journal of Pharmacology and Experimental Therapeutics*, 282, 1486–1502.

Chapelsky, M.C., Nicholas, A., Jorkasky, D.K., Lundberg, D.E., Knox, S.H., and Audet, P.R. (1992). Pharmacodynamics of SK&F105657 in healthy male subjects. *Clinical Pharmacology and Therapeutics*, 51, 154.

Coltman, C.A., Thompson, I.M. and Feigl, P. (1999). Prostate cancer prevention trial (PCPT) update. *European Urology*, 35, 544–547.

Cooper, K.L., McKiernan, J.M., and Kaplan, S.A. (1999). α-Adrenoceptor antagonists in the treatment of benign prostatic hyperplasia. *Drugs*, 57, 9–17.

Frye, S.V. (1996). Inhibitors of 5α-reductase. *Current Pharmaceutical Design*, 2, 59–84.

Graul, A., Silvestre, J., and Castaner, J. (1999). Dutasteride. *Drugs of the Future*, 24, 246–324.

Hirosumi, J., Nakayama, O., Chida, N., Inami, M., Fagan, T., Sawada, K., Shigematsu, S., Kojo, H., Notsu, Y., and Okuhara, M. (1995). FK143, a novel nonsteroidal inhibitor of steroid 5a-reductase: (2) *In vivo* effects on rat and dog prostate. *Journal of Steroid Biochemistry and Molecular Biology*, 52, 365–373.

Hobbs, S., Hermann, D.J., Gabriel, T., Wilson, B., Morrill, R., and Clark, R.V. (1998). Marked suppression of dihydrotestosterone in men by novel 5 alpha reductase inhibitor. *Fertility and Sterility*, 70 (Suppl. 3), 4555.

Holt, D.A., Yamashita, D.S., Konialian-Beck, A.L., Luengo, J.I., Abell, A.D., Bergsam, D.J., Brandt, M., and Levy, M.A. (1995). Benzophenone and indolecarboxylic acids: Potent type 2 specific inhibitors of human steroid 5α-reductase. *Journal of Medicinal Chemistry*, 38, 13–15.

Levy, M.A., Brandt, M., Sheedy, K.M., Dinh, J.T., Holt, D.A., Garrison, L.M., Bergsma, D.J., and Metcalf, B.W. (1994). Epristeride is a selective and specific uncompetitive inhibitor of human steroid 5α-reductase isoform 2. *Journal of Steroid Biochemistry and Molecular Biology*, 48, 197–206.

Li, X., Chen, C., Singh, M., and Labrie, F. (1995). The enzyme and inhibitors of 4-ene-3-oxosteroid 5α-oxidoreductase. *Steroids*, 60, 430–441.

McCellan, K.J. and Markham, A. (1999). Finasteride: a review of its use in male pattern hair loss. *Drugs*, 57, 111–126.

McConnell, J.D., Wilson, J.D., George, F.W., Geller, J., Pappas, F., and Stoner, E. (1992). Finasteride, an inhibitor of 5 alpha reductase suppresses prostatic dihydrotestosterone in men with benign prostatic hyperplasia. *The Journal of Clinical Endocrinology & Metabolism*, 74, 505–508.

McNulty, A.M., Audia, J.E., Bemis, K.G., Goode, R.L., Rocco, V.P., and Neubauer, B.L. (2000). Kinetic analysis of LY320236: competitive inhibitor of type I and noncompetitive inhibitor of type II human steroid 5α-reductase. *Journal of Steroid Biochemistry and Molecular Biology*, 72, 13–21.

Norman, R.W., Coakes, K.E., Wright, A.S., and Rittmaster, R. (1993). Androgen metabolism in men receiving finasteride before prostatectomy. *Journal of Urology*, 150, 1736–1739.

Picard, F., Barassin, S., Mokhtarian, A., and Hartmann, R.W. (2003). Synthesis and evaluation of 2`-substituted 4-(4`-carboxy- or 4`-carboxymethylbenzylidene)-N-acylpiperidines: Highly potent and *in vivo* active steroid 5α-reductase type 2 inhibitors. *Journal of Medicinal Chemistry*, 45, 3406–3417.

Picard, F., Schulz, T., and Hartmann, R.W. (2002). 5-Phenyl substituted 1-methyl-2-pyridones and 4-substituted biphenyl-4`-carboxylic acids. Synthesis and evaluation as inhibitors of steroid-5 alpha-reductase type 1 and 2. *Bioorganic Medicinal Chemistry*, 10, 437–448.

Rasmusson, G.H., Johnston, D.B., and Arth, G.E. (1983). 4-Aza-17-substituted 5α-androstan-3-one reductase inhibitors. International patent application U.S. 4,377,584, March 22.

Rasmusson, G.H., Johnston, D.B., Reinhold, D.F., Utne, T., and Jobson, R.B. (1980). Preparation of 4-aza-17-substituted-5α-androstan-3-ones useful as 5α-reductase inhibitors. International patent application U.S. 4,220,775, September 2.

Steiner, F. (1996). Clinical pharmacokinetics and pharmacodynamics of finasteride. *Clinical Pharmacokinetics*, 30, 16–27.

Zaccheo, T., Giudici, D., and DiSalle, E. (1997). Effect of turosteride, a 5 alpha reductase inhibitor on the Dunning R3327 rat prostatic carcinoma. *Prostate*, 30, 85–91.

Prostate Cancer and CYP 17

Akhtar, M., Corina, D., Miller, S., Shyadehi, A.Z., and Wright, N. (1994). Mechanism of the acyl-carbon cleavage and related reactions catalyzed by multifunctional P-450s: studies on cytochrome P450 17. *Biochemistry*, 33, 4410–4418.

De Coster, R., Wouters, W., and Bruynseels, J. (1996). P 450-dependent enzymes as targets for prostate cancer therapy. *Journal of Steroid Biochemistry and Molecular Biology*, 56, 133–143.

Hartmann, R.W., Ehmer, P.B., Haidar, S., Hector, M., Jose, J., Klein, C.D.P., Seidel, S., Sergejew, T.F., Wachall, B.G., Wächter, G.A., and Zhuang, Y. (2002). Inhibition of CYP 17, a new strategy for the treatment of prostate cancer. *Archive Der Pharmazie, Pharmaceutical and Medicinal Chemistry*, 33, 119–128.

Huggins, C. and Hodges, C.V. (1941). Studies of prostate cancer. The effects of castration, of estrogen and androgen injections on serum phosphatases in metastatic carcinoma of prostate. *Cancer Research*, 1, 293–307.

Nakajin, S., Shively, J.E., Yuan, P.M., and Hall, P.F. (1981). Microsomal cytochrome P450 from neonatal pig testis: Two enzymatic activities (17 α-hydroxylase and C17, 20-lyase) associated with one protein. *Biochemistry*, 20, 4037–4042.

Njar, V.C.O. and Brodie, A.M.H. (1999). Inhibitors of 17α-hydroxylase/17, 20-lyase (CYP 17): potential agents for the treatment of prostate cancer. *Current Pharmaceutical Design*, 5, 163–180.

Inhibitors of CYP 17

Angelastro, M.R., Laughlin, M.E., Schatzman, G.L., Bey, P., and Blohm, R. (1989). 17 (cyclopropylamino)androsta-5-en-3-ol, as a selective mechanism-based inhibitor of cytochrome P450 17 (steroid 17α-hydroxylase/C 17-20-lyase). *Biochemical and Biophysical Research Communications*, 162, 1571–1577.

Ayub, M. and Levell, M. (1987). Inhibition of testicular 17 alpha-hydroxylase and 17, 20-lyase but not 3 beta-hydroxysteroid oxidoreductase by ketoconazole and other imidazole drugs. *Journal of Steroid Biochemistry*, 28, 521–531.

Barrie, S.E., Haynes, B.P., Potter, G.A., Chan, F.C.Y., Coddard, P.M., Dowsett, M., and Jarman, M. (1996). Biochemistry and pharmacokinetics of potent nonsteroidal cytochrome P450 17α inhibitors. *Journal of Steroid Biochemistry and Molecular Biology*, 60, 347–351.

Haidar, S., Ehmer, P.B., Barassin, S., Batzl-Hartmann, C., and Hartmann, R.W. (2003). Effects of novel 17α-hydroxylase/C17, 20-lyase (P450 17, CYP 17) inhibitors on androgen biosynthesis *in vitro* and *in vivo*. *Journal of Steroid Biochemistry and Molecular Biology*, 84, 555–562.

Hartmann, R.W., Hector, M., Haidar, S., Ehmer, P.B., Reichert, W., and Jose, J. (2000). Synthesis and evaluation of novel steroidal oxime inhibitors of P 450 17 (17α-hydroxlase/C 17-20-lyase) and 5α-reductase types 1 and 2. *Journal of Medicinal Chemistry*, 43, 4266–4277.

Hartmann, R.W., Wachall, B.G., Yoshihama, M., Nakakoshi, M., Nomoto, S., and Ikeda, Y. (2002). Novel dihydronaphthalene compounds and processes of producing the same. U.S. patent US 2002/0032211 A1.

Higashi, Y., Omura, M., Suzuki, K., Inano, H., and Oshima, H. (1987). Ketoconazole as a possible universal inhibitor of cytochrome P450 dependent enzymes: its mode of inhibition. *Endocrinologia Japonica*, 34, 8105–8115.

Ideyama, Y., Kudoh, M., Tanimoto, K., Susaki, Y., Nanya, T., Nakahara, T., Ishikawa, H., Yoden, T., Okada, M., Fujikura, T., Akaza, H., and Shikama, H. (1998). Novel nonsteroidal inhibitor of cytochrome P450 17 (17α-hydroxylase-C17-20-lyase), YM116, decreased prostatic weights by reducing serum concentrations of testosterone and adrenal androgens in rats. *Prostate*, 37, 10–18.

Ideyama, Y., Kudoh, M., Tanimoto, K., Susaki, Y., Nanya, T., Nakahara, T., Ishikawa, H., Fujikura, T., Akaza, H., and Shikama, H. (1999). YM 116, 2-(1H-Imidazole-4-ylmethyl)-9H-carbazole, decreases adrenal androgen synthesis by inhibiting C17-20 lyase activity in NCI-H295 human adrenocortical carcinoma cells. *Japan Journal of Pharmacology*, 79, 213–220.

Klus, G.T., Nakamura, J., Li, J-S., Ling, Y-Z., Son, C., Kemppainen, A., Wilson, E.M., and Brodie, A.M.H. (1996). Growth inhibition of human prostate cells *in vitro* by novel inhibitors of androgen synthesis. *Cancer Research*, 56, 4956–4964.

Neubauer, B.L., Best, K.L., Blohm, T.R., Gast, C., Goode, R.L., Hirsch, K.S., Laughlin, M.E., Petrow, V., Smalstig, E.B., Stamm, N.B., Toomey, R.E., and Hoover, D. (1993). LY207320 (6-Methylene-4-pregnene-3,20-dione) inhibits testosterone biosynthesis, androgen uptake, 5α-reductase, and produces prostatic regression in male rats. *Prostate*, 23, 181–199.

Nnane, P., Long, B.J., Ling, Y-Z., Grigoryev, D.N., and Brodie, A.M. (2000). Anti-tumor effects and pharmacokinetic profile of 17-(5-isoxazolyl) androsta-4, 16-dien-3-one (L-39) in mice: an inhibitor of androgen synthesis. *British Journal of Cancer*, 83, 74–82.

Potter, G.A., Barrie, E.S., Jarman, M., and Rowlands, M.G. (1995). Novel steroidal inhibitors of human cytochrome P450 17α: potential agents for the treatment of prostatic cancer. *Journal of Medicinal Chemistry*, 38, 2463–2471.

Trachtenberg, J., Halpern, N., and Pont, A. (1984). Ketoconazole: A novel and rapid treatment for advanced prostatic cancer. *Journal of Urology*, 132, 61–63.

Wachall, B.W., Hector, M., Zhuang, Y., and Hartmann, R.W. (1999). Imidazole substituted biphenyls — a new class of highly potent and *in vivo* active inhibitors of P450 17 as potential therapeutics for treatment of prostate cancer. *Bioorganic and Medicinal Chemistry*, 7, 1913–1924.

Zhuang, Y., Wachall, B.W., and Hartmann, R.W. (2000). Novel imidazolyl and triazolyl biphenyl compounds: synthesis and evaluation as nonsteroidal inhibitors of human 17α-hydroxylase-C17, 20-lyase (P450 17). *Bioorganic and Medicinal Chemistry*, 8, 1245–1252.

5.8 THROMBIN INHIBITOR EXAMPLES

Torsten Steinmetzer

5.8.1 INTRODUCTION

The trypsin-like serine protease thombin is the terminal key enzyme of the blood-clotting cascade and specifically activates a number of macromolecular substrates by cleavage of a peptide bond at the C-terminal site of the basic amino acid arginine (Arg). This process normally is localized at the site of a vascular injury and protects the body against uncontrolled blood loss.

The procoagulant thrombin substrates include a number of clotting factors such as fibrinogen (factor I), the zymogen forms of the transglutaminase factor XIII, and cofactors V and VIII. Their activation finally results in the formation and stabilization of a fibrin clot and stops bleeding. In addition, thrombin promotes coagulation and induces a number of cellular effects by proteolytic activation of PARs (protease activatable receptors), localized at the surface of blood and endothelial cells. This can lead to an enhanced rate of cell proliferation and the release or transport of numerous proteins and small molecules to the cell surface, which accelerate and increase platelet plug formation. Such activated platelets also amplify the coagulation reactions by providing a prothrombogenic scaffold at their membrane surfaces, and make an important contribution to actions that speed both wound healing and repair processes. Under normal conditions, the coagulation is carefully balanced with the fibrinolytic system, which is also a proteolytic cascade that removes the fibrin clot in conjunction with the repair process.

Abnormalities in any step of the coagulation and fibrinolysis can lead to bleeding, and especially to thrombosis, which is the major cause of cardiovascular diseases in the Western world. Therefore, it is assumed that a potent and specific orally available thrombin inhibitor with a wide therapeutic range would improve present strategies of anticoagulant therapy and be especially useful for prophylaxis of thrombotic diseases. At present, there is only limited clinical use of some parenteral preparations of thrombin inhibitors in acute situations, especially when the common antithrombotic drugs heparin, warfarin, and aspirin are not effective or are associated with side effects.

Thrombin circulates in the blood as the inactive zymogen prothrombin and is only activated after vascular injury at the final step of the coagulation cascade by the prothrombinase complex. This complex consists of an additional serine protease (factor Xa) and the cofactor Va (the index "a" indicates the activated form of the protease or cofactor), and is formed on a phospholipid surface in the presence of Ca^{2+} ions, e.g., on thrombocytes or endothelial cells. Activated thrombin is normally controlled and finally downregulated by endogenous inhibitors, mainly by the serpins (serine protease inhibitor) antithrombin and heparin cofactor II, or other proteins, such as α_2-macroglobulin. A second mechanism for the termination of the clotting cascade is initiated by thrombin itself; after clot formation, thrombin mediates an antithrombotic effect by proteolytically activating protein C after binding to the cell-surface protein thrombomodulin. Protein Ca in conjunction with the cofactor protein S inactivates the coagulation factors Va and VIIIa, which results in a break of the thrombin-generation reactions and terminates clotting.

In contrast, the thrombin substrates are permanently present in blood and, consequently, inhibitors, to be of therapeutic value, must be present in plasma at adequate concentrations to immediately neutralize the thrombin generated upon massive activation after vascular injury. To be effective in anticoagulation, the plasma concentration of a potent inhibitor (K_i in the low nanomolar range) should be at least in the range of 100 to 200 nM or 0.05 to 0.1 µg/ml, assuming an average molecular weight of ≈500 g/mol for a synthetic thrombin inhibitor.

Up to the present, a tremendous number of potent and selective thrombin inhibitors have been developed, some of them with very promising properties.

5.8.2 First Electrophilic Substrate Analog Inhibitors

The development of synthetic inhibitors and substrates of thrombin had already started in the 1960s, long before any knowledge of the 3-D structure of the enzyme's active site was available. These substrate-analog structures were designed to mimic fibrinopeptide A (FPA), which is cleaved from the Aα-chain of the thrombin substrate fibrinogen (Figure 5.12). Starting from the poor substrate Tosyl-Arg-methylester (**5.188**), it was realized that in addition to the P1-Arg and P2-Val, the P9-Phe was also important for thrombin affinity of FPA mimetics. The tripeptide derivative with the best inhibiting effect on the thrombin fibrinogen reaction was benzoyl-Phe-Val-Arg-OMe (**5.189**); this peptide sequence was transformed to benzoyl-Phe-Val-Arg-*p*-nitroanilide, the first chromogenic peptide substrate of thrombin.

Further modifications by using the P1'-P3' sequence of the fibrinogen-cleavage site and replacement of the substrate-like Arg-OMe group by a C-terminal arginal, the aldehyde analog of arginine, improved thrombin inhibition (**5.190**). A comprehensive structure–activity relationship study finally led to the discovery of DPhe as the optimal P3 amino acid (**5.191**). These DPhe-Pro-Arg derivatives, including the irreversible inhibitor H-DPhe-Pro-Arg-chloromethyl ketone (**5.192**, PPACK), were the first highly potent and relatively selective thrombin inhibitors discovered in the mid-1970s (Figure 5.12).

Later, it could be demonstrated by x-ray crystallography that the side chains of the P3-DPhe in such synthetic inhibitors and of P9-Phe in natural FPA occupy the

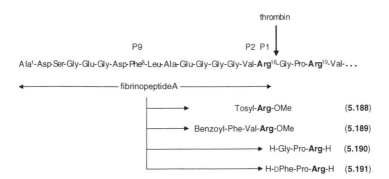

FIGURE 5.12 Schematic reaction of thrombin with the Aα-chain of fibrinogen — starting point for the development of the first synthetic substrate–analog thrombin inhibitors.

same so-called aryl-binding site formed by the hydrophobic thrombin residues Leu99, Ile174, and Trp215 (Figure 5.13A). Several additional binding interactions could be identified for each residue of compound **5.192** in complex with thrombin. The side chain of the catalytic Ser195 forms a covalent hemiketal bond with the P1 carbonyl group; in addition, His57, also part of the catalytic triad of serine proteases, is irreversibly alkylated by the methylene group of the inhibitor. The protonated guanidino group of the P1-Arginyl-ketone makes a double salt bridge with the side chain of Asp189 at the bottom of thrombin's S1 site.

The pyrrolidine ring of the P2-Pro is located in a hydrophobic S2 site below the amino acids Leu99, His57, and a thrombin-specific insertion loop formed of the amino acids Tyr60A, Pro60B, Pro60C, and Trp60D. The P2-Pro constrains the conformation of such tripeptide inhibitors and therefore strongly contributes to thrombin affinity.

In addition, several hydrogen bonds from the inhibitor backbone to the thrombin residues Gly216 and Ser214, which lead to the formation of a short antiparallel β-sheet between enzyme and inhibitor (Figure 5.13B), could be identified.

Inhibitors **5.191** to **5.192**, the close analog efegatran (**5.193**, a chemically more stable derivative of **5.191**), and ketone-based inhibitors **5.194** to **5.197** contain activated P1 carbonyl groups prone to the addition of nucleophiles that act as a trap for thrombin's catalytic Ser195 hydroxyl group. As a consequence of the formed hemiacetal or hemiketal structure (generated by the reaction of an aldehyde or ketone, respectively, with a hydroxyl group of an alcohol), the P1 carbonyl oxygen is transformed to a negatively charged oxyanion that is stabilized by two hydrogen bonds to the amide-NH of the residues Gly193 and Ser195. This binding mode resembles the putative transition state of the proteolytic mechanism of serine proteases (see Chapter 2) and generates highly potent inhibitors. A similar binding mode was observed for the remarkably potent boronic acid derivatives containing a C-terminal boroarginine (**5.198**, K_i 41 pM) or related ester analogs, which also form a covalent bond to thrombin's Ser195. However, the poor selectivity with respect to other trypsin-like enzymes, especially to the proteases of the fibrinolytic system, and their low oral bioavailability limited the clinical development of **5.198** and prompted for the first time the design of analogs with neutral P1 side chains. Exchange of boroArg at P1 with the pinacol ester of methoxypropylboroGly in derivative TRI-50b (**5.199**) enhanced both oral bioavailability and selectivity of inhibition. Derivative **5.199**, lacking the basic guanidino function of Arg that was thought to be essential for binding to Asp189 in the P1 specificity pocket of trypsin-like proteases, is still a potent thrombin inhibitor with a nanomolar K_i value (7 nM).

(**5.192**) PPACK   (**5.193**) efegatran

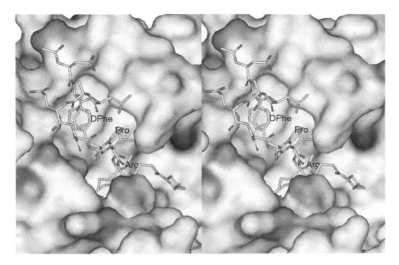

FIGURE 5.13A (See color insert following page 176.) Stereo view of the active site region of thrombin in complex with PPACK (**5.192**, green carbon atoms) superimposed with the structure of a proteolytically stable FPA mimetic Acetyl-Asp-Phe-Leu-Ala-Glu-Gly-Gly-Gly-Val-Argψ(CO–CH$_2$)Gly-Pro-OH, in which the NH of the P1'-Gly has been replaced by a CH$_2$ group (gray carbon atoms). Thrombin is represented as a solid surface with the colors showing the electrostatic surface potential (negatively and positively charged regions are shown in red and blue with different intensities, respectively). The side chains of P3-ᴅPhe, P2-Pro, and P1-Arg in inhibitor **5.192** and P9-Phe, P2-Val, and P1-Arg in the FPA mimetic occupy similar binding sites on thrombin. The structures were generated from PDB entries 1ppb for **5.192** and 1ucy for the FPA mimetic, both obtained from the Protein Data Bank (www.rcsb.org/pdb/).

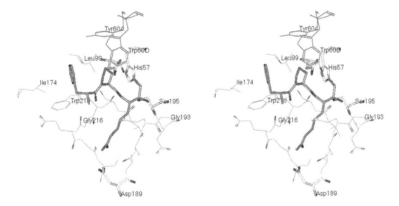

FIGURE 5.13B (See color insert following page 176.) Stereo view of the active site region of PPACK-thrombin. The residues of the catalytic triad (Ser195, His57, Asp102) and Asp189 at the bottom of the S1-binding pocket are shown as sticks with yellow carbon atoms; all other thrombin residues are shown as lines. Tyr60A and Trp60D of the thrombin-specific insertion loop are colored in green. The irreversible inhibitor PPACK (**5.192**) forms a covalent, hemiketal-like bond to the thrombin residue Ser195 and alkylates His57. Selected thrombin residues are labeled.

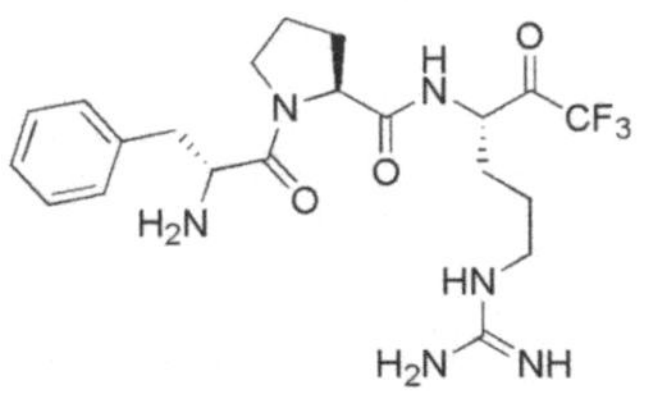

(5.194)

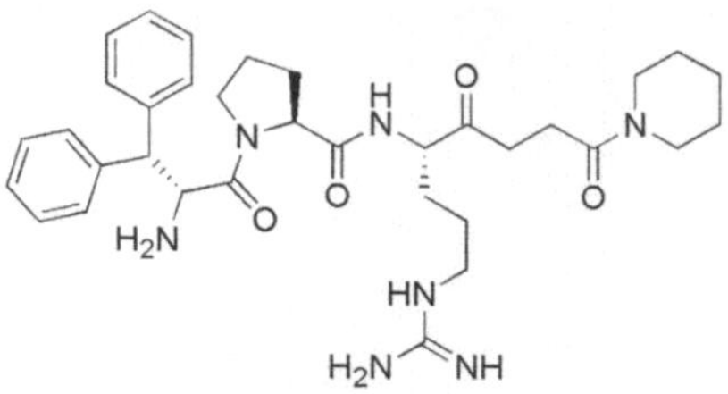

(5.195)

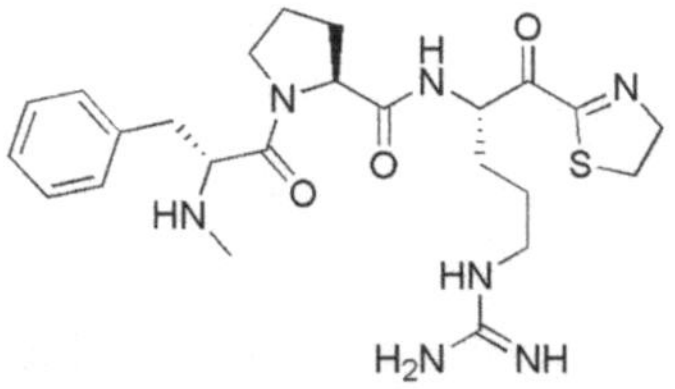

(5.196)

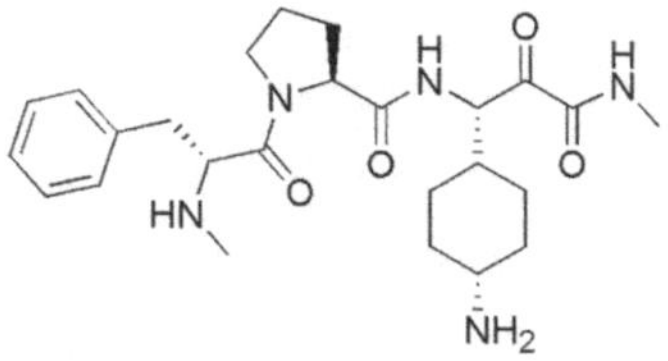

(5.197)

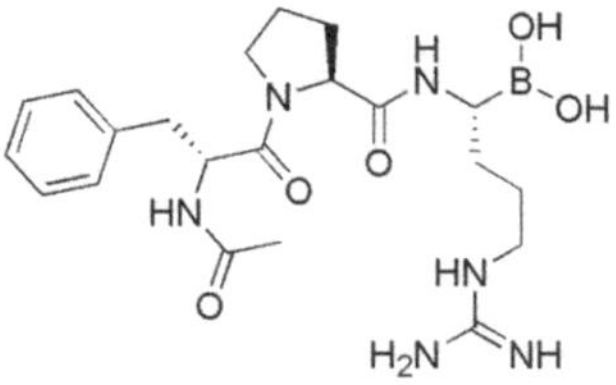

(5.198) DUP-714

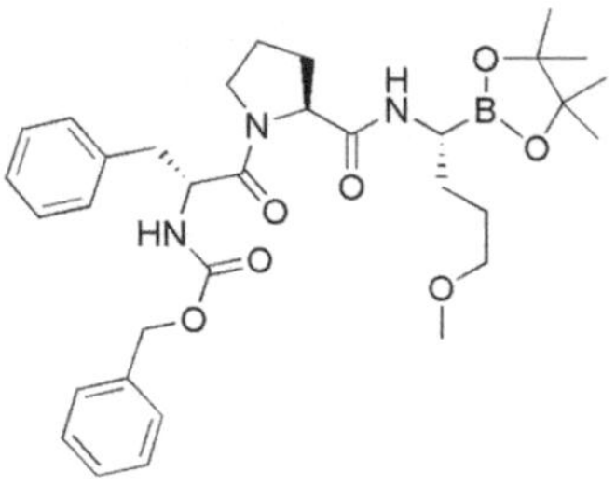

(5.199) TRI-50b

However, such transition-state analogs are only of limited use for drug design because their electrophilic carbonyl group can react nonspecifically with a variety of nucleophiles in blood, and therefore they often have a very limited half-life. Compared to non-transition-state analogs, in many cases these inhibitors are more difficult to synthesize and have the disadvantage that their P1 residues are prone to racemization. Therefore, recent effort was focused on the development of nonelectrophilic thrombin inhibitors with improved pharmacokinetic properties, especially prolonged half-life, stability, and oral bioavailability.

5.8.3 Nonelectrophilic Thrombin Inhibitors

First leads (**5.200** to **5.202**) were obtained in the early 1980s, and can be roughly separated into three groups still containing at least one amino acid: (1) tripeptide analogs with decarboxylated P1 residues derived from H-ᴅPhe-Pro-agmatine (**5.200**), (2) secondary amides of sulfonylated arginine based on argatroban (**5.201**), and (3) benzamidine derivatives of the NAPAP-type (**5.202**). In recent years, some nonpeptidic inhibitors with very different structural types have been identified, which can be summarized in a fourth group. For all leads, similar strategies were employed to improve affinity, selectivity, oral bioavailability, and half-life and to lower the toxicity of the synthetic inhibitors, which include:

- Reduction in basicity of the P1 residue by incorporation of Arg mimetics with lowered pK_a values or by employing a prodrug strategy to improve oral bioavailability
- Replacement of peptide segments by conformationally constrained peptide mimetics to reduce the number of peptide bonds, and improve affinity, stability, and oral bioavailability
- Incorporation of carboxyl groups, including their ester prodrug forms, to enhance hydrophilicity and half-life by avoiding the rapid clearance of the inhibitors from the blood circulation via a hepatobiliary elimination, typically found for very hydrophobic compounds.

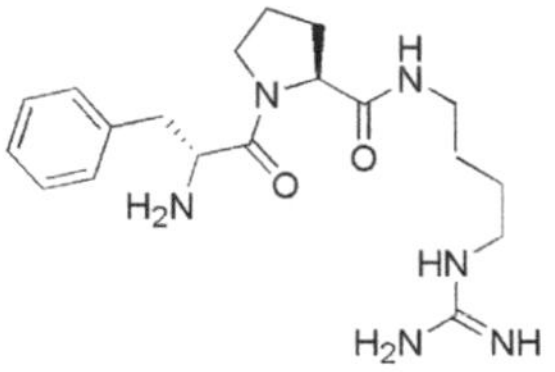

(**5.200**)	(**5.201**) argatroban	(**5.202**) NAPAP

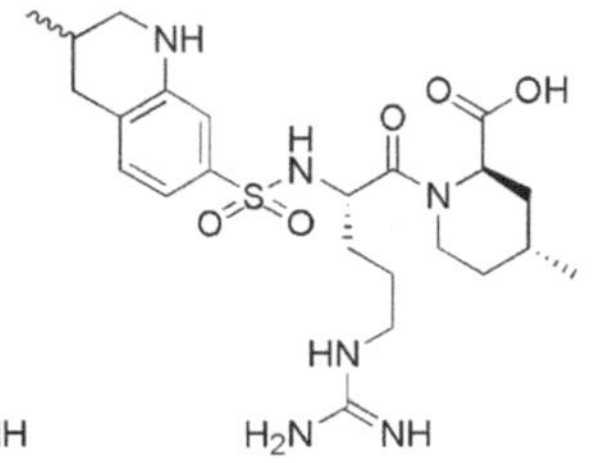

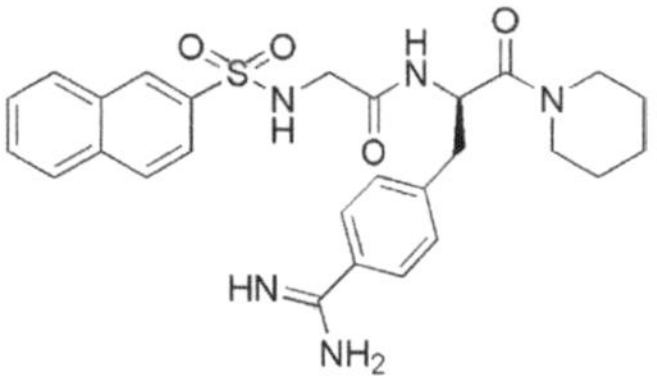

5.8.3.1 H-ᴅPhe-Pro-Agmatine Analogs

Because the agmatine-derivative (**5.200**, K_i 0.18 μM) was less potent than the aldehyde efegatran (**5.193**, K_i 1.8 nM), it was not further developed for several years. However, starting with the noragmatine-based inogatran (**5.203**, K_i 15 nM), a growing interest in this inhibitor class arose. The hydrogenated cyclohexyl side chain of inogatran fits better into the aryl-binding side of thrombin and significantly enhances the affinity compared to the ᴅPhe analogs. In addition, the N-terminal carboxymethyl group improved the hydrophilicity, which leads to a prolonged half-life. Due to their strong basic P1 residues, inogatran (pK_a of the guanidine ≈13) and the close analog melagatran (**5.204**, K_i 2 nM) containing a 4-amidinobenzylamide (pK_a of the amidine ≈11.5) showed only negligible oral bioavailability at below 5%. This problem could be solved in the case of ximelagatran, a double-prodrug derivative of melagatran that contains a noncharged hydroxyamidino moiety (pK_a value 5.2) and an ethyl-ester-protected carboxyl group.

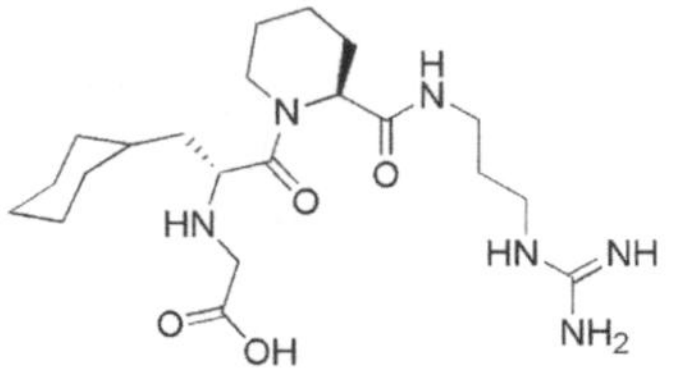

(**5.203**) inogatran

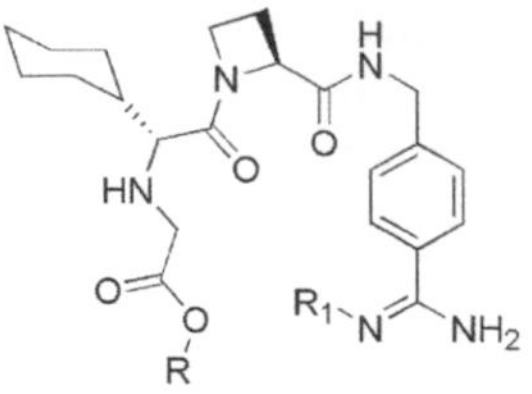

(**5.204**) (R = H, R₁ = H) melagatran

(**5.205**) (R = CH₂-CH₃, R₁ = OH) ximelagatran

Ximelagatran shows a relatively constant bioavailability of between 20 and 25% in humans. The inactive ximelagatran is easily converted to melagatran after absorption of the drug; this process proceeds via hydrolysis of the ethylester bond and reduction of the hydroxyamidine by a NADH-dependent liver microsomal benzamidoxime reductase.

Very recently, with ximelagatran (**5.205**) a first orally available thrombin inhibitor has been approved in Europe for the prevention of venous thromboembolic events in major orthopedic surgery after successful clinical Phase III. However, increased concentrations of liver transaminases were recently reported as side effects in a small percentage of the patients after long-term treatment using either melagatran or ximelagatran. This problem is probably due to liver damage, which may partially limit the use of these drugs and requires monitoring.

Close melagatran analogs (**5.206**, **5.207**) have been developed by replacement of the strong basic P1 group by residues that are only partially charged at physiological pH (pK_a ≈ 6 to 8) or even by residues without any basic P1 moiety (**5.208**). However, despite strong *in vitro* potencies (e.g., **5.207**, K_i 0.82 nM; **5.208**, K_i 0.74 nM), these analogs had poor activity in thrombosis models *in vivo*. Due to their strong lipophilicity and a nearly complete plasma–protein binding, loss in activity occurred.

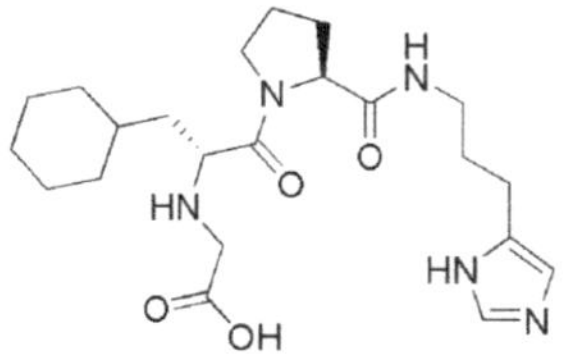

(**5.206**) 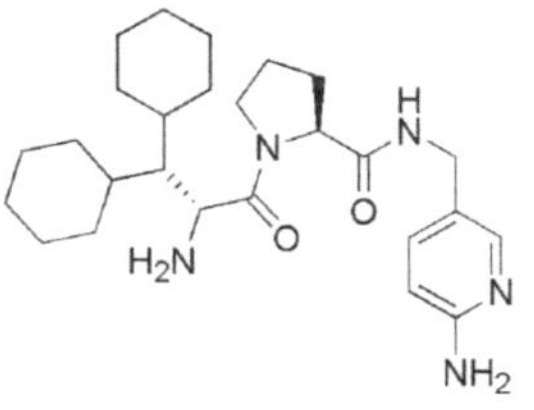   (**5.207**) 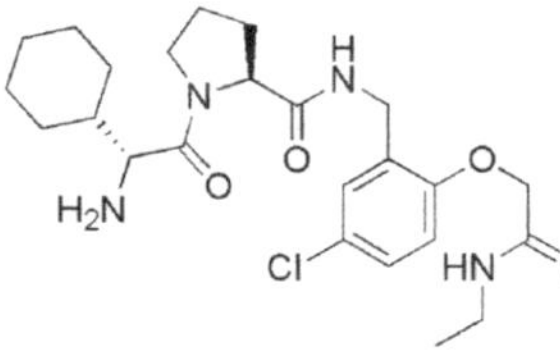   (**5.208**)

The next step was the replacement of the P3-P2 segment by nonpeptidic templates still containing the dipeptide backbone, as shown with the aminopyridinone and the aminopyrazinone acetamide inhibitors (**5.209**, **5.210**). Because of its excellent properties (K_i 0.8 nM; high selectivity; t$_{1/2}$ 231 min; bioavailability F = 60% in rhesus monkeys), the compound **5.210** was the subject of extensive metabolism studies, and several oxidation and conjugation products were identified. As a consequence, a metabolism-directed optimization process was performed, which resulted in a series of sufficiently stable and orally bioavailable compounds containing weakly basic pyridine moieties. The achiral compound **5.211** is still very potent (K_i 5.2 nM), despite the lack of any binding partner for Asp189 at the bottom of thrombin's S1 site, and exhibits good oral bioavailability in the African green monkey (F = 66%) with a half-life of 1.6 h after intravenous treatment, and was recently selected for clinical testing.

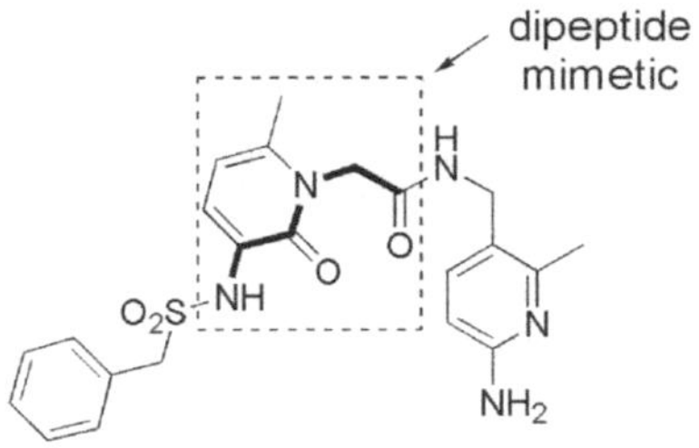

(5.209) 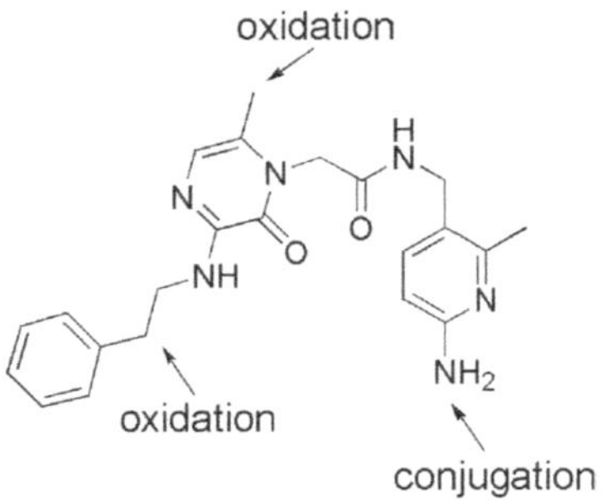   (5.210)

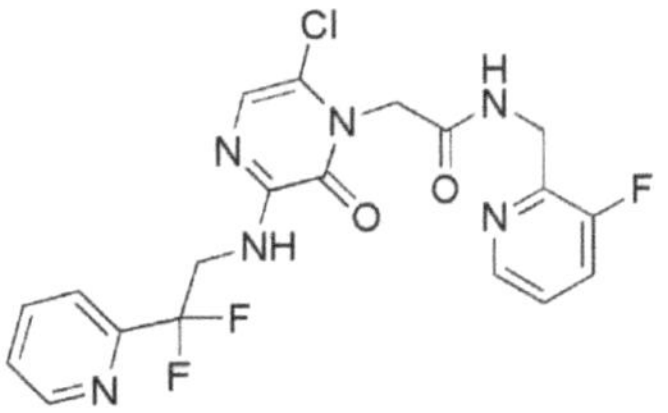

(5.211)

5.8.3.2 Secondary Amides of Sulfonylated Arginine

This type of thrombin inhibitor has been derived from the sulfonylated arginine-methylester (**5.188**). A key step toward lower toxicity within this substance class was the incorporation of an additional carboxyl group at the C-terminal amide group. The resulting argatroban (**5.201**) was the first synthetic thrombin inhibitor in clinical use so far. However, the use of argatroban is very limited because it is not sufficiently absorbed after oral administration and has to compete with a series of cheaper parenteral antithrombotics, mainly low-molecular-weight heparin mimetics. Close analogs of argatroban have been designed to improve oral bioavailability. One strategy was to lower the pK_a value of the P1-Arg by replacement of the guanidino group. This could be achieved for the canavanin-based inhibitor **5.212**, the 2-ami-nopyridine compound **5.213**, and the 4-amidrazonophenylalanine derivative **5.214**, with pK_a values of 7.01, 6.9, and 8.9, respectively. Also, inhibitor **5.215**, containing a noncharged P1 benzthiazolylalanine, showed a good overall profile including reasonable oral bioavailability when dosed to dogs (F 28%), despite the relatively high molecular weight of 825 Da.

UK-156406 (**5.216**, K_i 0.46 nM) contains a benzamidine moiety in the P1 position and an unsaturated P1' portion to eliminate the second chiral center present in argatroban. Data were presented that **5.216** may be potentially useful in the treatment of lung fibrosis in which activation of protease activatable receptors has been implicated. This was one of the first reports on therapeutic *in vivo* effects of thrombin inhibitors not related to blood clotting.

Promising data were also obtained for the weakly basic aminopyridine-derived inhibitor SSR182289A (**5.217**), which inhibits thrombin highly selectively with a K_i value of 31 nM, other trypsin-like proteases not being influenced. In dogs, a dose-related increase in clotting times was observed after oral dosing, whereas maximum anticoagulant effects were observed 2 h after administration. The measured clotting times were still markedly elevated 8 h after administration of either 3 or 5 mg/kg of inhibitor **5.217**.

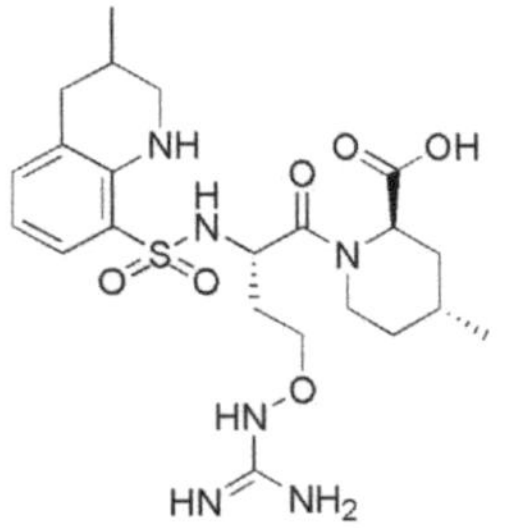

(5.212) 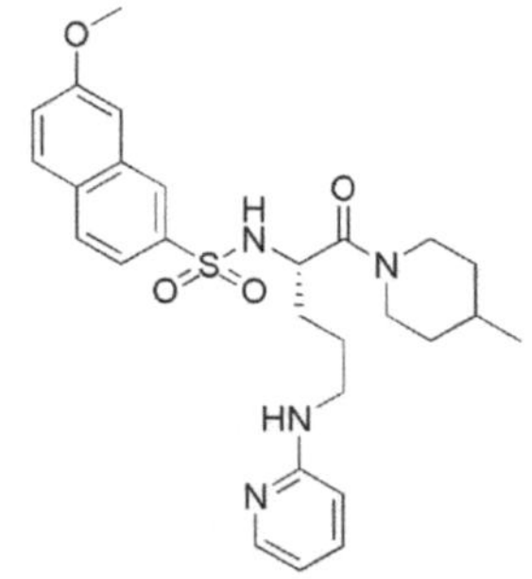   (5.213) 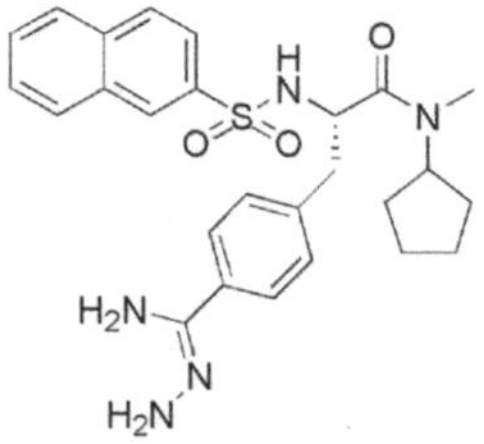   (5.214)

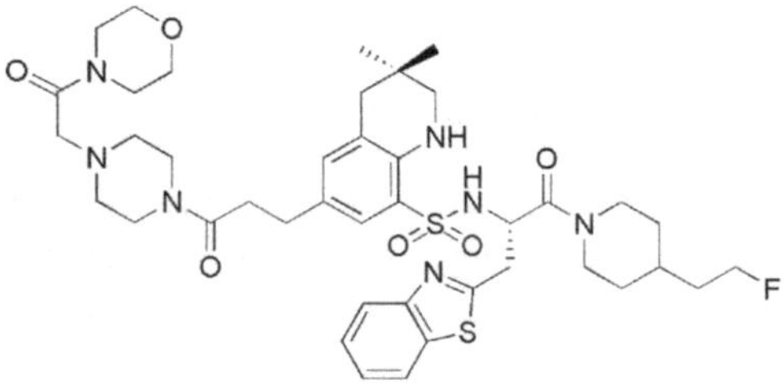

(5.215) 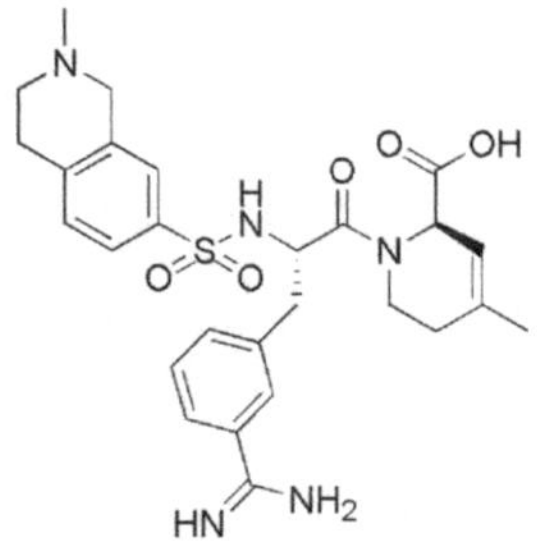   (5.216)

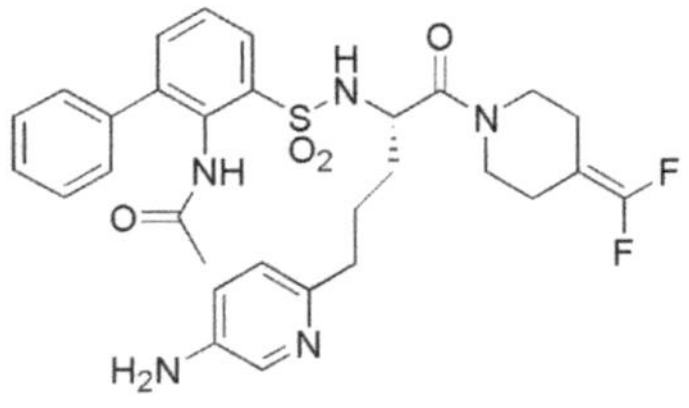

(5.217)

5.8.3.3   Benzamidine Derivatives of the NAPAP Type

Benzamidine is a nonspecific inhibitor of trypsin-like serine proteases (K_i for thrombin 0.3 mM). The ketone-derived 4-amidinophenylpyruvic acid (**5.218**) is nearly 100 times more potent (K_i 6.5 µM) and interacts under hemiketal formation with Ser195. Inhibitor **5.218** has outstanding oral bioavailability (up to 80%) and shows anticoagulant and antithrombotic effects *in vivo*.

The problems of poor selectivity and affinity found for such simple benzamidines were overcome by transferring them into nonnatural amino acids like 3- and 4-amidinophenylalanine. This enabled the incorporation of additional residues at their amino- and carboxy-terminal site, which was important to improve affinity and selectivity. The first analogs were $N\alpha$-tosylated piperidides of 3- and 4-amidinophenylalanine (**5.219**, **5.220**), which can also be considered as close analogs of argatroban (**5.201**). The incorporation of a glycine (Gly) spacer lead to NAPAP (**5.202**), which was the first thrombin inhibitor with a low nanomolar inhibition constant (K_i 2.1 nM for the D-enantiomer). Replacement of Gly in NAPAP by aspartic acid (Asp) containing a free carboxyl group and simultaneous modification of the N-terminal sulfonyl residue resulted in CRC-220 (**5.221**, K_i 2.4 nM). Both analogs, (**5.202**) and (**5.221**), were rapidly eliminated via an active hepatobiliary route using an ATP-dependent organic anion transporter 1 (OATP 1). This problem could be solved by the attachment of polyethylene glycols (PEG) to the side chain of Asp. The elimination half-life of (**5.222**) was prolonged to several hours with longer PEG chains having an average molecular weight of approximately 10 kDa. However, such analogs have the disadvantage that they will never be orally available and can only be used parenterally.

Also, in this class of inhibitors, some analogs with reduced basicity were developed, as shown with inhibitor **5.223** containing a P1 β-[6-(1-aminoisoquinolinyl)alanine] having a pK_a value of 7.5 (K_i 1.3 nM).

An outstanding selectivity was found for napsagatran (**5.224**), which contains S-3-(amidomethyl)-amidinopiperidine as a P1 residue attached to an Asp-side-chain carboxyl group. Despite high potency (K_i 0.3 nM), napsagatran has no improved pharmacokinetic properties and its clinical development was stopped in Phase II.

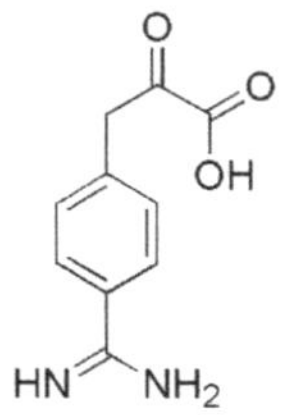

(**5.218**)

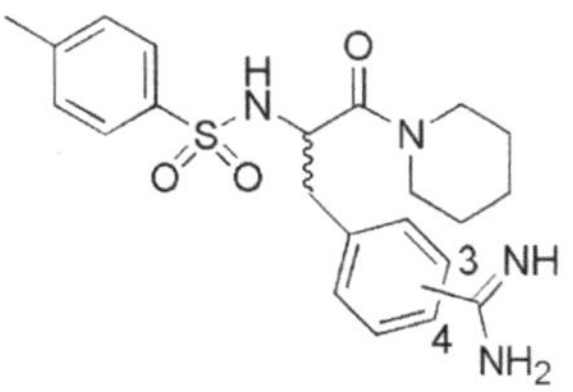

(**5.219**) amidine in 3-position
(**5.220**) amidine in 4-position

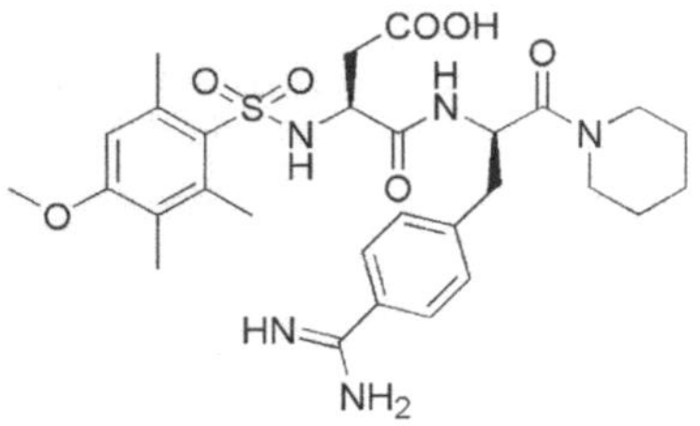

(5.221) CRC-220

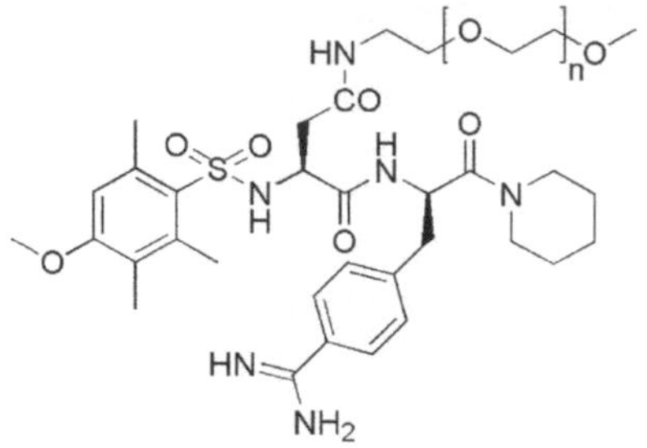

(5.222)

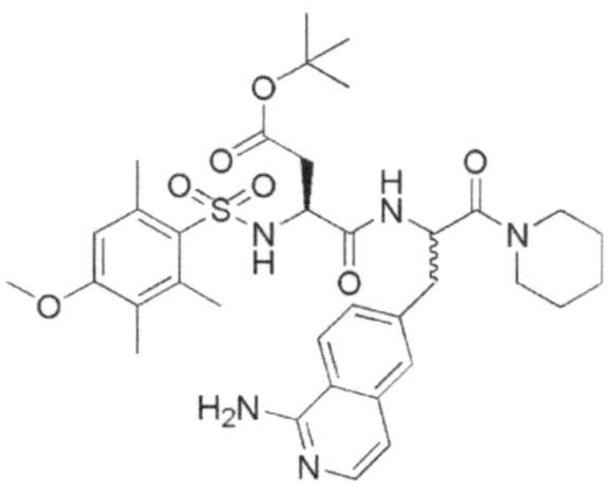

(5.223) Org 37476

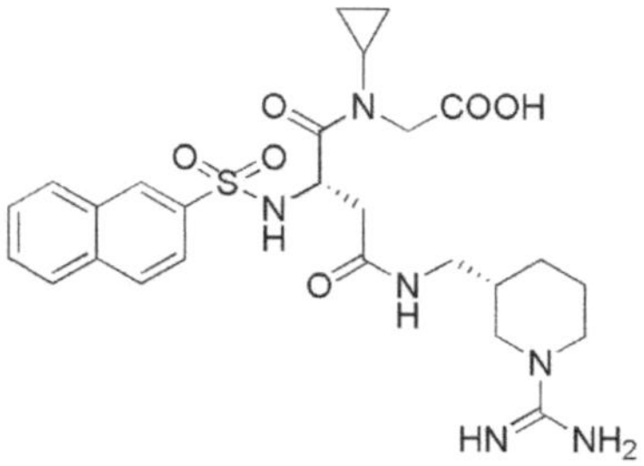

(5.224) napsagatran

5.8.3.4   Nonpeptidic Thrombin Inhibitors

Most of the known thrombin inhibitors were developed by a laborious stepwise optimization of first lead compounds. Since the beginning of the 1990s, a number of x-ray structures for thrombin in complex with small-molecule inhibitors became available. This information enabled the search for new lead inhibitors by an *in silico* screening of 3-D compound databases, using different types of docking strategies. An additional method is the *de novo* design, in which groups that fit into the binding pockets of thrombin were identified in 3-D segment databases, and opportunities were proposed concerning the connection of these residues. Compounds **5.225** and **5.226** are examples that could be identified by such strategies; both analogs inhibit thrombin in the nanomolar range with inhibition constants of 58 and 95 nM, respectively.

An additional possibility is the high-throughput screening of large databases of compounds, which are available in pharmaceutical companies. In this way, the nonpeptidic 2,3-disubstituted benzothiophene (**5.227**) was identified. This compound had originally been patented as an antifertility agent and showed only a weak thrombin inhibition with a K_i of 373 nM. The affinity could be enhanced by introduction of a hydroxyl group on the benzothiophene ring. Further optimization resulted in analogs like **5.228** with greatly improved potency (K_i 0.3 nM). A similar strategy was employed with a series of 4-aminopyridine derivatives. First leads were found by broad screening and inhibit thrombin at concentrations in the micromolar range. By lead optimization, several potent and selective nonpeptidic inhibitors were developed, e.g., compound **5.229** with a K_i value of 4 nM.

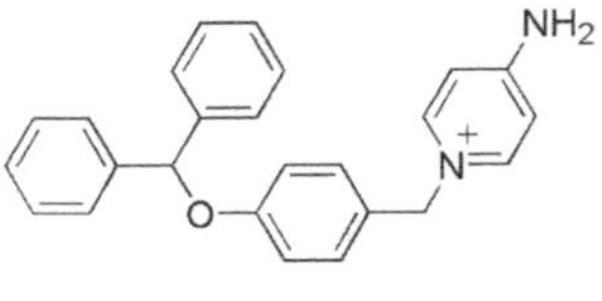

(5.225)

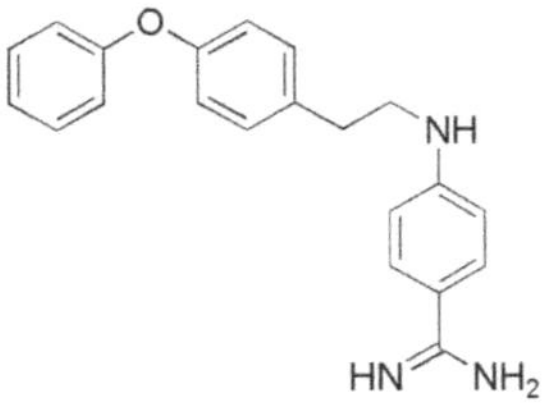

(5.226)

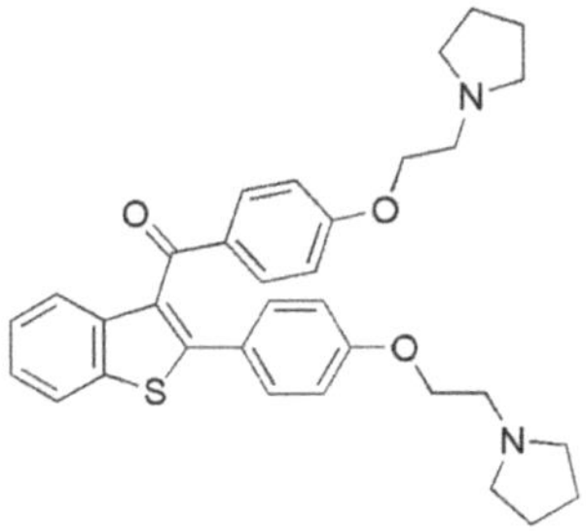

(5.227)

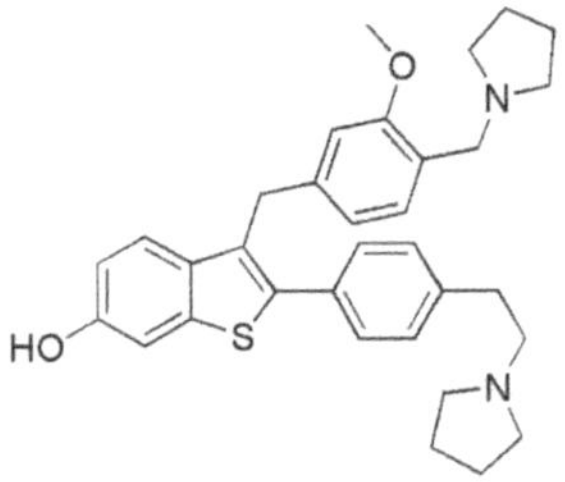

(5.228)

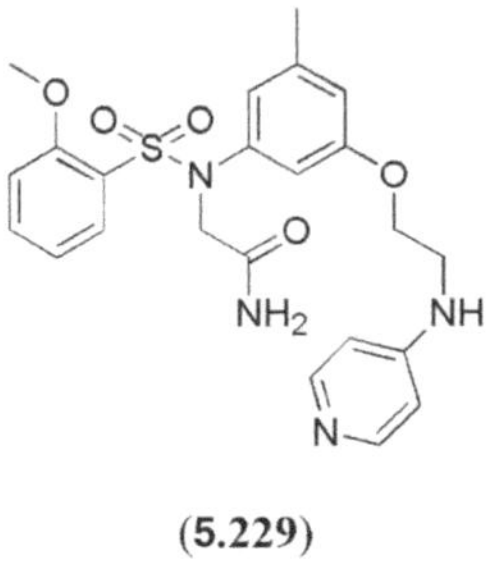

(5.229)

5.8.4 BIVALENT INHIBITORS

In addition to the active site, thrombin contains two positively charged exosites with a high percentage of the basic amino acids Arg and lysine (Lys) (also named anion binding sites I and II). The anion binding site I is important for the recognition of naturally occurring thrombin substrates such as fibrinogen or the thrombin receptor. Such substrates contain an acidic amino acid sequence that is electrostatically attracted by Exosite I and orients the cleavable peptide bond with high selectivity within the active site of thrombin. The anion binding site II is involved in the binding of heparin and therefore named also as the heparin binding site. Many bloodsucking animals contain highly potent thrombin inhibitors with exceptional selectivity because of a unique bivalent binding mode. These inhibitors bind to the active site and block one of the exosites at the same time. The best understood example is hirudin, a small 65-residue polypeptide originally isolated from the medicinal leech, which inhibits thrombin with a K_i of 22 f*M*. At the same time, the C-terminal tail of hirudin (residues 55 to 65) binds to Exosite I, and the N-terminal core domain blocks the active site of thrombin. Recently, two recombinant variants of hirudin (K_i 230 f*M*) have been approved for the prevention of deep venous thrombosis after hip replacement surgery and as heparin replacement for patients with heparin-induced thrombocytopenia.

Based on hirudin, a class of synthetic peptides has been designed. Hirulog (**5.230**), the first analog of these bifunctional inhibitors, binds highly selectively to thrombin (K_i 1.9 n*M*) and has been approved recently under the name Angiomax™ for clinical use as an antithrombotic during angioplasty.

The DPhe-Pro-Arg-Pro sequence of hirulog blocks the active site of thrombin and the C-terminal part binds to its anion binding site I, both inhibitor segments being connected by a pentaglycine linker. The proline in the P1′ position of hirulog is important to improve its stability; nevertheless, hirulog is slowly cleaved by thrombin and therefore has to be considered a poor thrombin substrate.

H-DPhe-Pro-Arg-Pro-$(Gly)_5$-Asn-Gly-Asp-Phe-Glu-Glu-Ile-Pro-Glu-Glu-Tyr-Leu-OH

(**5.230**) hirulog

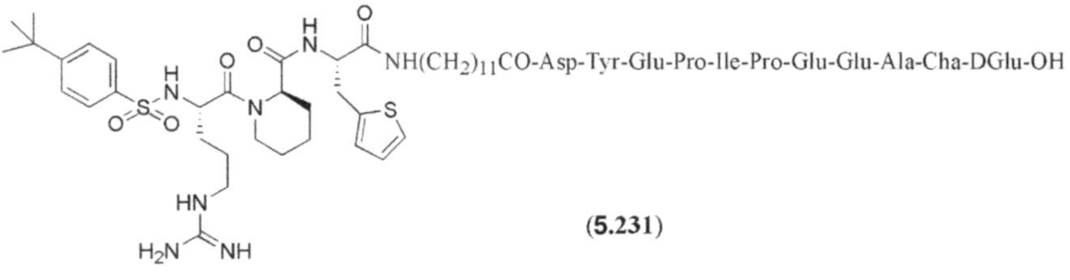
NH(CH2)11CO-Asp-Tyr-Glu-Pro-Ile-Pro-Glu-Glu-Ala-Cha-DGlu-OH
S
N
HN
H2N
NH

(**5.231**)

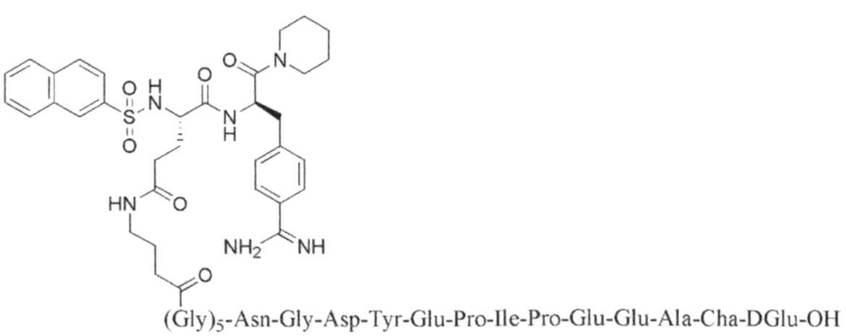
N
HN
NH2
NH
(Gly)5-Asn-Gly-Asp-Tyr-Glu-Pro-Ile-Pro-Glu-Glu-Ala-Cha-DGlu-OH

(**5.232**)

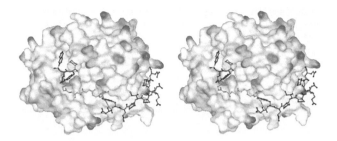

FIGURE 5.14 (See color insert following page 176.) Stereo view of the bivalent inhibitor **5.232** in complex with thrombin generated from the PDB entry 1qur. Thrombin is represented as a transparent solid surface with the colors showing the electrostatic surface potential (for an explanation see Figure 5.13). The carbon atoms of the inhibitor's active site segment are colored in magenta, the linker is shown in green, and the hirudin-like segment is colored with gray carbons.

Several highly potent analogs of hirulog have been developed, which contain proteolytically stable, active, site-directed inhibitor segments, e.g., those derived from argatroban (**5.231**, K_i 17 fM) or from NAPAP (**5.232**, K_i 290 fM). The structure of inhibitor **5.232** in complex with thrombin obtained by x-ray crystallography is shown in Figure 5.14.

Although these newer analogs were even more effective as antithrombotics in animal models compared to the approved hirulog (**5.230**), none of them reached clinical development.

A very large number of low molecular weight, synthetic thrombin inhibitors have been described during the past years. Although by stringent lead optimization some of them have sufficient oral bioavailability, it might be too early to say that the ideal antithrombotic has finally been discovered. Several compounds are presently in clinical development, and with double-prodrug ximelagatran (**5.205**) a first orally available thrombin inhibitor has been recently approved in Europe. Other compounds, such as **5.211** and **5.217**, show promising *in vitro* and *in vivo* characteristics with improved pharmacokinetic properties.

FURTHER READING

Coburn, C.A. (2001). Small-molecule direct thrombin inhibitors: 1997–2000. *Expert Opinion on Therapeutic Patents*, 11, 721–738.

Gustafsson, D., Bylund, R., Antonsson, T., Nilsson, I., Nyström, J.E., Eriksson, U., Bredberg, U., Teger-Nilsson, A.C. (2004). A new oral anticoagulant: the 50-year challenge. *Nature Reviews Drug Discovery*, 3, 649–659.

Hauptmann, J. and Stürzebecher, J. (1999). Synthetic inhibitors of thrombin and factor Xa: From bench to bedside. *Thrombosis Research*, 93, 203–241.

Maraganore, J.M., Bourdon, P., Jablonski, J., Ramachandran, K.L., and Fenton II, J.W. (1990). Design and characterization of hirulogs: A novel class of bivalent peptide inhibitors of thrombin. *Biochemistry*, 29, 7095–7101.

Rewinkel, J.B.M. and Adang, A.E.P. (1999). Strategies and progress towards the ideal orally active thrombin inhibitor. *Current Pharmaceutical Design*, 5, 1043–1075.

Steinmetzer, T. and Stürzebecher, J. (2004). Progress in the development of synthetic thrombin inhibitors as new orally active anticoagulants. *Current Medicinal Chemistry,* 11, 2297–2321.

Stubbs, M.T. and Bode, W. (1993). A player of many parts: The spotlight falls on thrombin's structure. *Thrombosis Research,* 69, 1–58.

5.9 HIV-1 PROTEASE DRUG DEVELOPMENT EXAMPLES

Paul J. Ala and Chong-Hwan Chang

5.9.1 INTRODUCTION

The human immunodeficiency virus (HIV) was identified as the causal agent of the acquired immunodeficiency syndrome (AIDS) in the early 1980s, and the Food and Drug Administration (FDA) approved the first anti-HIV drug in 1987. Currently, there are 18 drugs that target two viral enzymes, reverse transcriptase (RT) and HIV protease (PR). RT is responsible for reverse-transcribing the single-stranded viral RNA into double-stranded DNA, which is then integrated into the host genome by HIV integrase. PR, which is the topic of this section, plays a pivotal role in the maturation step of viral particles by processing the polyprotein gene products of *gag* and *gag-pol* into active structural and replicative proteins. Inactivating the protease by mutating its catalytic aspartates or by administering synthetic competitive inhibitors produces immature, noninfectious viral particles. This proof of concept, coupled with the availability of crystal structures of PR in complex with inhibitors, effectively launched the HIV protease rational drug discovery area. In just 4 years (from 1995 to 1999), the FDA approved the use of five PR inhibitors: saquinavir (Invirase and Fortovase; Hoffmann-LaRoche), ritonavir (Norvir; Abbott Laboratories), indinavir (Crixivan; Merck and Co.), nelfinavir (Viracept; Pfizer), and amprenavir (Agenerase; GlaxoSmithKline). Lopinavir (Kaletra; Abbott's second generation drug) was approved in 2000 and atazanavir (Reyataz; Bristol Myers Squibb) in 2003. Currently, there are seven approved inhibitors and four more in clinical trials. In general, PR inhibitors can be divided into two classes (Figure 5.15).

Class I inhibitors are flexible, linear molecules that retain varying degrees of peptidic character. This class can be further subdivided into four groups based on the presence of unique substituents and design concepts. For example, Group A compounds contain a hydroxyethylamine isostere and a Pro-Ile (P1'-P2') mimic (decahydroisoquinoline-*tert*-butylamide or piperazine-*tert*-butylamide). Group B compounds contain a hydroxyethylamine isostere and a sulfonamide substituent at the C-terminal side of the scissile bond. Group C inhibitors contain a hydroxyethylhydrazine isostere, and Group D compounds are pseudosymmetric molecules produced by duplicating the N-terminal side of the scissile bond. Class II inhibitors are nonpeptidic molecules that contain a cyclic core, which interacts with the catalytic aspartates and displaces the structural water molecule (Figure 5.16). A common feature of these successful drug discovery programs is the utilization of structural information to discover and/or optimize lead compounds. In general, leads were generated through rational design or random screening and then optimized using a combination of structure–activity relationship studies and iterative structure-based

I Peptidomimetic Inhibitors

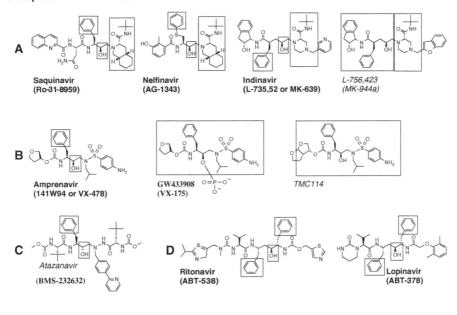

II Non-Peptidic Inhibitors

FIGURE 5.15 Classes of HIV-1 protease inhibitors. Drugs approved by the FDA (bold letters) and in clinical trials (italic letters) can be classified as either linear, flexible peptidomimetic (Class I) or rigid, nonpeptidic (Class II) inhibitors. Structural similarities are identified by rectangles. For example, all compounds contain at least one benzyl group at P1 and a hydroxyl between P1 and P1′; Class IA inhibitors contain similar Pro-Ile replacements; and Class IB compounds are nearly identical to amprenavir.

approaches, which relied on information obtained from crystal structures of protease–inhibitor complexes and sophisticated modeling exercises.

5.9.2 LEAD DISCOVERY

All PR leads were designed, except for indinavir and tipranavir, which were discovered empirically by screening a library of renin inhibitors and a diverse set of compounds, respectively. Although it is difficult to determine exactly what information was used during the design process, important information was extracted from (1) HIV protease's enzymatic mechanism of action, (2) composition of the PR clip sites in the *gag* and *gag-pol* polyprotein precursors, and (3) crystal structures of the apoprotein and inhibitor complexes (Table 5.3).

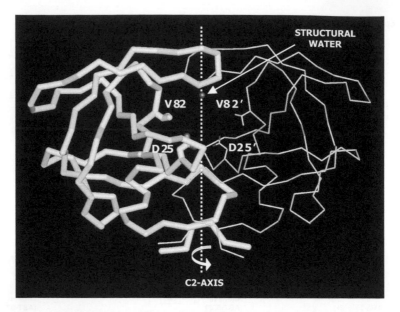

FIGURE 5.16 (See color insert following page 176.) Structural features of HIV-1 protease that influence inhibitor design. HIV-1 protease is a C2-symmetric homodimer and its active site is located at the dimer interface. The catalytic aspartates (D25 and D25′) are located at the base of the pocket and interact with the hydroxyl group adjacent to the scissile bond. The structural water molecule (red sphere) at the top of the pocket bridges the flaps and peptido-mimetic inhibitors. Val 82 and 82′, which are located in the S1 and S1′ subsites, are involved in key hydrophobic interactions with the P1 and P1′ substituents of most inhibitors. The bound inhibitor has been omitted for clarity.

5.9.2.1 Mechanism of Action

Elucidating the mechanism of action of aspartyl proteases was pivotal to the design of potent inhibitors. Ever since a water molecule was found between the catalytic aspartates in the crystal structures of penicillopepsin, endothiapepsin, and rhizopus-pepsin, the reaction has been thought to proceed via a high-energy tetrahedral transition state formed by adding an activated water molecule to the amide carbonyl of the scissile bond. Consequently, inhibitors of renin and pepsin were designed by replacing the scissile bonds of substrates with nonhydrolyzable isosteres (such as hydroxyethyl, hydroxymethyl, and hydroxyethylamine). The addition of a hydroxyl group, which displaces the catalytic water, to reduced-isostere inhibitors of porcine pepsin increased potency approximately 3000 times. This gain in binding affinity is primarily due to favorable entropic energy associated with the release of the tightly bound water molecule into bulk solvent. This interaction energy is believed to be one of the most significant forces holding the complex together. Similarly, potent peptidomimetic inhibitors of PR were rapidly obtained simply by designing peptide isosteres centered on the scissile bonds of PR clip sites. All PR peptidomimetic leads were therefore designed based on concepts developed during the renin and pepsin drug discovery programs.

TABLE 5.3
Information Used to Design HIV-1 Protease Inhibitors

FDA-Approved Drugs and Inhibitors in Clinical Trials	Lead Discovery					Lead Optimization		
	Substrate Information		Structural Information		Screen Library of Compounds	Prior Knowledge	SAR	Molecular Modeling
	Mimic Transition State	Mimic Protease Clip Site	C2 Symmetry	Displace Structural Water				
Saquinavir	x	x	—	—	—	—	x	—
Nelfinavir	x	x	—	—	—	Saquinavir	x	x
Indinavir	—	—	—	—	x	Saquinavir	x	x
L-756,423	—	—	—	—	—	Indinavir	x	—
Amprenavir	x	x	—	—	—	—	x	x
GW433908	—	—	—	—	—	Amprenavir	x	—
TMC114	—	—	—	—	—	Amprenavir	x	—
Atazanavir	x	x	x	—	—	—	x	x
Ritonavir	x	x	x	—	—	—	x	x
Lopinavir	—	—	—	—	—	Ritonavir	x	x
Mozenavir	—	—	x	x	—	—	x	x
Tipranavir	—	—	—	—	x	—	x	x

Note: Information and technologies used to develop inhibitors are indicated by the symbol x.

5.9.2.2   HIV-1 Protease Cleavage Sites

HIV-1 protease cleaves the viral polyprotein precursors at sites that contain a phenyl-alanine and/or proline or both at P1 and P1', respectively. The Phe-Pro sequence is particularly interesting because mammalian proteases do not normally cleave peptide bonds that have a proline at the C-terminal side of the scissile bond. Inhibitors that mimic this dipeptide should therefore have a higher specificity for viral proteases. Classes IA and IB inhibitors not only mimic the dipeptide but also mimic the P2 (Asn) and P2' (Ile) residues of the RT-PR clip site. For example, saquinavir contains Asn and Phe at P2 and P1, respectively, and a decahydroisoquinoline-*tert*-butyl-amine, which mimics the Pro-Ile dipeptide at P1'-P2'; and amprenavir contains a Phe at P1 and an Asn mimic at P2. All other inhibitors mimic the clip site by placing benzyl groups at P1 or P1' or both.

5.9.2.3   Structural Information

The crystal structures of apo-PR and PR in complex with a reduced substrate analog, acetyl-pepstatin, and a hydroxyethylamine isostere were determined in December and February 1989, and August and November 1990, respectively. These structures revealed (1) that the protein is indeed a member of the aspartyl protease family, (2) that the catalytic water molecule is displaced by the hydroxyl group of the central statine unit, and (3) the presence of a structural water molecule between the flaps and inhibitors. This information was extremely important because it validated the use of design concepts borrowed from renin and pepsin drug discovery programs and spurred on the *de novo* design of nonpeptidic inhibitors. Currently, there are almost 200 crystal structures of PR-inhibitor complexes in the public domain. The availability of structural information at the early stages of the drug discovery process is one of the main reasons for the successes of these programs. The important structural features that influenced the design of lead compounds are described in the following text (and illustrated in Figure 5.16).

- The apostructure of PR revealed that the protein is a homodimer, in which each polypeptide chain of 99 amino acids contributes equally to the active site. The twofold (C2) symmetry axis lies between the catalytic aspartates, dividing the active site into two equal halves (Figure 5.16). Because mammalian aspartyl proteases possess asymmetric active sites, it was believed that symmetric inhibitors would have greater specificity for viral proteases. This information was incorporated into the design of ritonavir, atazanavir, and mozenavir. For example, ritonavir was designed to complement the C2 symmetry of the PR active site by duplicating the N-terminal side of the RT-PR clip site; and substituents were attached symmetrically around the hydroxyethylhydrazine isostere and seven-member cyclic urea ring of atazanavir and mozenavir, respectively.
- Structures of PR-inhibitor complexes revealed the existence of a water bridge between the flaps of the protein and the carbonyl oxygens of the P2 and P1' substituents of the inhibitors. This information was incorpo-

rated into the design process in two different ways. In the case of most peptidomimetic inhibitors, the spatial arrangement between the hydrogen-bonding acceptors (carbonyl oxygens) was optimized to enhance inter-actions with the structural water molecule. In the case of nonpeptidic inhibitors, a carbonyl oxygen was used to displace the structural water and interact directly with the flaps. Displacing the structural water is believed to be energetically beneficial, as described earlier for the cata-lytic water molecule.

- Crystal structures also provided the necessary information to design non-peptidic leads *de novo*. For example, knowledge of the spatial arrange-ment, size and hydrophobic character of the subsites, and location of the catalytic aspartates were used to construct a 3-D pharmacophore model (based on intramolecular distances between S1, S1′, and the catalytic aspartates) that was used to discover the precursor of mozenavir. The key features of mozenavir are (1) C2 symmetry, (2) direct interaction with the flaps, (3) a rigid, cyclic compound that is preorganized for highly com-plementary interactions with the protease, i.e., its conformational entropy penalty associated with binding is lower than that for flexible ligands, and (4) the optimal stereochemistry needed to project the P1, P1′, P2, and P2′ benzyl groups into their complementary subsites.

5.9.3 Lead Optimization

HIV-1 protease inhibitors were optimized through iterative cycles of synthesis, biological testing, and structural information gathering. Initially, the design emphasis was focused on optimizing *in vitro* potency but quickly shifted to improving phar-macological properties because the effectiveness of lead compounds was hampered by low water solubility, poor oral bioavailability, and susceptibility to proteolytic enzymes due to their peptide-like nature. Traditionally, it has been very challenging to overcome the undesirable properties of peptidomimetic inhibitors. Structural insights, which played an important role in lead discovery, were now being used to identify regions of the inhibitors that could be modified to improve oral bioavail-ability without significantly affecting activity. The primary selection criterion was aqueous solubility, and secondary criteria were: (1) potency against PR, (2) activity in cell-based assays in the presence of serum proteins, (3) concentration in plasma after oral administration to animals, and (4) the onset of resistance.

Saquinavir was the first inhibitor approved by the FDA to combat the progression of HIV and the only inhibitor to be designed without (or very little) PR structural information. One of the most significant modifications that led to the discovery of saquinavir was the addition of the decahydroisoquinoline (DIQ) group at P1′. This substituent limits the conformational freedom of the inhibitor and simultaneously improves aqueous solubility and potency. The pharmacokinetic properties of all other leads were optimized with the benefit of information discovered during the saquinavir program and the availability of extensive structural information. For example, the solubilities of nelfinavir and a precursor to indinavir were improved by incorporating saquinavir's DIQ substituent; the crystal structure of PR in complex with A-77003 (a precursor of ritonavir) revealed that P3 and P3′ could be replaced with polar substituents in an attempt to improve solubility; the structure of ritonavir guided the design of lopinavir, which was designed specifically to inhibit ritonavir-selected mutants by de-emphasizing interactions with Val82 and Ile84; the structure of CGP-53820 (a precursor of atazanavir) revealed that larger P1 groups could interact favorably with the protease; and the structure of phenprocoumon (a precursor of tipranavir) identified a need for increased flexibility in order to properly project substituents into the S1 and S2 subsites. The detailed steps involved in the design of each PR inhibitor are summarized in Table 5.4.

Regardless of the diverse methods used to discover and optimize inhibitors, all approved and experimental drugs have very similar electrostatic character (Figure 5.17) and bound conformations (Figure 5.18). In general, all inhibitors have a central polar region made up of oxygen atoms, which either displace the structural water or hydrogen bond to it, and a mixture of hydrophobic and hydrophilic substituents at P2, P2′, P3, and P3′.

These similarities are not surprising because they were designed against the same target, i.e., to hydrogen bond to the structural water molecule and catalytic aspartates as well as interact with hydrophobic residues in the various subsites. The greatest structural similarities are found within Class I compounds, for example, Class IB inhibitors differ only at P2, Class IA inhibitors differ at P2, P3, and P3′, and Class IC compounds, which contain the most diverse members, have significant overlap with Class IB compounds (Figure 5.18). Even nonpeptidic inhibitors, which contain novel templates, have similar bound conformations to Class I inhibitors. Finally, the greatest diversity among all inhibitors is found at P3 and P3′ because these substituents were primarily designed to modulate pharmacokinetic properties and not potency (Figure 5.18).

5.9.4 Drug Resistance

It is generally accepted that resistant variants have reduced sensitivities to protease inhibitors as a result of mutations in the active site of PR. Mutations have been shown to directly interfere with inhibitor binding by increasing steric contacts or reducing van der Waals interactions between the protein and inhibitors. Although the activity of first-generation inhibitors (saquinavir, ritonavir, indinavir, amprenavir, and nelfinavir) against drug-resistant mutants was continually monitored during their development, they were primarily designed to inhibit wild-type PR. Very little has changed for second-generation inhibitors. Except for lopinavir, recently developed inhibitors have been designed to combat resistance by improving pharmacokinetic properties and increasing potency against the wild-type protease. This strategy is based on the assumption that a highly effective dose will reduce the viral load and prevent the virus from replicating and mutating. The use of these indirect methods, however, highlights the inability to design inhibitors that do not select for resistant variants.

All PR inhibitors have similar resistance profiles because they interact with the same residues in the active site (Figure 5.18). The use of multiple PR inhibitors is therefore not a viable therapy on its own because cross-resistant mutants will eventually emerge. However, if needed, one should select PR inhibitors from different

TABLE 5.4
Rational Designs of HIV-1 Protease Inhibitors

Saquinavir

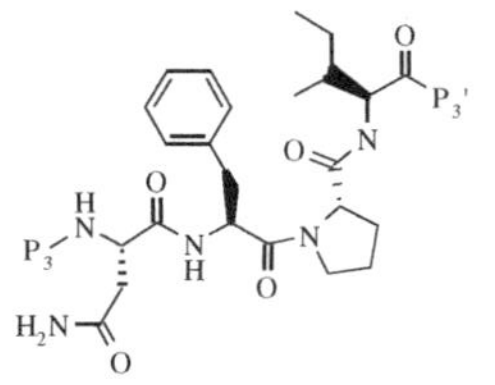

HIV-1 protease cleaves the *gag* and *gag-pol* polyproteins at eight different sites. The cleavage sequence between PR and RT was chosen as the starting point for the design of selective peptidomimetic inhibitors because (1) it contains the consensus protease-cleavage sequence Asn-Phe-Pro, where the amide bond between Phe and Pro is the scissile bond, and (2) the Phe-Pro sequence is not susceptible to cleavage by mammalian peptidases.

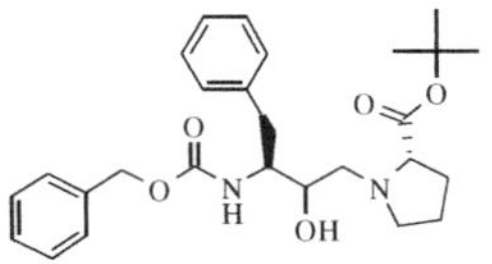

Various transition-state mimics were designed based on the approach used to design inhibitors against other aspartyl proteases, such as renin. One of the first PR inhibitors was a simple, protected Phe-Pro dipeptide where the scissile bond C(O)N was replaced with the hydroxyethylamine moiety CH(OH)CH$_2$N (IC$_{50}$ = 6500 nM). Note that the C-terminal blocking group *tert*-butoxycarbonyl (BOC) is an isoleucine mimic.

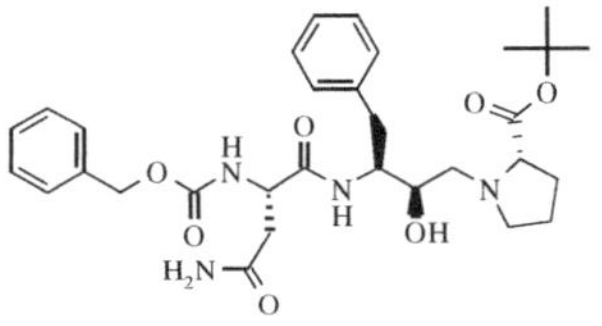

Potency of the lead was increased 45-fold (IC$_{50}$ =140 nM) by (1) extending the amino terminus with an asparagine, a residue of the consensus clip site, and (2) discovering the preference for R stereochemistry at the hydroxyl-bearing carbon atom.

P3-Asn-Phe-Pro-Ile-P3'
Cleavage site between HIV-1
Protease and Reverse transcriptase

TABLE 5.4 (Continued)
Rational Designs of HIV-1 Protease Inhibitors

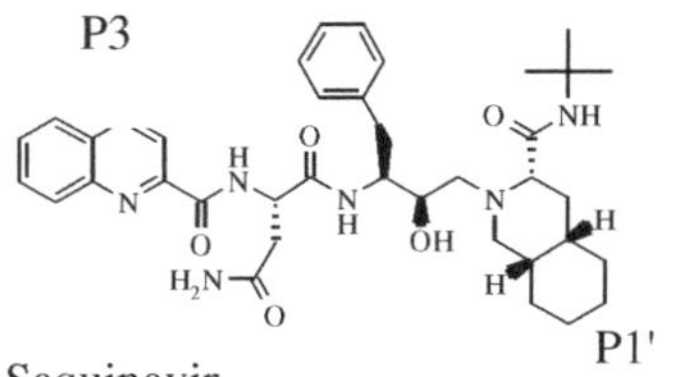

A potent ($IC_{50} < 0.4$ nM) and selective inhibitor was discovered after systematically modifying all substituents. No alternatives were found for P1 and P2 substituents. However, significant increases in potency were observed when P3 and P1′ were replaced with a quinoline and decahydroisoquinoline (DIQ), respectively. These groups fill the S3 and S1′ subsites of the protease and have extensive van der Waals interactions with the protein. Finally, only a limited search for P3′ substituents was made because modeling results suggested that inhibitors containing the DIQ group would preclude extension of the C-terminus.

P3

P1′

Saquinavir
(Ro–31-8959)

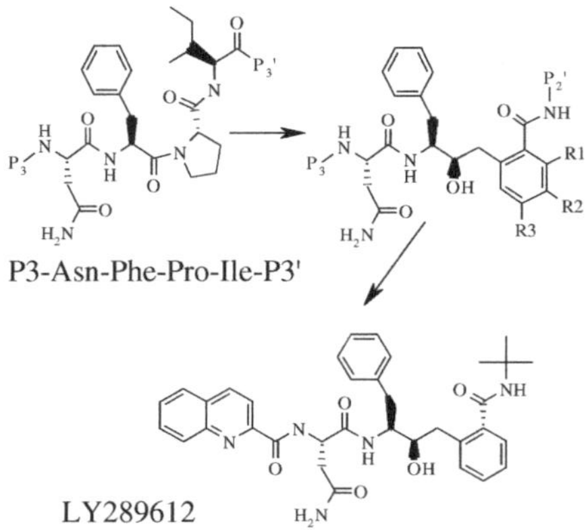

The lead compound LY289612 was designed by replacing the nonhydrolyzable hydroxyethylamine unit of saquinavir with a hydroxyethylbenzamide moiety. This substituent was chosen because derivatives are readily accessible. A systematic analysis of the effects of benzamide substitution on enzyme activity revealed that the unsubstituted ring (LY289612) is the most potent enzyme inhibitor of the series ($IC_{50} = 1.4$ nM). Substitutions at R1, R2, and R3 affect the freedom of rotation of the amide carbonyl, which forms an important hydrogen bond with the structural water molecule. Unfortunately, this compound is poorly bioavailable; subsequent cycles of drug design were thus focused on increasing potency and reducing its molecular weight and peptidic nature.

Nelfinavir

P3-Asn-Phe-Pro-Ile-P3′

LY289612

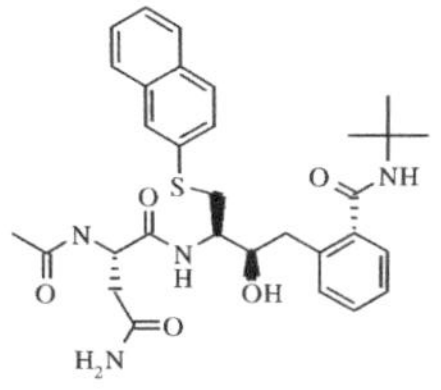

The crystal structure of PR in complex with LY289612 revealed that the P1 and P3 substituents are only 3.2 Å apart. Based on this information, attempts to enlarge P1 and truncate P3 led to the design of LY297135. Structural analysis correctly predicted the bound conformation of the S-naphthyl substituent, which spans S1 and S3. Although this truncated compound is a potent inhibitor (IC_{50} = 1.1 nM), its antiviral activity in whole cell assays is about 10 times poorer than that of LY289612.

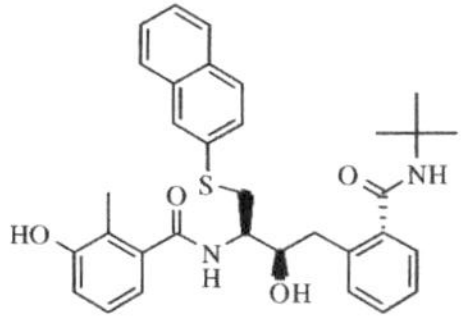

New substituents were selected to replace P2-P3 to further reduce the peptidic nature of LY297135. Based on the crystal structures of the latter two compounds in complex with PR, bicyclic rings (where the second ring is a heterocycle) were first selected as replacements. The bicyclic ring systems were subsequently replaced with a single ring because the most active bicycles contained a saturated heterocycle, where the heteroatom forms a hydrogen bond with the amide carbonyl oxygen of Gly 48. Optimizing this series led to the design of AG1254, which contains a m-hydroxy-o-methyl phenyl replacement. Overall, the initial goal of designing a compound with decreased molecular weight and comparable *in vitro* activity (K_i = 3 nM) to LY289612 was achieved.

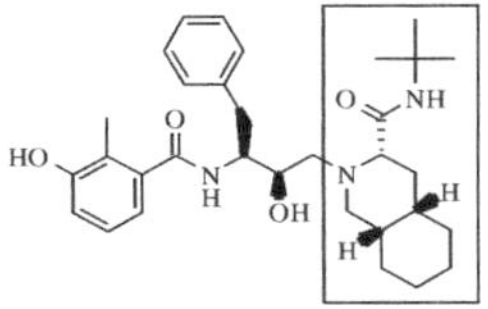

Attempts to improve antiviral activity led to the hypothesis that incorporation of the m-hydroxyl-o-methyl phenyl P2 group of AG1254 into the hydroxyethylamine isostere of saquinavir would improve antiviral activity. This led to the design of the hybrid compound AG1310, which is slightly less active in the enzyme assay (K_i = 21 nM) but 97 times more active in the whole cell assay compared to AG1254.

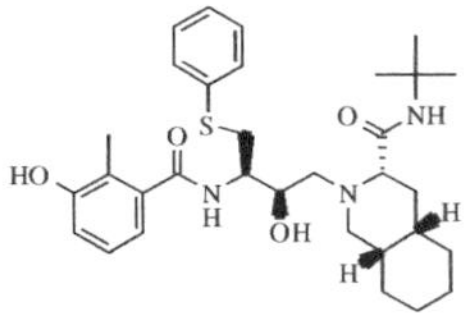

Optimization of P1 was guided by previous studies demonstrating that S-aryl substituents could effectively span the S1 and S3 subsites. AG1343, the S-phenyl analog of AG1310, has a K_i value of 2 nM and is twice as effective in the antiviral whole cell assay compared to AG1310.

LY297135

AG1254

AG1310

Nelfinavir (AG1343)

TABLE 5.4 (Continued)
Rational Designs of HIV-1 Protease Inhibitors

Indinavir and L-756,423

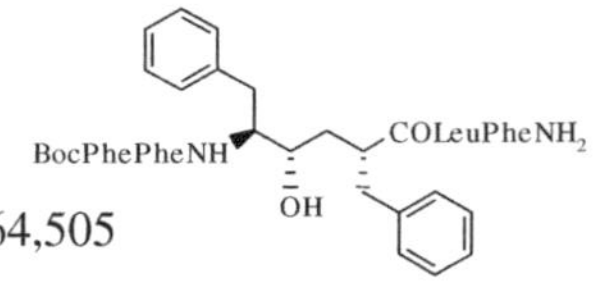

The lead compound L-364,505 was identified by screening a collection of renin inhibitors against PR. It contains a hydroxyethylene moiety, and it has an IC_{50} of 1 nM.

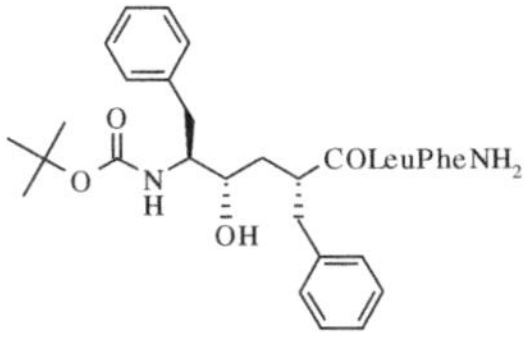

N-terminal truncation increased and reduced potency against PR ($IC_{50} = 0.6$ nM) and renin, respectively. Despite its high potency it is poorly absorbed and vulnerable to degradative enzymes.

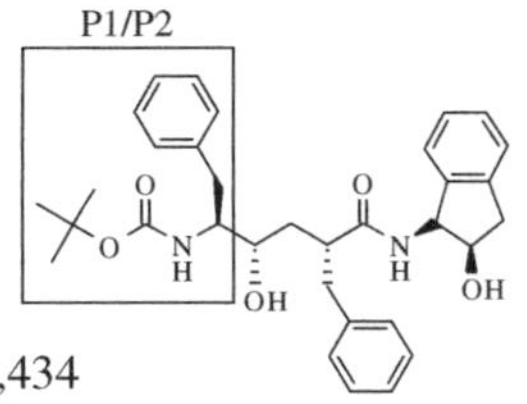

Efforts to reduce molecular weight and peptidic nature resulted in replacing the Leu-Phe C-terminal dipeptide with a benzyl amide ($IC_{50} = 111$ nM). Potency was then increased by conformationally constraining the substituent by forming an aminoindan amide ($IC_{50} = 19$ nM). A significant increase in potency ($IC_{50} = 0.3$ nM), however, was not observed until an alcohol group (which hydrogen bonds to Gly27' and Asp29') was added at 2-position cis to the nitrogen on the indan ring.

L-364,505

L-685,434

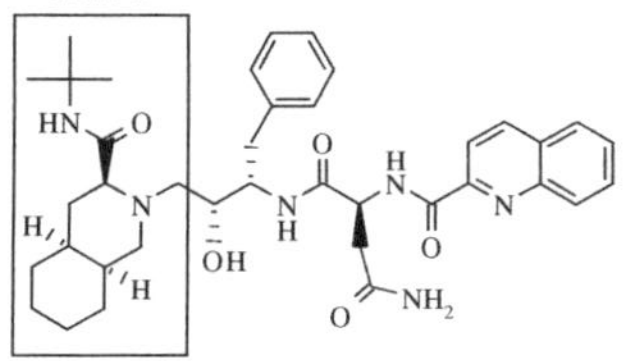

Attempts to increase aqueous solubility and oral bioavailability in animal models were guided by molecular modeling and crystal structures of PR-inhibitor complexes. A superposition of the bound conformations of L-685,434 and saquinavir suggested that the basic decahydroisoquinoline *tert*-butylamide (P1'/P2') of saquinavir could replace the P1 and P2 groups of L-685,434 and improve bioavailability. Although the resulting hybrid compound is a less potent inhibitor (IC$_{50}$ = 7.8 n*M*), it possesses favorable pharmacokinetic properties.

To regain potency, the decahydroisoquinoline was replaced with piperazine, which possesses a second nitrogen that can be derivatized with a P3 substituent. Attempts to modify the physical properties of the piperazine anologs while maintaining potency led to L-735,524 (arrows indicate sites of indinavir metabolism).

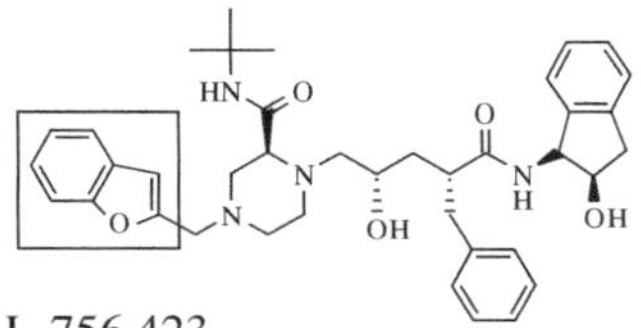

The pharmacokinetic properties of a series of indinavir analogs was improved by trying to (1) block the metabolism associated with the pyridyl moiety, (2) minimize protein binding, (3) increase solubility in water, (4) not inhibit P450 isozymes, and most important, (5) increase plasma concentrations C$_{max}$ and C$_{8h}$ (concentration after 8 h). The metabolically labile 3-pyridylmethyl was first derivatized at several positions and then eventually replaced with a lipophilic heterocycle.

Saquinavir

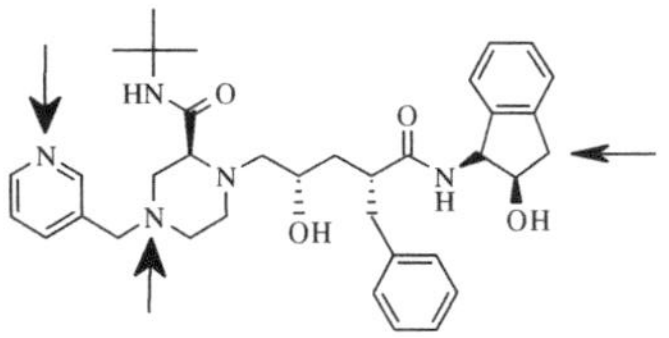

Indinavir
(L-735,524 or MK-639)

L-756,423
(MK-944a)

TABLE 5.4 (Continued)
Rational Designs of HIV-1 Protease Inhibitors

Amprenavir, GW4333908, and TMC114

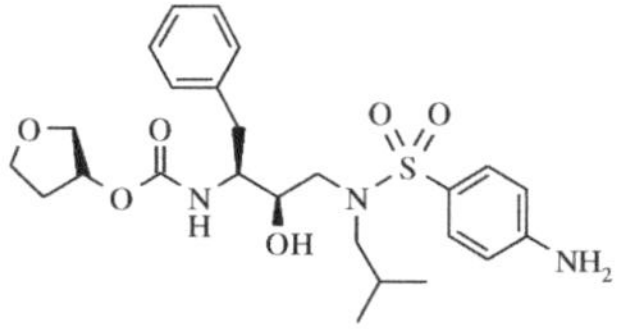

Amprenavir was designed based on the following concepts: (1) reduce the energy strain of the bound conformation while maintaining optimal interactions with the protease, (2) incorporate the P1 phenyl group of the cleavage site, (3) replace the P2 asparagine side chain with the conformationally restricted mimic tetrahydrofuran, and (4) assume low-molecular-weight molecules (<500 Da) are more likely to be bioavailable. Amprenavir is one of the smallest and most potent ($K_i = 0.6$ nM) inhibitors of HIV-1 protease.

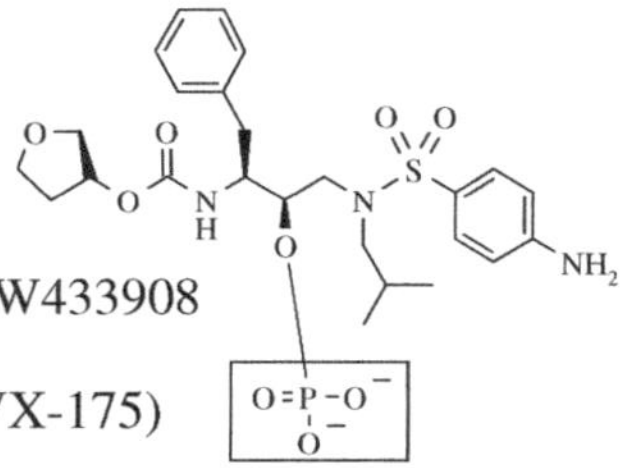

A calcium phosphate ester prodrug (which is dephosphorylated during absorption through the gut epithelium) of amprenavir was designed to increase solubility and ultimately reduce the current pill count of eight capsules (150 mg each) to two capsules daily.

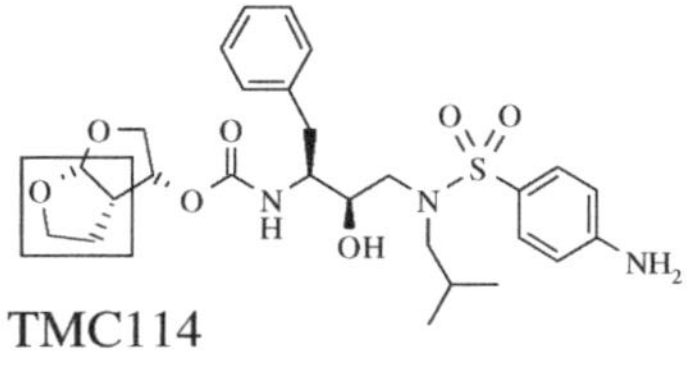

Various P2 analogs of amprenavir were designed based on the assumption that potency can be increased by incorporating stereochemically defined and conformationally constrained cyclic ethers. The introduction of a bis-tetrahydrofuryloxyurethane produced TMC114 ($K_i = 2.1$ nM). This compound is being developed by Tibotec-Virco.

Amprenavir
(141W94 or VX-478)

GW433908
(VX-175)

TMC114

A pseudosymmetric hydroxyethyl hydrazine isostere was selected as a starting point because it complements the C2 symmetry of the PR active site, and it is synthetically accessible to a broad range of substituents by acylation of the amino and hydrazine groups.

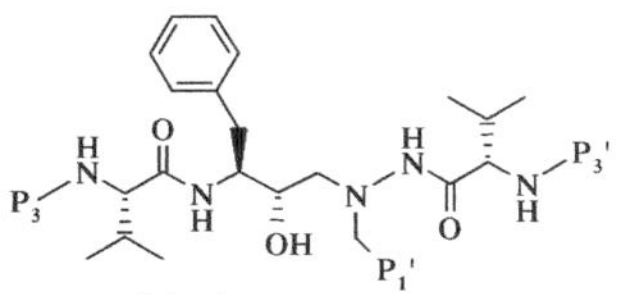

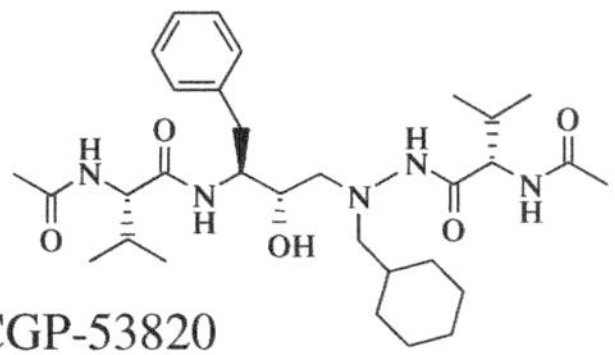

A systematic exploration of P3 and P3' resulted in a 1700-fold increase in potency when *tert*-butoxycarbonyl was replaced with acetyl groups; however, no futher increases in potency were observed by projecting substituents deeper into the S3 and S3' subsites. Further improvements were achieved by varying P1'. CGP-53820, a potent enzyme inhibitor ($IC_{50} = 8.5$ nM) with moderate activity in a whole cell assay, was identified by attaching a cyclohexyl moiety at P1'.

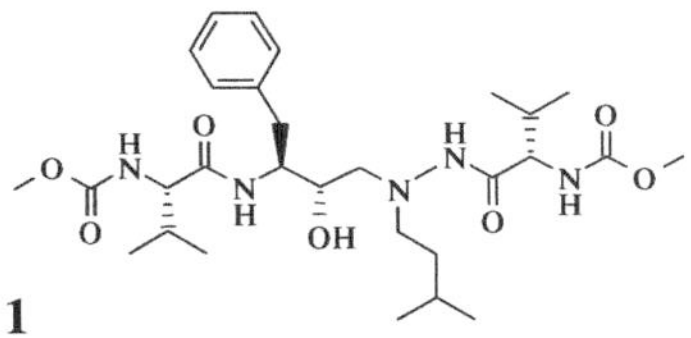

An extensive structure–activity study was undertaken in parallel to optimize the pharmacokinetic profile and antiviral activity of CGP-53820. First, a series of P1' analogs was synthesized based on the crystal structure of PR in complex with CGP-53820. The structure revealed that the S1' subsite could accommodate lipophilic substituents of diverse size. Replacing the cyclohexyl with an isobutyl produced compound 1, which is highly effective in cellular assays but is not bioavailable. Second, a series of P2-P3 and P2'-P3' analogs restricted to either acyl residues or carbamates was synthesized based on the desire to retain an important hydrogen bond between the carbonyl oxygen of the acyl residues and the amide nitrogens of Asp 29 and 29' of the protein. Replacing the P3 and P3' acetyl groups of CGP-53820 with ethyl carbamates produced compound 2, which is orally well absorbed but has weak antiviral activity.

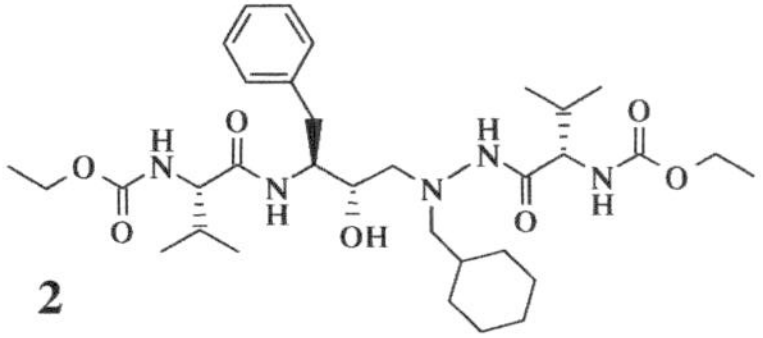

Atazanavir

aza-peptide isostere

CGP-53820

1

2

TABLE 5.4 (Continued)
Rational Designs of HIV-1 Protease Inhibitors

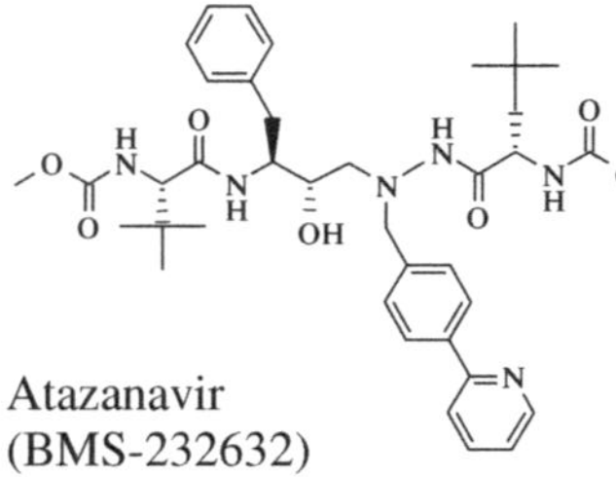

In an effort to combine good oral bioavailability and potent antiviral activity, CGP-53820 was redesigned using larger P1′ substituents. This idea was based on information obtained from the crystal structure of PR in complex with CGP-53820, which revealed that large P1′ substituents can interact favorably with the enzyme. A systematic approach was therefore used to design a series of bicyclic P1′ derivatives. The addition of a pyridylphenyl substituent increased potency and bioavailability. Further improvements were observed when the P2 and P2′ valines were replaced with *tert*-leucines and the P3 and P3′ acetyls were replaced with carbamates (as in compound I). Although atazanavir is not the most potent (IC$_{50}$ = 26 nM) aza-peptide isostere, it possesses the best combination of antiviral activity and oral bioavailability, and it is the first drug approved for a single daily dose.

The lead compound A-75925 was designed based on the concept that symmetric compounds have greater selectivity for homodimeric PR and superior pharmacokinetic properties than peptidomimetics because they are less peptide-like. The tetrahedral C2 symmetric core of the inhibitor was created by duplicating the P1 side of the cleavage site, based on its greater importance in previous renin inhibitor binding studies. The symmetric additions of valine and Cbz (carbobenzyloxy, a common N-terminal blocking group for amino acids) increased potency approximately 100-fold (IC$_{50}$ = 0.22 nM); however, this compound is highly lipophilic. Given the early successes in this program, the major goal quickly shifted from optimizing inhibitor binding to improving aqueous solubility.

Atazanavir
(BMS-232632)
(CGP-73547)

Ritonavir and Lopinavir

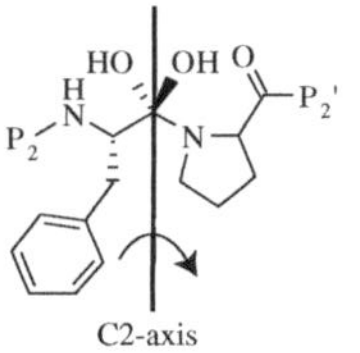

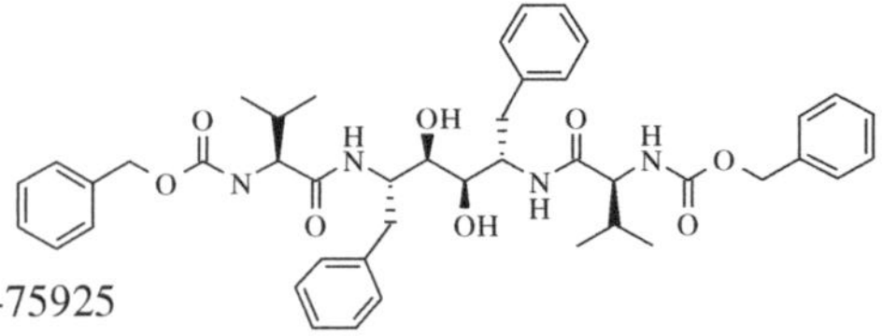

A-75925

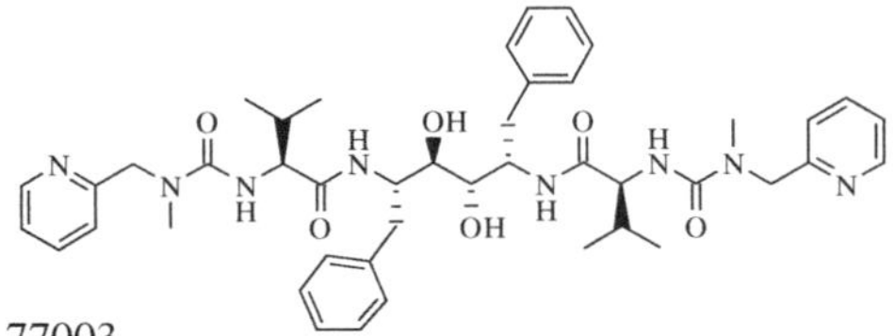

Crystal structures of PR-inhibitor complexes revealed that the phenyl of Cbz could be replaced with a more polar group because the S3 subsite is solvent exposed. Pyridyl analogs have good activity (IC$_{50}$ < 1 nM) and improved aqueous solubility but they are poorly absorbed and vulnerable to degradative enzymes.

Although diol analogs are usually more potent inhibitors than monools, they have poorer aqueous solubility, probably due to higher desolvation energy. In addition, SAR studies revealed that the right side of the inhibitor could be truncated without significant loss of activity (IC$_{50}$ < 1 nM). This inhibitor is unfortunately cleared rapidly through N-oxidation of the pyridyl groups.

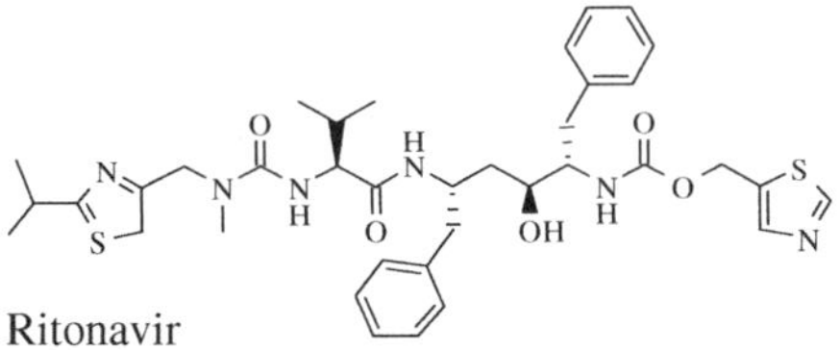

Efforts to increase the stability of A-80987 by reducing the rate of oxidative metabolism by steric and electronic modifications to the pyridyl did not increase bioavailability. However, replacement with other six- and five-membered heterocycles eventually produced ritonavir, which has excellent pharmacokinetic properties.

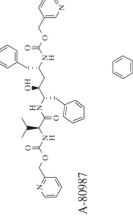

A-77003

A-80987

Ritonavir
ABT-538

TABLE 5.4 (Continued)
Rational Designs of HIV-1 Protease Inhibitors

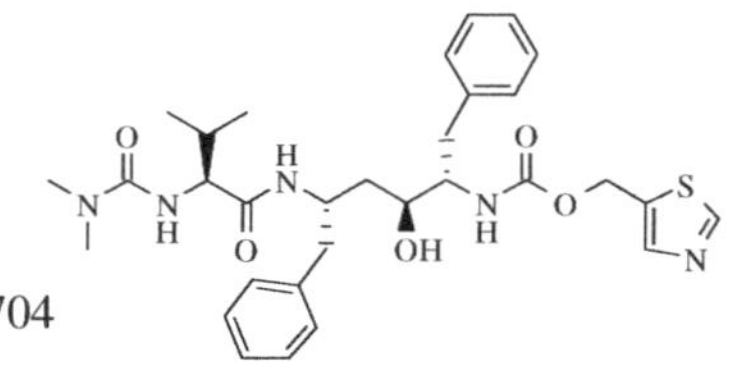

The therapeutic benefit of ritonavir is limited by rapid selection of Val82-resistant mutants, which are 12 to 52 times less sensitive to the inhibitor. New selection criteria were therefore implemented based on the assumption that the onset of resistance can be delayed by increasing the concentration of unbound drug in human plasma. This was accomplished by (1) screening analogs in the presence of human serum to assess the effects of protein binding, (2) evaluating the pharmacokinetic properties following coadministration of ritonavir, as the latter has been shown to elevate the plasma concentrations of other inhibitors by inhibiting cytochrome P450 3A4-mediated metabolism. In addition, inhibitors were designed to have fewer interactions with Val82.

A-155704

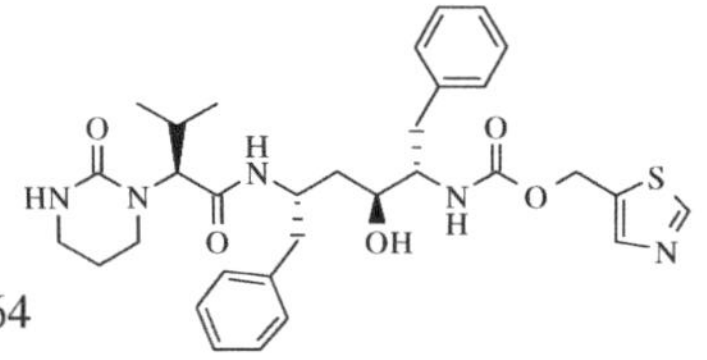

A truncated analog of ritonavir (A-155704) was designed based on the crystal structure of PR in complex with ritonavir. Eliminating the P3 isopropylthiazolyl group of ritonavir (which interacts with Val82) resulted in a 30-fold loss of anti-HIV activity, but only a fourfold difference in the presence of 50% human serum. Conformationally constraining the urea group produced A-155564, which is 2.5 times less active than ritonavir in the absence of human serum but 1.6 times more active in the presence of 50% human serum.

A-155564

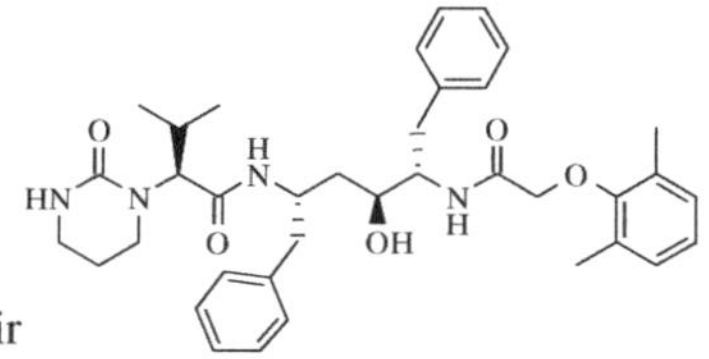

Replacement of the P2′ substituent with a dimethylphenoxyacetyl group produced lopinavir, which is 3.5 times more active than ritonavir in the absence of human serum and 10 times more active in the presence of 50% human serum. In addition, the activity of lopinavir is only reduced 6 to 13 times against various Val82 mutants, compared to a reduction of 17 to 41 times for ritonavir against the same mutants.

Lopinavir
ABT-378

Mozenavir

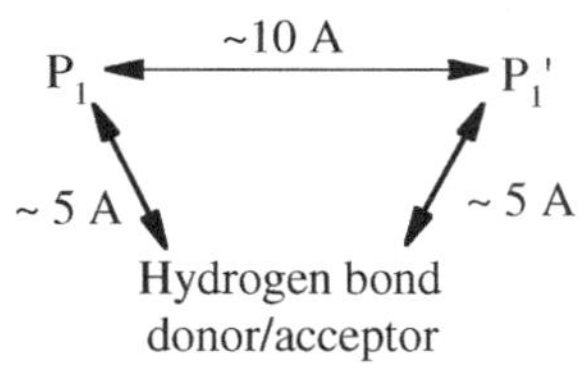

A 3-D pharmacophore model relating P1, P1′, and the central hydroxyl of peptidomimetic inhibitors was generated based on various crystal structures of PR–inhibitor complexes and a model of a C2 symmetric diol docked in the active site.

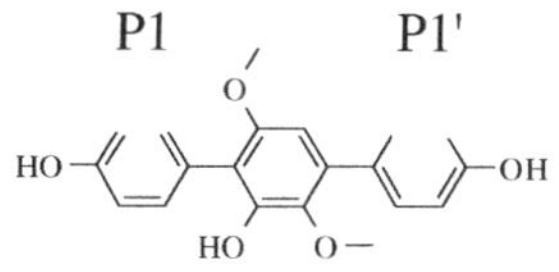

The pharmacophore was used to search a subset of the Cambridge Structural Database. The hit obtained from the 3-D search suggested that a six-membered ring can properly position (1) a structural water mimic, which hydrogen bonds to the flaps of the protein, and (2) a catalytic water mimic, which interacts with the catalytic aspartates.

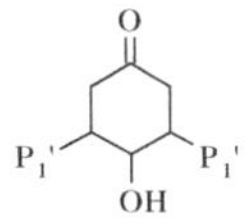

A cyclohexanone ring was chosen the initial synthetic scaffold, as modeling studies showed it is better suited to project substituents into the various subsites of the protein, displace the structural water, and interact with the catalytic aspartates.

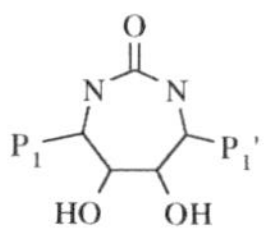

The cyclohexanone was enlarged to a seven-membered ring to incorporate diol functionality, and the ketone was replaced with a urea group. These modifications were designed to increase interactions with the catalytic aspartates and strengthen hydrogen bonds to the flaps, respectively.

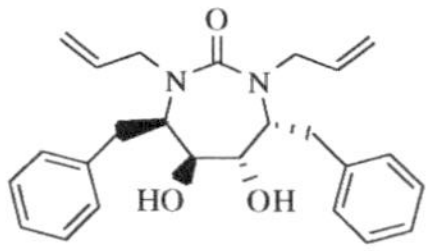

The conformation and stereochemistry of the cyclic urea ring needed for optimal complementarity between the P1/P1′/P2/P2′ and the corresponding subsites were predicted from modeling studies. Finally, substituents were added symmetrically to optimize interactions with the C2-symmetric active site, and benzyl groups were added to mimic the P1 and P1′ groups of the protease clip site (K_i = 4.7 nM).

TABLE 5.4 (Continued)
Rational Designs of HIV-1 Protease Inhibitors

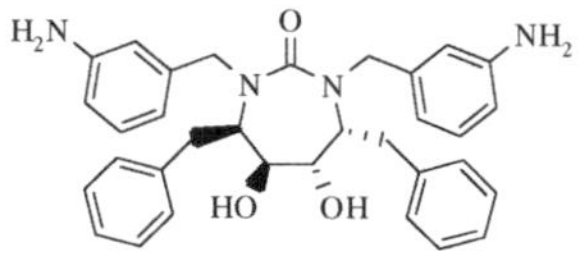

Mozenavir
(DMP450)

Tipranavir

Warfarin

Phenprocoumon

Potency and solubility were further improved by substituting the P2 and P2′ allyls with weakly basic meta-aminobenzyl groups. This decision was based on structural information that showed that the S2 and S2′ subsites can accommodate large hydrophobic substituents as well as hydrogen bond donors and acceptors ($K_i = 0.8$ nM)

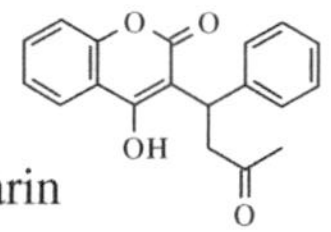

A fluorescence-based high-throughput screen of a broad set of 5000 compounds from Pharmacia Upjohn's chemical collection uncovered a weak inhibitor of PR (IC$_{50}$ = 30 μM) called warfarin, an anticoagulant drug already approved for use in humans.

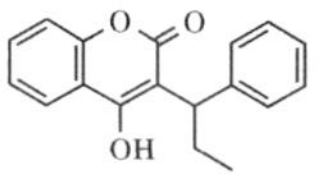

A focused screen around the 4-hydroxycoumarin template uncovered phenprocoumon ($K_i = 1$ μM). The hydroxycoumarin template is a promising lead because it is small, shows weak antiviral activity, and most important, it has also been approved for use in humans.

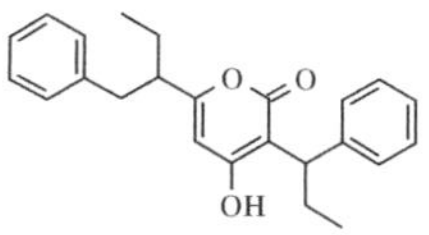

Lead optimization was performed using iterative cycles of structure-based design. The structure of PR in complex with phenprocoumon revealed that increased flexibility, by removing the planarity of the 4-hydroxycoumarin ring system, was needed to project substituents into the S1 and S2 subsites. Benzyl and ethyl groups were then added to C-6 of the 4-hydroxypyrone ring, producing PNU-96988 (K_i =38 nM).

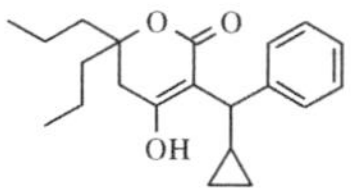

A series of dihydropyrone analogs was designed to further increase flexibility. One of the most potent compounds contains two propyl groups attached to C-6 (K_i =15 nM).

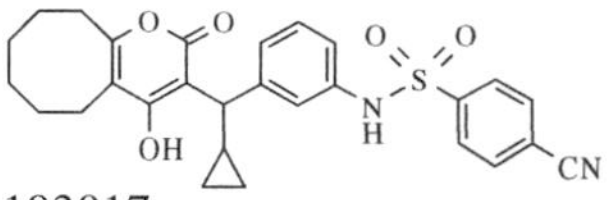

Another approach was to keep the pyrone template but attach a flexible cyclooctyl ring to C-5 and C-6 of the pyrone. Optimization of this series was based on the observation that the meta position of the C-3 phenyl of phenprocoumon binds close to the binding site of the P2 histidine of a peptidomimetic inhibitor, which hydrogen bonds to the protease. In an attempt to mimic these interactions and project a hydrophobic substituent into the S3 subsite, a carboxamide linker was attached at the meta position of the C-3 phenyl. A sulfonamide linker was later used to further increase hydrogen bonding potential and eventually produce PNU-103017 (K_i = 1 nM).

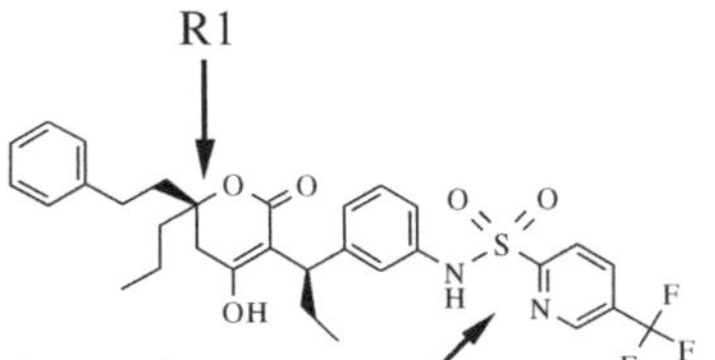

Sequential modifications at R1 (C-6) and R2 (sulfonamine substituent) of the dihydropyrone template, using previously identified substituents, produced tipranavir (K_i = 8 pM).

PNU-96988

PNU-103017

R1

R2

Tipranavir
(PNU-140690)

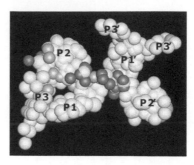

FIGURE 5.17 Common electrostatic characteristics of all HIV-1 protease inhibitors. A super-position (based on the alpha carbon atoms of the protein) of the bound conformations of PR inhibitors (saquinavir, nelfinavir, indinavir, L-756,423, amprenavir, ritonavir, lopinavir, CGP-73547, mozenavir, and tipranavir) reveals that most inhibitors have (1) a polar central core that hydrogen bonds to the catalytic aspartates and structural water molecule, (2) hydrophobic groups at P1 and P1', and (3) a mixture of hydrophilic and hydrophobic groups at P2, P2', P3, and P3', reflecting the increased solvent exposure of these subsites.

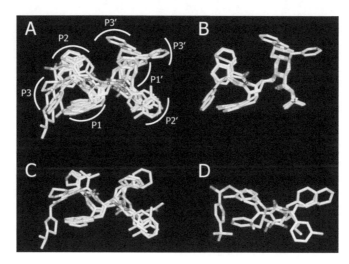

FIGURE 5.18 (See color insert following page 176.) HIV-1 protease inhibitors interact with the same residues in the active site of the protein. A: Superposition of the bound conformations of saquinavir, nelfinavir, indinavir, L-756,423, amprenavir, ritonavir, lopinavir, CGP-73547, mozenavir, and tipranavir clearly reveals that they bind in the same subsites. B: Class IA compounds only differ at P2, P3, and P3'. C: Similar binding modes are observed for classes IB and IC; D: Class II inhibitors appear to be the most diverse, except for the central core that displaces the structural water and places a hydroxyl between the catalytic aspartates.

classes and subclasses. Currently, the most effective therapy used to combat AIDS is still a combination of PR and RT inhibitors. The major benefit of simultaneously inhibiting multiple targets is a delay in the onset of resistance. The long-term benefit of these inhibitors will rely on our ability to (1) increase our understanding of the selection process so that we can predict resistance profiles, and (2) inhibit additional targets in the viral life cycle, such as HIV integrase and the cell entry receptors gp120 and gp41, which is the target for the recently approved Fuzeon.

FURTHER READING

Saquinavir

Pettit, G.R., Herald, C.L., Boyd, M.R., Leet, J.E., Dufresne, C. et al. (1991). Novel binding mode of highly potent HIV-proteinase inhibitors incorporating the (R)-hydroxyethylamine isostere. *Journal of Medicinal Chemistry*, 34, 3340–3342.

Roberts, N.A., Martin, J.A., Kinchington, D., Broadhurst, A.V., Craig, J.C., Duncan, I.B. et al. (1990). Rational design of peptide-based HIV proteinase inhibitors. *Science*, 248, 358–361.

Nelfinavir

Kaldor, S.W., Appelt, K., Fritz, J.E., Hammond, M., Crowell, T.A. et al. (1995). A systematic study of P_1-P_3 spanning sidechains for the inhibition of HIV-1 protease. *Bioorganic and Medicinal Chemistry Letters*, 5, 715–720.

Kaldor, S.W., Hammond, M., Dressman, B.A., Fritz, J.E., Crowell, T.A. et al. (1994). New dipeptide isosteres useful for the inhibition of HIV-1 protease. *Bioorganic and Medicinal Chemistry Letters*, 4, 1385–1390.

Kaldor, S.W., Kalish, V.J., Davies, J.F., Shetty, B.V., Fritz, J.E. et al. (1997). Viracept (nelfinavir mesylate, AG1343): A potent, orally bioavailable inhibitor of HIV-1 protease. *Journal of Medicinal Chemistry*, 40, 3979–3985.

Kalish, V.J., Tatlock, J.H., Davies, J.F., Kaldor, S.W., Dressman, B.A. et al. (1995). Structure-based drug design of nonpeptidic P_2 substituents for HIV-1 protease inhibitors. *Bioorganic and Medicinal Chemistry Letters*, 5, 727–732.

Indinavir

Chen, Z., Li, Y., Chen, E., Hall, D.L., Darke, P.L. et al. (1994). Crystal structure at 1.9-Å resolution of human immunodeficiency virus (HIV) II protease complexed with L-735,524, an orally bioavailable inhibitor of the HIV protease. *Journal of Biological Chemistry*, 269, 26344–26348.

Dorsey, B.D., Levin, R.B., McDaniel, S.L., Vacca, J.P., Guare, J.P. et al. (1994). L-735,524: The design of a potent and orally bioavailable HIV protease inhibitor. *Journal of Medicinal Chemistry*, 37, 3443–3451.

Lyle, T.A., Wiscount, C.M., Guare, J.P., Thompson, W.J., Anderson, P.S. et al. (1991). Benzocycloalkyl amines as novel C-termini for HIV protease inhibitors. *Journal of Medicinal Chemistry*, 34, 1228–1230.

Vacca, J.P., Guare, J.P., deSolms, S.J., Sanders, W.N., Giuliani, E.A. et al. (1991). L-687,908, a potent hydroxyethylene-containing HIV protease inhibitor. *Journal of Medicinal Chemistry*, 34, 1225–1228.

MK-944a

Dorsey, B.D., McDonough, C., McDaniel, S.L., Levin, R.B., Newton, C.L. et al. (2000). Identification of MK-944a: a second clinical candidate from the hydroxyethylamine-pentanamide isostere series of HIV protease inhibitors. *Journal of Medicinal Chemistry*, 43, 3386–3399.

Amprenavir

Kim, E.E, Baker, C.T., Dwyer, M.D., Murcko, M.A., Rao, B.G. et al. (1995). Crystal structure of HIV-1 protease in complex with VX478, a potent and orally bioavailable inhibitor of the enzyme. *Journal of the American Chemical Society*, 117, 1181–1182.

GW433908

Falcoz, C., Jenkins, J.M., Bye, C., Hardman, T.C., Kenney, K.B. et al. (2002). Pharmacokinetics of GW433908, a prodrug of amprenavir, in healthy male volunteers. *Journal of Clinical Pharmacology*, 42, 887–898.

Gatell, J.M. (2001). From amprenavir to GW433908. *Journal of HIV Therapy*, 6, 95–99.

TMC114

Ghosh, A.K., Kincaid, J.F., Cho, W., Walters, D.E., Krishnan, K. et al. (1995). Potent HIV protease inhibitors incorporating high-affinity P_2-ligands and (*R*)-(hydroxyethylamino) sulfonamide isostere. *Bioorganic and Medicinal Chemistry Letters*, 8, 687–690.

Atazanavir

Bold, G., Fässler, A., Capraro, H.-G., Cozens, R., Klimkait, T. et al. (1998). New aza-dipeptide analogues as potent and orally absorbed HIV-1 protease inhibitors; candidate for clinical development. *Journal of Medicinal Chemistry*, 41, 3387–3401.

Fässler, A., Bold, G., Capraro, H.-G., Cozens, R., Mestan, J. et al. (1996). Aza-peptide analogs as potent human immunodeficiency virus type-1 protease inhibitors with oral bioavailability. *Journal of Medicinal Chemistry*, 39, 3203–3216.

Fässler, A., Rösel, J., Grötter, M., Tintelnot-Blomley, M., Alteri, E. et al. (1993). Novel pseudosymmetric inhibitors of HIV-1 protease. *Bioorganic and Medicinal Chemistry Letters*, 3, 2837–2842.

Priestle, J.P., Fässler, A., Rösel, J., Tintelnot-Blomley, M., Strop, P. et al. (1995). Comparative analysis of the x-ray structures of HIV-1 and HIV-2 proteases in complex with CGP 53820, a novel pseudosymmetric inhibitor. *Structure*, 3, 381–389.

Ritonavir

Erickson, J., Neidhart, D.J., VanDrie, J., Kempf, D.J., Wang, X.C. et al. (1990). Design, activity, and 2.8 Å crystal structure of a C_2 symmetric inhibitor complexed to HIV-1 protease. *Science*, 249, 527–533.

Kempf, D.J., Cadacovi, L., Wang, X.C., Kohlbrenner, W.E., Wideburg, N.E. et al. (1993). Symmetry-based inhibitors of HIV protease. Structure-activity studies of acylated 2,4-diamino-1,5-diphenyl-3-hydroxypentane and 2,5-diamino-1,6-diphenylhexane-3,4-diol. *Journal of Medicinal Chemistry*, 36, 320–330.

Kempf, D.J., Marsh, K.C., Cadacovi Fino, L., Bryant, P., Craig-Kennard, A. et al. (1994). Design of orally bioavailable, symmetry-based inhibitors of HIV protease. *Bioorganic and Medicinal Chemistry Letters*, 2, 847–858.

Kempf, D.J., Marsh, K.C., Denissen, J.F., McDonald, E., Vasavanonda, S. et al. (1995). *Proceedings of the National Academy of Sciences U.S.A.*, 92, 2484–2488.

Kempf, D.J., Marsh, K.C., Paul, D.A., Knigge, M.F., Norbeck, D.W. et al. (1991). Antiviral and pharmacokinetic properties of C_2 symmetric inhibitors of the human immunodeficiency virus type 1 protease. *Antimicrobial Agents and Chemotherapy*, 35, 2209–2214.

Kempf, D.J., Norbeck, D.W., Cadacovi, L., Wang, X.C., Kohlbrenner, W.E. et al. (1990). Structure-based, C_2 symmetric inhibitors of HIV protease. *Journal of Medicinal Chemistry*, 33, 2687–2689.

Kempf, D.J., Sham, H.L., Marsh, K.C., Flentge, C.A., Betebenner, D. et al. (1998). Discovery of ritonavir, a potent inhibitor of HIV protease with high oral bioavailability and clinical efficacy. *Journal of Medicinal Chemistry*, 41, 602–617.

Lopinavir

Sham, H.L., Kempf, D.J., Molla, A., Marsh, K.C., Kumar, G.N. et al. (1998). ABT-378, a highly potent inhibitor of the human immunodeficiency virus protease. *Antimicrobial Agents and Chemotherapy*, 42, 3218–3224.

Stoll, V., Qin, W., Stewart, K.D., Jakob, C., Park, C. et al. (1992). X-ray crystallographic structure of ABT-378 (lopinavir) bound to HIV-1 protease. *Bioorganic and Medicinal Chemistry*, 10, 2803–2806.

Mozenavir

Lam, P., Jadhav, P.K., Eyermann, C.J., Hodge, C.N., Lee, Y. et al. (1994). Rational design of potent, bioavailable, nonpeptide cyclic ureas as HIV protease inhibitors. *Science*, 263, 380–384.

Tipranavir

Thaisrivonds, S., Tomich, P.K., Watenpaugh, K.D., Chong, K.T., Howe, W.J. et al. (1994). Structure-based design of HIV protease inhibitors: 4-hydroxycoumarins and 4-hydroxy-2-pyrones as non-peptidic inhibitors. *Journal of Medicinal Chemistry*, 37, 3200–3204.

Thaisrivonds, S., Watenpaugh, K.D., Howe, W.J. Tomich, P.K., Dolak, L.A. et al. (1995). Structure-based design of novel HIV protease inhibitors: carboxamine-containing 4-hydroxycoumarins and 4-hydroxy-2-pyrones as potent nonpeptidic inhibitors. *Journal of Medicinal Chemistry*, 38, 3624–3637.

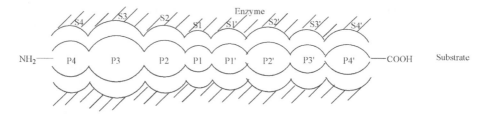

FIGURE 5.19 Standard nomenclature for substrates' binding subsites (Sn ... Sn') and inhibitor-binding pockets in proteases (Pn – Pn'). The substrate will be hydrolyzed between amino acid residue P1 and P1'.

5.10 METALLOPROTEINASE–COLLAGENASE INHIBITOR EXAMPLES

Claudiu T. Supuran and Andrea Scozzafava

5.10.1 INTRODUCTION

Proteases (PRs), also termed proteinases or peptidases, constitute one of the largest functional groups of proteins, with more than 560 members described.[1,2] By hydrolyzing one of the most important chemical bonds present in biomolecules, i.e., the peptide bond, PRs perform crucial functions in organisms found all over the phylogenetic tree, from viruses, bacteria, protozoa, metazoa and fungi to plants and animals. Numerous practical applications of such enzymes in biotechnology, and the understanding that PRs are important targets for drug design, ultimately fueled much research in this field.[1,2] Five catalytic types of PRs have been recognized so far, in which serine, threonine, cysteine, or aspartic groups as well as metal ions play a primary role in catalysis. The first three types of PRs are catalytically very different from the aspartic and metallo-PRs, mainly because the nucleophile of the catalytic site is part of an amino acid in the former, whereas it is an activated water molecule in the latter group of enzymes. Thus, acyl enzyme intermediates are formed only in the reactions of the Ser or Thr or Cys PRs, and only these peptidases can readily act as transferases. The terminology used in describing the specificity of PRs depends on a model in which the catalytic site is considered to be flanked on one or both sides by specificity subsites for the substrate, each able to accommodate the side chain of a single amino acid residue, as originally proposed by Berger and Schechter[3] and adopted thereafter by most researchers in this field. These sites are numbered from the catalytic site: S1, S2, ... Sn toward the N-terminus of the substrate, and S1', S2', ... Sn' toward the C-terminus. The residues they accommodate are numbered P1, P2, ... Pn, and P1', P2', ... Pn', respectively, as shown in Figure 5.19.

5.10.2 METALLOPROTEINASES

The extracellular matrix (ECM) plays a critical role in the structure and integrity of various tissue types in higher vertebrates.[4,5] ECM turnover is involved in important physiological and physiopathological events, such as embryonic development, blastocyst implantation, nerve growth, ovulation, morphogenesis, angiogenesis, tissue resorption and remodeling (such as wound healing), bone remodeling, apoptosis, cancer invasion and metastasis, arthritis, atherosclerosis, aneurysm, breakdown of blood–brain barrier, periodontal disease, skin ulceration, corneal ulceration, gastric ulcer, and liver fibrosis, among others.[4–6] The matrix metalloproteinases (MMPs), a family of zinc-containing endopeptidases (also called matrixins), were shown to play a central role in these processes.[4–6]

At least 20 members of this enzyme family, all sharing significant sequence homology, have been reported (Table 5.5).[1,6] They can be subdivided (considering the macromolecular substrate requirements) into (1) collagenases (MMP-1, -8, -13, and -18); (2) gelatinases (MMP-2 and -9); (3) stromelysins (MMP-3, -10, and -11); and (4) membrane-type MMPs (MT-MMPs; MMP-14, -15, -16, and -17). Recently, some new members of the family have been discovered, but little is known currently regarding their properties, substrate specificity, and inhibition (Table 5.5).[6]

MMPs possess a modular structure consisting of:

1. An N-terminal signal peptide sequence (*Pre*, see Figure 5.20).
2. A propeptide sequence (*Pro*) that has the role of conferring latency on the enzyme. In fact, this domain contains a conserved Cys residue that is coordinated to the catalytic Zn(II) ion, inhibiting in this way the autolysis of these highly active enzymes. MMPs require the removal of this pro-domain to acquire catalytic activity.
3. The catalytic domain (of about 170 amino acid residues), which contains a highly conserved zinc-binding motif consisting of three histidine residues and a conserved glutamate, important in catalysis. The Zn(II) binding motif is HE*XX*H*XX*G*XX*H (where *X* can be any amino acid residue). Stromelysin 3 and the MT-MMPs also contain a furin-recognition sequence between the propeptide sequence and the catalytic domain.
4. A variable, C-terminal domain. In matrylisin this domain is missing, whereas for other MMPs (such as the collagenases), it is essential for the recognition of macromolecular substrates. MT-MMPs also contain a transmembrane region within the C-terminal domain, which serves to anchor the enzyme to the cell membrane, whereas the N-terminal part of the molecule protrudes into the extracellular space.
5. Several metal ions, with different functions. All MMPs contain two Zn(II) and from two to three Ca(II) ions. One of the zinc ions, coordinated by the histidines belonging to the binding motif mentioned earlier, is critical for catalysis because the water coordinated to it as the fourth ligand in the quasi-tetrahedral geometry of Zn(II) acts as the nucleophile during the proteolytic process. The other zinc ion and the calcium ions have a structural role, probably in stabilizing the enzyme from autocleavage.[1,6]

In general, MMPs are secreted as zymogens, which are inactive, latent proenzymes.[1] These proforms need activation in order to give fully active proteases. Extracellular activation is generally a two-step process: an initial cleavage by an

TABLE 5.5
Vertebrate MMPs, Their Molecular Weights, Substrates, and Preferred Scissile Amide Bonds

Protein	MMP	MW (kDa)	Principal Substrates	Preferred Scissile Amide Bonds
Collagenase 1	MMP-1	52	Fibrillar and nonfibrillar collagens (types I, II, III, VI, and X), gelatins	Gly-Ile
Gelatinase A	MMP-2	72	Basement membrane and nonfibrillar collagens (types IV, V, VII, X), fibronectin, elastin	Ala-Met
Stromelysin 1	MMP-3	57	Proteoglycan, laminin, fibronectin, collagen (types III, IV, V, IX); gelatins; pro-MMP-1	Gly-Leu
Matrilysin	MMP-7	28	Fibronectins, gelatins, proteoglycan	Ala-Ile
Collagenase 2	MMP-8	64	Fibrillar collagens (types I, II, III)	Gly-Leu; Gly-Ile
Gelatinase B	MMP-9	92	Basement membrane collagens (types IV, V), gelatins	Gly-Ile; Gly-Leu
Stromelysin 2	MMP-10	54	Fibronectins, collagen (types III, IV) Gelatins, pro-MMP-1	Gly-Leu
Stromelysin 3	MMP-11	45	Serpin	Ala-Met
Macrophage elastase	MMP-12	53	Elastin	Ala-Leu; Tyr-Leu
Collagenase 3	MMP-13	51.5	Fibrillar collagens (types I, II, III), gelatins	Gly-Ile
MT1-MMP	MMP-14	66	Pro-72 kDa gelatinase	Not determined
MT2-MMP	MMP-15	61	Not determined	Not determined
MT3-MMP	MMP-16	55	Pro-72 kDa gelatinase	Not determined
MT4-MMP	MMP-17	58	Not determined	Ala-Gly
Collagenase 4 (*Xenopus*)	MMP-18	53	Not determined	Gly-Ile
RASI 1	MMP-19	?	Gelatin	Not determined
Enamelysin	MMP-20	?	Amelogenin (dentine), gelatin	Not determined
XMMP (*Xenopus*)	MMP-21	?	Not determined	Not determined
CMMP (chicken)	MMP-22	?	Not determined	Not determined
(No trivial name)	MMP-23	?	Not determined	Not determined

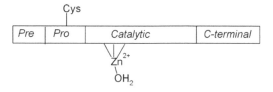

FIGURE 5.20 Schematic representation of the structure of a typical MMP enzyme. *Pre* represents the N-terminal signal peptide and *Pro* the propeptide sequence, followed by the catalytic domain and the C-terminal domain.

activator protease of an exposed susceptible loop in the propeptide domain (the so-called bait region), leading to the destabilization of the propeptide-binding interactions and disruption of the coordination of the conserved Cys residue to Zn(II); this is then followed by a final cleavage, usually assisted by another MMP, with the release of the amino terminus of the mature enzyme.[1]

Due to their ubiquitous spread in many tissues where they play critical physiological functions, MMPs have recently become interesting targets for drug design in the search for novel types of anticancer, antiarthritis, or other pharmacological agents useful in the management of osteoporosis, restenosis, aortic aneurysm, glomerulonephritis, or multiple sclerosis among others.[4,5]

In MMPs the catalytic Zn(II) ion is coordinated by three histidines, with the fourth ligand being a water molecule or hydroxide ion, which is the nucleophile intervening in the catalytic cycle of these enzymes (Figure 5.21).[5,7,8,9] In MMPs, the zinc-bound water molecule interacts with the carboxylate moiety of the conserved glutamate (Glu198 in MMP-8), forming two hydrogen bonds with it[7,8] and so generating a very effective nucleophile, which will attack the amide scissile bond.

The proteolytic mechanism of MMPs involves the binding of the substrate, with its scissile carbonyl moiety weakly coordinated to the catalytic Zn(II) ion (Figure 5.21a), followed by nucleophilic attack of the zinc-bound (and hydrogen-bonded glutamate) water molecule (Figure 5.21b) on this carbon atom. The water molecule donates a proton to the carboxylate moiety of Glu198, which transfers it to the nitrogen atom of the scissile amide bond (Figure 5.21c). The Glu198 residue then shuttles the second remaining proton of the water to the nitrogen of the scissile amide bond, resulting in peptide-bond cleavage (Figure 5.21d). During these processes, the Zn(II) ion stabilizes the developing negative charge on the carbon atom of the scissile amide bond, whereas a conserved alanine residue (Ala161 in MMP-8) helps to stabilize the positive charge at the nitrogen atom of the scissile amide.[5,7–8]

5.10.3 INHIBITION

As for other metalloenzymes, inhibition of MMPs is correlated with binding of the inhibitor molecule to the catalytic metal ion, with or without substitution of the metal-bound water molecule. Thus, MMP inhibitors (MMPIs) must contain a zinc-binding function attached to a framework that interacts with the binding regions of the protease.[4,5] The usual MMPIs of peptidic nature generally belong to the so-called right-hand-side inhibitors, in that they bind in the "primed" subsites shown in Figure 5.19.[4] Depending on the zinc-binding functions contained in their molecules, MMPIs belong to several chemical classes, such as carboxylates, hydroxamates, thiols, phosphorus-based inhibitors, sulfodiimines, etc.[4,5] The strongest inhibitors are the hydroxamates, and only such compounds will be discussed here. Many of the MMPIs were derived by replacing the scissile peptide bond with such a zinc-binding function (eventually followed by a methylene moiety) in such a way that the zinc-binding moiety is available for coordination to the catalytic Zn(II) ion.

The interaction of the catalytic domain of several MMPs with some inhibitors has been recently investigated by means of x-ray crystallography, NMR, and homology modeling (Figure 5.22).[7,10]

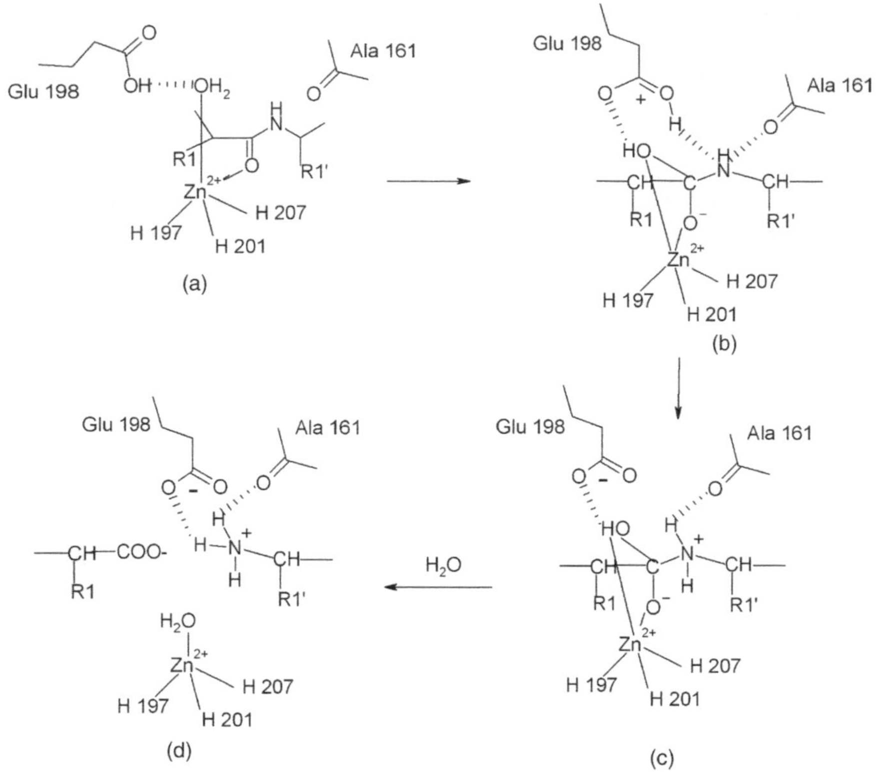

FIGURE 5.21 Catalytic mechanism of MMPs (exemplified for one of the best-studied cases, MMP-8). (Adapted from Lovejoy, B., Hassell, A.M., Luther, M.A., Weigl, D., and Jordan, S.R. (1994). *Biochemistry*, 33, 8207–8217.)

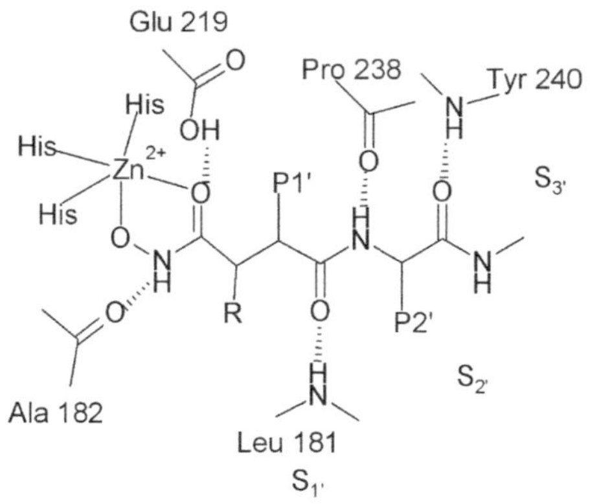

FIGURE 5.22 Binding of a hydroxamate inhibitor to MMP-7, as determined by x-ray crystallography. The Zn(II) ligand and hydrogen-bond interactions in the enzyme–inhibitor adduct are shown (Adapted from Grams, F., Crimmin, M., Hinnes, L., Huxley, P., Pieper, M., Tschesche, H., and Bode, W. (1995). *Biochemistry*, 34, 14012–14020).

Hydroxamates bind bidentately to the catalytic Zn(II) ion of the enzyme, which in this way acquires a distorted trigonal bipyramidal geometry.[7,10] The hydroxamate anion forms a short and strong hydrogen bond with the carboxylate moiety of Glu219 that is oriented toward the unprimed binding regions. The NH hydroxamate also forms a hydrogen bond with the carbonyl oxygen of Ala182. Thus, several strong interactions are achieved at the zinc site without any significant unfavorable contacts.

As with many other proteases, the main approach to the identification of synthetic, potent MMPIs was the substrate-based design of peptide-like compounds, derived from information on the amino acid sequence at the cleavage site.[2] Both right-hand-side as well as left-hand-side inhibitors were investigated initially, but because the compounds of the first type acted as much stronger inhibitors (compared with the other type), they were subsequently the most investigated for different types of pharmacological applications.[4] Thus, mainly this type of MMPI will be discussed in detail here, although a few left-hand-side inhibitors important for drug design are also mentioned. Hydroxamates are by far the most investigated class of MMPIs, and thousands of structural variants containing the CONHOH moiety have been synthesized and assayed as inhibitors of MMPs and other types of metalloenzymes. Two main classes of such MMPIs have been reported: (1) the succinyl hydroxamates (and their derivatives) and (2) the sulfonamide-based inhibitors.

For the first class of MMPIs, it was observed that the presence of a P_1 substituent (α- to the hydroxamate moiety) in this type of compound confers broad-spectrum activity against a variety of MMPs.[4] Thus, two important MMPIs, batimastat (**5.233**) and marimastat (**5.234**), were discovered by scientists from British Biotech Pharmaceuticals.[4] Batimastat possesses a thienylthiomethyl α-substituent, whereas in marimastat this is an OH group. These compounds showed very good *in vivo* activity in several disease models, but batimastat is not orally bioavailable, unlike marimastat, probably due to the increased water solubility induced by the presence of the hydrophilic OH moiety.

(5.233) batimastat   **(5.234)** marimastat

IC_{50} (nM)

	MMP-1	MMP-2	MMP-3	MMP-8	MMP-9	MMP-14
(5.233)	10	4	20	10	1	3
(5.234)	5	6	200	2	3	1.8

Further developments in this field involved variations of the α-substituent and the P_1 to P_3 moieties in order to obtain stronger or more selective inhibitors or both.

In the second class of MMPIs, sulfonylated amino acid hydroxamates were recently discovered to act as efficient MMPIs.[4,5] The first compounds from this class to be developed for clinical trials, of types (**5.235**; CGS 27023A) and (**5.236**; CGS 25966), possess the following structural features: (1) an isopropyl substituent α- to the hydroxamic acid moiety, considered to slow down metabolism of the zinc-binding function. It probably binds within the S_1 subsite; (2) a bulkier pyridylmethyl or benzyl moiety substituting the amino nitrogen atom and probably binding within the S_2 pocket; and (3) the arylsulfonyl group occupies (but does not fill) the specificity S_1 pocket.[4] CGS 27023A is a potent inhibitor of MMP-12, an enzyme that seems to be implicated in the development of emphysema that results from chronic inhalation of cigarette smoke.[5] A related compound from Agouron (**5.237**) also recently entered clinical studies, showing a range of pharmacological activities in animals, and inhibiting tumor growth in models of human glioma, human colon carcinoma, Lewis lung carcinoma, and human non-small-cell lung carcinoma.[5]

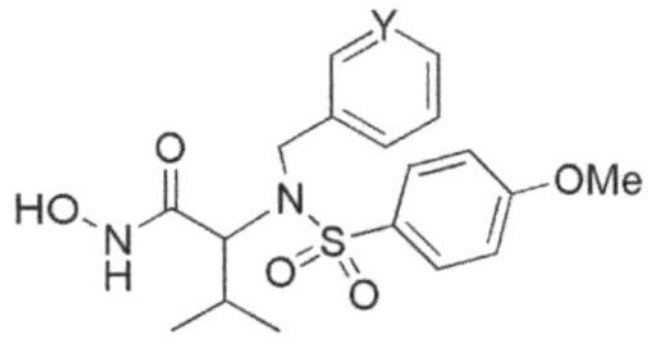

(5.235) Y = N
(5.236) Y = CH

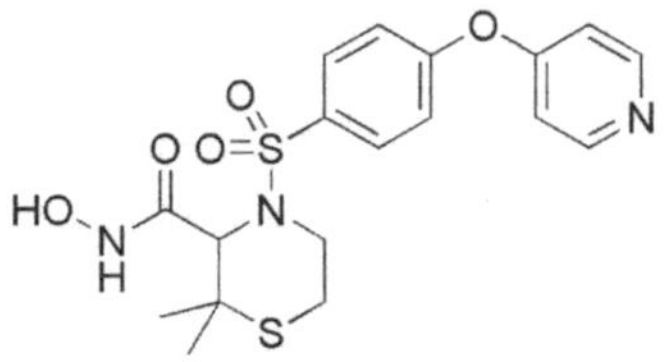

(5.237)

K_I (nM)

	MMP-1	MMP-2	MMP-3	MMP-7	MMP-9	MMP-13
(5.235)	33	20	43	–	8	–
(5.236)	–	–	92	–	–	–
(5.237)	8	0.08	0.27	54	–	0.038

(-) means that the compound was not tested

A large number of structurally related arylsulfonyl hydroxamates derived from glycine, L-alanine, L-valine, and L-leucine, possessing *N*-benzyl- or *N*-benzyl-substituted moieties, with nanomolar affinities for MMP-1, MMP-2, MMP-8, and MMP-9, were also reported, together with their diverse structural variants of types **5.238–5.247**.[11,12]

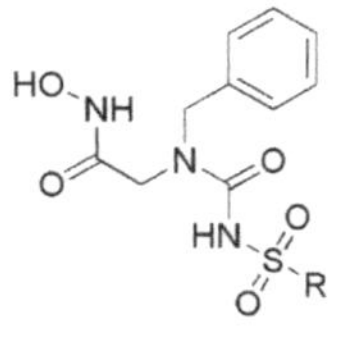

(5.238-4.241) 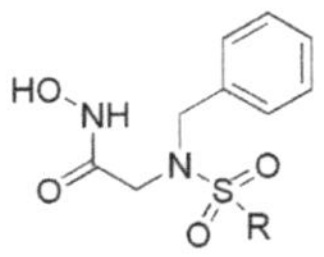   **(5.242-4.243)** 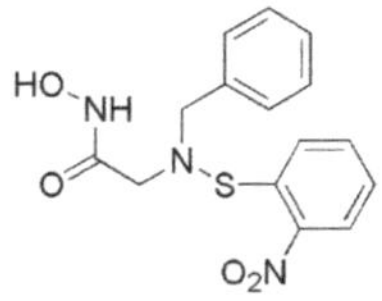   **(5.244)**

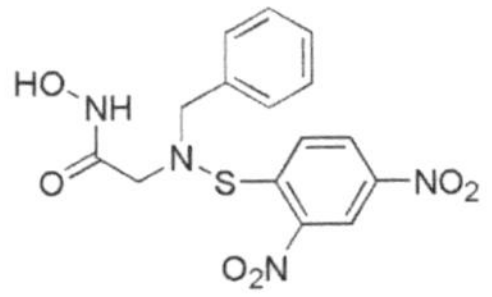

(5.245) 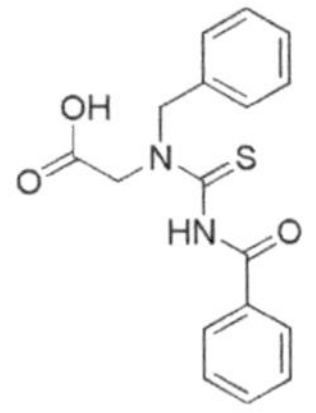   (5.246) 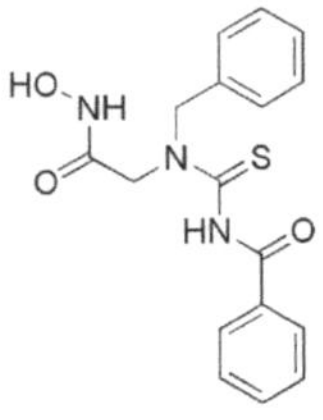   (5.247)

	R	K_I (nM) (against ChC)
(5.238)	4-F-C$_6$H$_4$	18
(5.239)	4-Cl-C$_6$H$_4$	15
(5.240)	4-Me-C$_6$H$_4$	15
(5.241)	2-Me-C$_6$H$_4$	14
(5.242)	3-CF$_3$-C$_6$H$_4$	6
(5.243)	C$_6$F$_5$	6
(5.244)	-	12
(5.245)	-	10
(5.246)	-	500
(5.247)	-	5

Some of these MMPIs were also shown to act as inhibitors of other enzymes that degrade ECM, such as the bacterial collagenases isolated from *Clostridium histolyticum* (ChC).[5,9] This collagenase (EC 3.4.24.3) is a 116-kDa protein that is able to hydrolyze triple-helical regions of collagen under physiological conditions, as well as an entire range of synthetic peptide substrates. In fact, the crude homogenate of *Clostridium histolyticum*, which contains several distinct collagenase isozymes, is the most efficient system known for the degradation of connective tissue, being also involved in the pathogenicity of this and related clostridia, such as *C. perfringens*, which causes human gas gangrene and food poisoning, among other diseases. Typically, these bacteria (and their collagenases) cause so much damage so quickly that antibiotics are ineffective. Similar to the vertebrate MMPs, ChC incorporates the conserved HExxH zinc-binding motif, which in this specific case is His[415]ExxH, with the two histidines (His415 and His419) acting as Zn(II) ligands, whereas the third ligand seems to be Glu447, and a water molecule or hydroxide ion acts as a nucleophile in the hydrolytic scission. Similar to the MMPs, ChC is also a multiunit protein, consisting of four segments, S1, S2a, S2b, and S3, with S1 incorporating the catalytic domain.[5] The sulfonylated, sulfenylated, or arylsulfonylureido-derivatized amino acid hydroxamates mentioned earlier, of type (5.238–5.247), were proved to possess nanomolar affinity for the Type II ChC (the most abundant and active isozyme).[9,11,12] Some of the most active inhibitors and their K_i data are shown above.

MMPIs of the types discussed earlier were investigated recently in several animal models of human disease, mainly cancer and arthritis, and promising pharmacological effects have been observed in many cases. Thus, as the controlled degradation of ECM is crucial for growth, invasive capacity, metastasis, and angiogenesis in human tumors, inhibition of some of the enzymes involved, such as the

MMPs, can lead to the introduction of novel anticancer therapies based on inhibitors of these proteases.[2,4]

REFERENCES

1. Barrett, A.J., Rawlings, N.D., and Woessner, J.F., Eds. (1998). *Handbook of Proteolytic Enzymes*. London: Academic Press (CD-ROM), and references cited therein.
2. Smith, H.J. and Simons, C., Eds. (2002). *Proteinase and Peptidase Inhibition — Recent Potential Targets for Drug Development*. London: Taylor & Francis.
3. Berger, A. and Schechter, I. (1970). Mapping the active site of papain with the aid of peptide substrates and inhibitors. *Philosophical Transactions of the Royal Society of London*, 257, 249–264.
4. Whittaker, M., Floyd, C.D., Brown, P., and Gearing, A.J.H. (1999). Design and therapeutic application of matrix metalloproteinase inhibitors. *Chemical Reviews*, 99, 2735–2776.
5. Supuran, C.T. and Scozzafava, A. (2002). Matrix metalloproteinsases (MMPs), in *Proteinase and Peptidase Inhibition — Recent Potential Targets for Drug Development*. Smith, H.J. and Simons, C., Eds., 35–61. London: Taylor & Francis.
6. Nagase, H. and Woessner, J.F. Jr. (1999). Matrix metalloproteinases. *Journal of Biological Chemistry*, 274, 21491–21494.
7. Grams, F., Crimmin, M., Hinnes, L., Huxley, P., Pieper, M., Tschesche, H., and Bode, W. (1995). Structure determination and analysis of human neutrophil collagenase complexed with a hydroxamate inhibitor. *Biochemistry*, 34, 14012–14020.
8. Lovejoy, B., Hassell, A.M., Luther, M.A., Weigl, D., and Jordan, S.R. (1994). Crystal structures of recombinant 19-kDa human fibroblast collagenase complexed to itself. *Biochemistry*, 33, 8207–8217.
9. Supuran, C.T., Briganti, F., Mincione, G., and Scozzafava, A. (2000). Protease inhibitors: synthesis of L-alanine hydroxamate sulfonylated derivatives as inhibitors of *Clostridium histolyticum* collagenase. *Journal of Enzyme Inhibition*, 15, 111–128.
10. Brandstetter, H., Engh, R.A., Graf von Roedern, E., Moroder, L., Huber, R., Bode, W. et al. (1998). Structure of malonic acid-based inhibitors bound to human neutrophil collagenase. A new binding mode explains apparently anomalous data. *Protein Science*, 7, 1303–1309.
11. Scozzafava, A. and Supuran, C.T. (2000a). Protease inhibitors. Part 9. Synthesis of *Clostridium histolyticum* collagenase inhibitors incorporating sulfonyl-L-alanine hydroxamate moieties. *Bioorganic and Medicinal Chemistry Letters*, 10, 499–502.
12. Scozzafava, A. and Supuran, C.T. (2000b). Protease inhibitors. Synthesis of potent matrix metalloproteinase and bacterial collagenase inhibitors incorporating N-4-nitrobenzylsulfonyl glycine hydroxamate moieties. *Journal of Medicinal Chemistry*, 43, 1858–1865.

Index

T - #0377 - 071024 - C4 - 234/156/15 - PB - 9780367393571 - Gloss Lamination